HARC

R0017635044

REF
QD
505
.A52
cop.1

Anderson, John Russell

Structure of metallic catalysts

DATE DUE

REFERENCE

REF
QD
505
.A52
cop.1

FORM 125 M

SCIENCE DIVISION

The Chicago Public Library

JUL 2 6 1976

Received

Structure of Metallic Catalysts

Structure of Metallic Catalysts

J. R. Anderson

1975

Academic Press
London   New York   San Francisco
A Subsidiary of Harcourt Brace Jovanovich, Publishers

ACADEMIC PRESS INC. (LONDON) LTD.
24/28 Oval Road,
London NW1

United States Edition published by
ACADEMIC PRESS INC.
111 Fifth Avenue
New York, New York 10003

Copyright © 1975 by
ACADEMIC PRESS INC. (LONDON) LTD.

All Rights Reserved
No part of this book may be reproduced in any form by photostat, microfilm, or any other means, without written permission from the publishers

Library of Congress Catalog Card Number: 74-17426
ISBN: 0-12-057150-1

Printed in Great Britain by
Unwin Brothers Limited
The Gresham Press, Old Woking, Surrey
A member of the Staples Printing Group

Preface

Metals are used as catalysts for so many technically important processes that the need to understand their structure as fully as possible hardly requires emphasis. The scope of this book is, in the main, confined to catalysts where a metal is present as a distinct and separate metallic phase. Many catalysts consist, of course, of chemical compounds containing a metal in a combined form; these are not discussed in the following chapters except occasionally in passing.

This is not a book on catalytic reaction chemistry. I have been concerned with the structure of the catalyst itself. Even so, the literature is very large. Much of it is, however, highly empirical and there undoubtedly remains a great deal which is buried in confidential industrial literature and practice. This book relies almost entirely on information available in the open scientific and technical literature, and a deliberate choice has been made to try to limit the material to catalyst archetypes, to descriptions which have some general or correlative validity, and to behaviour which can be understood in terms of the known chemistry and physics of the substances involved. I have avoided trying to compile an exhaustive catalogue of the multitudinous catalyst varieties which are listed in the literature. So many of these are only known via the proprietary literature that such a catalogue would, in any case, be of very doubtful value. What is possible, and what I have attempted to do, is to strike a workable balance between generality and speciality.

The book is essentially qualitative and descriptive. In part this is dictated by the nature of the subject, but I have also kept it in mind as a deliberate objective so that the book will be of maximum benefit to those who are coming to the subject without an extensive specialist background. I hope it will enable people of this sort to equip themselves relatively painlessly. At the same time I hope it will serve as a reference work for practitioners with greater experience in the subject.

Model catalysts are important because they offer a better prospect of controlling variables such as surface topography and composition than do normal technical catalysts. On the whole, with model catalysts it has proved easier to control contamination than detailed surface structure, and one of the problems is that the structure of a metal surface after it has functioned as a catalyst may well not be the same as it was to begin with. At any event, in the subsequent chapters I have devoted a good deal of attention to model metallic catalysts, since these form a yardstick against which technical catalysts can be judged.

Even when a catalyst support is viewed in a passive role, it will be clear that its structure and surface chemistry are likely to be important in controlling the ultimate performance of a supported catalyst. For instance, a number of factors including the pore structure, surface area and the nature of the surface, influence the degree to which a metal is dispersed when it is introduced onto the support, while the pore structure will influence the accessibility of the supported metal to the reactant. However, a material which is used as a support for the metallic phase may have catalytic activity in its own right. The dual function platinum/silica–alumina or platinum/zeolite catalysts for hydrocarbon reforming are well-known examples. For these reasons the chapter on support materials deals not only with structure, but also summarizes at least some of the more generally important aspects of support surface chemistry.

Catalysts are materials, and all of the techniques which are available to the materials scientist are at least of potential value for catalyst characterization. A number of these investigational techniques are indicated in passing when describing catalyst structure. However, a detailed description of all of these techniques *per se* would have imposed such enormous length requirements on the book as to be out of the question. In the end, I tried to strike a compromise by devoting two chapters to an outline of techniques which are used for studying and measuring surface area, particle size, pore structure, surface composition and surface structure. For more general techniques of the type used for studying the bulk structure and composition of materials, the reader must be referred to the extensive literature in experimental methods in materials science.

There can be few books indeed that are written by an author in total isolation. I have had help of two sorts. A number of friends helped by critically reading various chapters. Their efforts resulted in numerous improvements. They know they have my thanks, but it is also appropriate that I should take this opportunity to thank them once again. Finally but by no means least, I must say that without encouragement and sympathetic support this book would probably never have been finished, and I can here only offer my profound gratitude. The people concerned know who they are: they shall only remain anonymous to emphasize that the residual imperfections are my responsibility alone.

CSIRO Division of Tribophysics
University of Melbourne
Parkville, Victoria, Australia
November, 1974

J. R. A.

Contents

Preface		v
Chapter 1	**Introduction to Metals and Catalysts**	1
	1. Metallic Structure	1
	2. Chemical Bonding at Metal Surfaces	8
	3. Metals as Catalysts	21
	References	28
Chapter 2	**Support Materials**	31
	1. Silica	39
	2. Glasses	45
	3. Alumina	46
	4. Chromia	54
	5. Titania	56
	6. Zirconia	61
	7. Magnesia	63
	8. Thoria	65
	9. Metal Sulphates, Phosphates, Chlorides and Carbonates	67
	10. Silica-Alumina and other Mixed Oxides	68
	11. Zeolites and Natural Clays	74
	12. Carbon	81
	13. Miscellaneous Supports	86
	14. Single Crystal Surfaces	89
	References	94
Chapter 3	**Massive Metal Catalysts**	101
	1. Ideal Surfaces	103
	2. Unsupported Bulk Metal	113
	3. Evaporated Metal Films	130
	4. Alloy Catalysts	143
	References	158
Chapter 4	**Dispersed Metal Catalysts**	163
	1. Supported Metal Catalysts	164
	2. Unsupported Metal Catalysts	218
	3. Dispersed Multimetallic Catalysts	231
	References	236

Chapter 5	**Structure and Properties of Small Metal Particles**	244
	1. Ideal Crystallographic Particles	246
	2. Observed Equilibrium Crystallite Structure	253
	3. Surface Atoms of Low Co-ordination	260
	4. Bimetallic Particles	263
	5. Properties of Small Metal Particles	266
	6. Metal Cluster Compounds	270
	7. Interaction of Metal Particles with a Non-Metallic Support	275
	8. Particle Growth	280
	References	286
Chapter 6	**Measurement Techniques: Surface Area, Particle Size and Pore Structure**	289
	1. Surface Area Measurement	290
	2. Particle Size	358
	3. Pore Structure	376
	References	387
Chapter 7	**Measurement Techniques: Surface Composition and Structure**	395
	1. Surface Structure	398
	2. Surface Topography	403
	3. Surface Composition	409
	References	440
Appendix 1		445
Appendix 2	**Illustrative Recipes for the Preparation of Metallic Catalysts**	451
	1. Platinum–Metals Powder	451
	2. Platinum–Metals Powder	452
	3. Nickel–Copper Powder	452
	4. Skeletal Nickel (Raney Nickel)	453
	5. Stabilized Porous Iron from Magnetite Fusion	455
	6. Cobalt Fischer-Tropsch Catalyst	455
	7. Platinum/Silica Gel (impregnation)	456
	8. Platinum/Aluminosilicate (coprecipitation)	456

9. Platinum/Silica Gel (adsorption) 456
10. Platinum/Zeolite (adsorption) 457
11. Nickel/Kieselguhr (impregnation) 457
12. Nickel/Magnesia (impregnation) 458
13. Nickel/Alumina (coprecipitation) 458
14. Palladium Colloid 458

Index 461

CHAPTER 1

Introduction to Metals and Catalysts

		page
1.	METALLIC STRUCTURE	1
2.	CHEMICAL BONDING AT METAL SURFACES	8
3.	METALS AS CATALYSTS	21

Technically important metallic catalysts always contain the metal more or less in a high state of dispersion so as to maximize as far as possible the surface area available for a given mass of metal. In many instances it is arranged for the metal particles to be distributed on a support. This offers a means of retaining the metallic particles in a state that is both stable towards agglomeration and is accessible to reactant. The support also makes possible the preparation of a catalyst which has convenient mechanical properties towards handling or incorporation as an element in a process system, and it can provide a material with suitable heat transfer properties.

However, technical catalysts are often not well enough defined for proper use in experimental work on reaction mechanisms, and various sorts of alternatives have been developed: metal wires, foils, single crystals and evaporated films are obvious examples.

1. Metallic Structure

Catalysts are chemicals which, under reaction conditions, form chemical bonds of varying degrees of stability with other atoms or groups. Although one may hope ultimately to explain the chemical behaviour of a metallic catalyst in purely quantum mechanical terms, this objective still remains largely beyond reach, and one is then left with the construction of essentially empirical correlations.

The transition metals are the ones which are of greatest importance as catalysts, and the physical properties of these elements show some general trends on moving across each of the transition series. Parameters for the more common metals are collected into Tables A1–A4 of Appendix 1: here we do no more than comment on some of the main features of the trends.

A number of physical properties of the metals are related in a general way to the cohesive energy: for instance, melting point, boiling point, interatomic distance and surface energy. With some minor variations, maxima occur in the heat of atomization, melting point and boiling point of the transition elements round about group VIB, while the surface energy and density show maxima, and the interatomic distance a minimum in the region of group VIII.

With only very few exceptions, all the metals fall into one of three crystal structures, f.c.c., b.c.c. or h.c.p. The atom arrangements of the conventional unit cells are shown in Figs 1.1–1.3. Some properties of these three structures are listed in Table 1.1.

TABLE 1.1 Some properties of f.c.c., b.c.c. and h.c.p. structures

	f.c.c.	b.c.c.	h.c.p.
number of nearest neighbours	12	8	12
number of second neighbours	6	6	6
second neighbour distance	$d\sqrt{2}$	$2d/\sqrt{3}$	$d\sqrt{2}$
number of atoms per unit cell	4	2	2
unit cell dimensions	cube side $=d\sqrt{2}$	cube side $=2d/\sqrt{3}$	$C = 2d\sqrt{2}/\sqrt{3}$ $(a = b = c) = d$
maximum proportion of space filled by hard spheres	74%	68%	74%

d is the nearest neighbour distance

A description of metallic structure and binding may be approached from two points of view—band theory (cf. ref. 1) and valence bond theory.

In the band theory approach, the cohesive energy of the metal is ascribed to the electrostatic interaction between the positively charged ions and the valence electrons which are able to move through the crystal in a periodic potential due to the ions. To a good approximation, electrons in the closed inner shells around each ion are considered to be localized. For every quantum state of an electron in the free atom there is a band of energies in the crystal, and the width of the band increases the more the atomic wave functions overlap. This behaviour is illustrated in Fig. 1.4 in a schematic manner for copper (Raimes[2]), and the general situation is similar for the transition metals. At 0 K the valence electrons occupy the

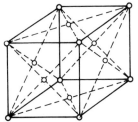

Fig. 1.1 f.c.c. structure.

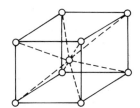

Fig. 1.2 b.c.c. structure.

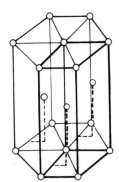

Fig. 1.3 h.c.p. structure.

lowest available energy levels: the maximum occupied at 0 K is the Fermi level, E_F. At a temperature >0 K there will always be some electrons with energies above the Fermi level, but this energy spread is always relatively small compared with E_F at ordinary temperatures.

The band widths in metals are such that bands frequently overlap: in particular in transition metals s and d bands overlap. This has the important consequence of affecting the degree of occupancy of the d band. For instance, ten electrons per atom are required to fill the $3d$ band: in the case of nickel there are ten $3d$ and $4s$ electrons, but because of band overlap some of the electrons go into the $3d$ band and some into the $4s$ band. With nickel, palladium and platinum there are, on the average, about 0·4–0·6 electron holes in the d band per atom, with cobalt 0·75 and with iron 0·95 holes. On moving out of the transition series to metals such as copper and silver, the d band is completely filled, but the s band remains half filled. The situation is illustrated schematically in Fig. 1.5. Because the d band is narrow, the density of states is very high. This has the consequence that with a transition metal where the d band is partly filled the density of states at the Fermi level is also very high. This approach is oversimplified in that it ignores orbital hybridization, but the results due to Wood[3] indicate that much of the general character of the model remains unchanged.

Since the position of the Fermi level changes on moving across the transition series while the general band structure remains the same, the density of states at the Fermi level also changes on moving across the transition series. Figures 1.6 and 1.7 show this for the first two transition series (Mott[4]).

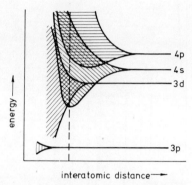

FIG. 1.4 The dependence of electron band widths upon interatomic distance in copper (schematic). The vertical broken line is drawn at roughly the interatomic distance of the normal solid metal. Reproduced with permission from Raimes, S. "The Wave Mechanics of Electrons in Metals" North-Holland, Amsterdam (1961).

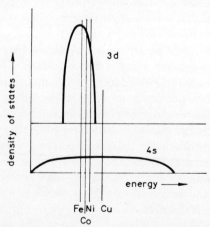

FIG. 1.5 Approximate representation of $3d$ and $4s$ band filling in iron, cobalt, nickel and copper. The bands are filled to the left of the vertical line in each case. After Mott, N. F. and Jones, H. "The Theory of the Properties of Metals and Alloys" Oxford University Press, Oxford (1936).

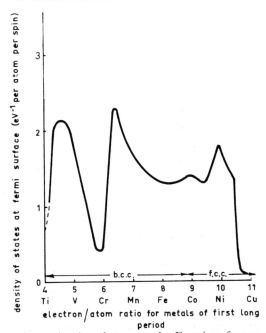

FIG. 1.6 Approximate density of states at the Fermi surface as a function of the electron/atom ratio for metals in the first long period and their alloys. Values for iron, cobalt and nickel refer to a paramagnetic state of the metal. Reproduced with permission from Hayward, D. O. *In* "Chemisorption and Reactions on Metallic Films" (J. R. Anderson, ed.) Academic Press, London, Vol. 1 (1971), p. 225.

The magnetic properties of the transition metals can be accounted for by the presence of unpaired electrons in the incompletely filled d band. Moreover, if one continues with the assumption of a rigid band model, it is possible to account for the magnetic properties of group VIII—group IB alloys in broad terms on the basis that s electrons from the group IB element fill the holes in the d band, and there will be an obvious critical concentration when this filling is just complete. However, this approach has serious problems because a rigid band model is badly inaccurate, and in group VIII—group IB alloys it is now clear that there is no common d band at all.

A ligand-field modification to the band theory of metals has been given by Trost[5] and Goodenough[6] in which the effect of the crystal field is to direct the valence electrons into the regions between nearest and next nearest neighbour atoms. This modification amounts to a transition between the band theory of metals and a valence bond approach. In an isolated atom in a cubic field, the five-fold degenerate d level splits into three-

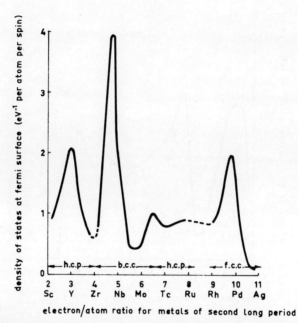

FIG. 1.7 Approximate density of states at the Fermi surface as a function of electron/atom ratio for metals in the second long period and their alloys. Reproduced with permission from Hayward, D. O. *In* "Chemisorption and Reactions on Metallic Films" (J. R. Anderson, ed.) Academic Press, London, Vol. 1 (1971), p. 225.

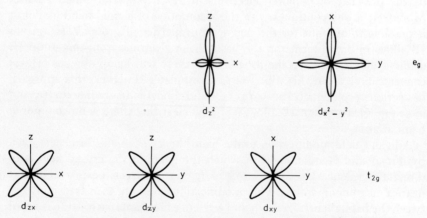

FIG. 1.8 Schematic representation of t_{2g} and e_g orbitals.

fold (t_{2g}) and two-fold (e_g) degenerate levels. The former contains the d_{xy}, d_{zy} and d_{zx} orbitals, the latter contain the $d_{x^2-y^2}$ and d_{z^2} orbitals. In a metal, the d band is split in an analogous fashion into sub-bands, the t_{2g} and e_g, of analogous orientations to the atomic orbitals.

In an f.c.c. metal there are twelve nearest neighbours to any particular atom, and nearest neighbour bonds may be formed by overlap using the twelve lobes of the t_{2g} orbitals. There are six next nearest neighbours, and bonds may be formed using the six lobes of the e_g orbitals (cf. Fig. 1.8), although one would expect next nearest neighbour bonding to be weak because of the reduced orbital overlap resulting from the greater internuclear distance.

An exclusively valence bond approach to metallic bonding has been formulated by Pauling[7,8]. It is assumed that each metal atom provides *dsp* hybrid orbitals, overlap between which leads to metallic bonding. It is assumed that the actual electronic configuration in the metal arises from resonance between all possible bonding arrangements, the number of orbitals used being in excess of the number of electrons; thus, there is on this model a necessary correlation between incompletely filled orbitals, the consequent large number of ways in which the electrons can be arranged, and an increased stability of the metallic crystal due to resonance.

Two types of *d* orbitals are distinguished: those that are involved in bonding and which form *dsp* hybrids, and those that remain as atomic *d* orbitals. It is further assumed that by virtue of resonance, the number of atomic *d* orbitals is not limited to an integral number. There are nine orbitals per atom, and it is assumed the magnetic properties of iron, cobalt and nickel are due to unpaired electrons in the atomic *d* orbitals. It is possible to account for the magnetic properties of these elements by assuming that there are 6 occupied bonding orbitals, 2·3 atomic *d* orbitals and 0·7 vacant bonding orbitals per atom, with the variable number of electrons affecting the occupancy of the atomic orbitals and hence the saturation magnetic moment per atom. This orbital–type proportion is assumed to hold for all the transition metals. As one moves from left to right across each of the transition series, electrons enter the bonding orbitals: since these are six in number, they are filled on reaching the metals in group VIB, and this agrees with maximum in cohesive energy observed for the metals in this group. There are, of course, twelve nearest neighbours in f.c.c. or h.c.p. metals, and eight for b.c.c. metals, so each of the nearest neighbour bonds cannot be a full electron pair bond: the bond orders are $\frac{1}{2}$ (f.c.c. and h.c.p.) and $\frac{3}{4}$ (b.c.c.).

It is also possible to compute the contribution which the *d* orbitals make to the *dsp* hybrids, and the values are listed in Table 1.2.

TABLE 1.2 Percentage d-character of the metallic bond in transition metals

Sc	Ti	V	Cr	Mn	Fe	Co	Ni	Cu
20	27	35	39	40	39·5	40	40	36
Y	Zr	Nb	Mo	Tc	Ru	Rh	Pd	Ag
19	31	39	43	46	50	50	46	36
La	Hf	Ta	W	Re	Os	Ir	Pt	Au
19	29	39	43	46	49	49	44	—

The percentage d-character of the metallic bond has often been correlated in an empirical way with catalytic or adsorption data. On an orbital model, it presumably reflects the linear extension of a hybrid dsp orbital in space. A valence bond approach has also been described by Altmann *et al.*[9]

2. Chemical Bonding at Metal Surfaces

A quantitative quantum mechanical account of chemisorption on metal surfaces is not available. Nevertheless, before proceeding to empirical "chemical" descriptions, it is useful to consider qualitatively some of the processes which occur.

Consider an adsorbing atom close to a metal surface and with the energy of the atomic valence states (e.g. $1s$ for a hydrogen atom) lying within the energy range of the valence band of the metal. At least within a restricted part of the valence band the Bloch wave functions near the surface (which decay exponentially away from the surface towards free space) will mix with the localized valence states of the adatom to produce wave functions which encompass both the Bloch states and the valence states of the adatom: that is, close to the adatom these wave functions have the character of the free atom valence states, and on moving towards the metal they transform continuously towards the Bloch functions for the metal surface. These have been termed virtual bound states (cf. Grimley[10]). Corresponding antibonding states are also created, but being of high energy remain unoccupied. In total, the electron population of the metallic valence states and the virtual bound states is increased by the number of valence electrons brought with the adsorbing atom.

On the other hand, if the energy of the atomic valence states lies outside the metallic valence band, virtual bound states cannot occur. If the atomic state were to lie sufficiently far below the Fermi level it would in principle be possible for an electron to be transferred to it from the Fermi level thus creating an adion, without the coulomb repulsion between the electrons in the ion raising the energy of the ionic state (after correction for an image charge interaction between the ion and the metal) to such an extent that the net binding energy becomes unfavourable. It seems, however, that atomic states seldom lie deep enough to make pure ionic adsorption possible. A compromise is struck in which the electrons, instead of being localized on the adatom, are distributed between it and one or more surface metal atoms to give quasi-conventional covalent bonding. It seems likely that the electrons will be distributed with metallic valence states near the top of the valence band where (for transition metals) the density of states is greatest.

On a valence bond model of metallic bonding, the formation of a chemical bond to a surface metal atom can be visualized as involving "dangling orbitals" of the surface metal atoms; that is, the metal atoms in the surface are assumed to have the same arrangement of orbitals as in the bulk, although not possessing their full complement of nearest neighbours. The "dangling orbitals" may be thought of in terms of either the Pauling or the Goodenough model. Thus, on the Pauling model there will be dsp dangling orbitals (bonding orbitals in the bulk), and these will be oriented towards the positions that would be occupied by metal atoms in the missing layer. There will also be dangling atomic d orbitals (assumed not involved in bulk bonding), and these point towards missing next nearest neighbours. Although not contributing to the cohesive energy of the crystal on Pauling's model, these orbitals could still be used for surface bonding. Alternatively, on Goodenough's model, it is the t_{2g} sub-band dangling orbitals which point towards the positions of the missing nearest neighbour atoms, and the e_g sub-band orbitals which point towards the missing next nearest neighbours. The geometries of the two models are, of course, identical: for every orbital lobe in Goodenough's scheme there is a corresponding one in Pauling's. The geometric arrangement of these surface orbitals for the low index crystal faces of f.c.c. metals is summarized in Fig. 1.9, following the representation due to Bond.[43]

It is clear that by making the assumption that the orbital geometry at the surface is the same as in the bulk, one is risking considerable error. A surface atom, unlike a bulk atom, will not be in a field of central symmetry. Indeed, there is certain to be a very strong field gradient normal to the surface. How this affects the orbital energy and geometry is not yet fully worked out. Nevertheless, the inadequacies of this model have recently become apparent from the work of Fassaert et al.[42] who considered hydrogen adsorption on nickel atom clusters of limited size using molecular orbital theory. They concluded that the main interaction takes place between the nickel $3d_{z^2}$ orbital, pointing towards the hydrogen atom, the $4s$ orbital, and the hydrogen $1s$ orbital. They also concluded that the idea of the d-orbitals which afford the e_g representation ($d_{x^2-y^2}$ and d_{z^2}) being non-bonding in the metal and thus suited for adsorption bonding is not

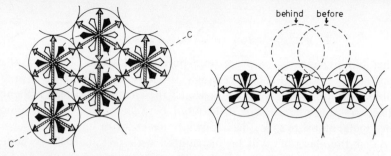

Fig. 1.9 Diagrammatic representation of the emergence of orbitals at the surface of a f.c.c. metal.
a. (100) face. Filled arrows; e_g orbitals in plane of paper: hatched arrows; t_{2g} orbitals in plane of paper: open arrows; t_{2g} orbitals emerging at 45° to plane of paper. The broken circle shows the position of an atom in the next layer above the surface layer. In both the plan and section, an e_g orbital emerges normal to the plane of the paper from each atom.

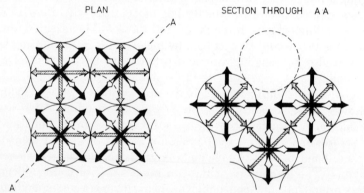

b. (111) face. Filled arrows; e_g orbitals emerging at an angle of 36°16′ to the plane of the paper: other arrows; as in a. Note: no orbitals emerge normal to the surface.

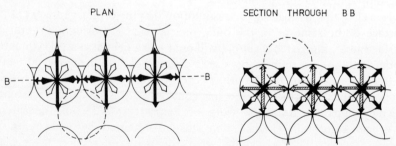

c. (110) face. Filled arrows; e_g orbitals in plane of paper or emerging at 45°, hatched arrows; t_{2g} orbitals in plane of paper: open arrows; t_{2g} orbitals emerging at 30° to plane of paper. In the plan, a t_{2g} orbital emerges normal to the paper from each atom: in the section, it is an e_g orbital. Reproduced with permission from Bond, G. C. *Discussions Faraday Soc.* **14**, 200 (1966).

substantiated. They also considered adsorption in crystallographic surface sites but concluded that in these positions hydrogen atoms are less strongly bound than when directly over a surface metal atom. This relative adsorption strength in these two positions is at variance with that proposed by Horiuti and Toya[45]. The trend of the results obtained by Fassaert et al. is therefore to emphasize the idea of hydrogen chemisorption by a diatomic chemical bond, and to de-emphasize the crystallographic implications of a model which seeks to locate adsorbed hydrogen atoms in surface sites of maximum co-ordination number. It must be remembered, however, that what is true for hydrogen chemisorption cannot be automatically extrapolated to other types of adsorbed species, and we refer to this question again in a later paragraph when dealing with adsorption of species such as oxygen and sulphur.

The covalent chemisorptive bond should be describable in similar parameters to those used for bonds in covalent molecules. There is a difficulty if an adatom is co-ordinated to more than one surface metal atom, and in this situation the heat of adsorption cannot be simply correlated with the energy of a *single* covalent bond between the adatom and the surface. Nevertheless, particularly when bearing in mind the comments of the previous paragraph, it is instructive to proceed *as though* chemisorption did only involve a single covalent bond with the surface. For instance, one may seek to express the energy of a covalent bond, D(M–H) between a surface metal atom and a chemisorbed hydrogen atom in terms of the energies of the two homoatom bonds D(M–M) and D(H–H). This has been done (cf. Eley[11], Stevenson[12]) using Pauling's correlation equation

$$D(M\text{--}H) = \tfrac{1}{2}\{D(M\text{--}M) + D(H\text{--}H)\} + (\chi_M - \chi_H)^2 \qquad (1.1)$$

where χ is the electronegativity. In equation 1.1, it is assumed that the bond energies and electronegativities are expressed in kJ mol^{-1}. If the bond energies are in kJ mol^{-1} but the electronegativities are in eV, the last term in the equation should be multiplied by 96·5.

There have been a number of attempts to evaluate the term $(\chi_M - \chi_H)^2$. Following the final form of the original treatment, χ_M is evaluated from

$$\chi_M = \phi/65 \qquad (1.2)$$

provided both the work-function, ϕ, and χ_M are expressed in kJ mol^{-1}. If ϕ is in eV, the right-hand side of equation 1.2 should be multiplied by 96·5. The factor 65 in equation 1.2 is an empirical scaling factor. One thus obtains

$$D(M\text{--}H) = \tfrac{1}{2}\{D(M\text{--}M) + D(H\text{--}H)\} + (\phi/65 - \chi_H)^2 \qquad (1.3)$$

The value of D(M–M) may be estimated from the heat of atomization of the metal

$$D(M-M) = \frac{-\Delta H_{at}}{n/2} \quad (1.4)$$

where n is the number of nearest neighbours possessed by an atom in the bulk of the metal. This puts all of the cohesive energy of a metal into nearest neighbour interactions, which is clearly inaccurate: nevertheless, for computational purposes, D(M–M) has generally been taken as $-\Delta H_{at}/6$ within the accuracy of the model, irrespective of the metallic crystal structure. It probably somewhat overestimates D(M–M) at the surface.

Since for dissociative chemisorption, the heat of adsorption is given by

$$-\Delta H_{ad} = 2D(M-H) - D(H-H) \quad (1.5)$$

substitution gives

$$-\Delta H_{ad} = -\Delta H_{at}/6 + 2(\phi/65 - \chi_H)^2 \quad (1.6)$$

assuming all quantities in kJ mol^{-1}. Using $\chi_H = 202 \cdot 5$ in these units, Table 1.3 compares some calculated and experimental values.

TABLE 1.3 Comparison of calculated and experimental heats of chemisorption of hydrogen

metal	$-\Delta H_{ad}$ (kJ mol^{-1})	
	calculated	observed (surface coverage → 0)
tantalum	210	188
tungsten	192	184
molybdenum	179	170
chromium	101	189
manganese	159	71
nickel	122	151
iron	134	142
rhodium	134	117
palladium	96	117

Although the calculated values are of the correct order, the general agreement is only semi-quantitative, and when applied to other adsorbates the situation is, if anything, worse. The difficulties are obvious. The model does not not necessarily assume the proper atomic symmetry of the adsorbate–adsorbent system, and the estimation of D(M–M) is of dubious accuracy.

The semi-empirical quantum mechanical treatment of Higuchi et al.[13] offers what is, in effect, an alternative method of calculating the effect of the electronegativity difference, and is on the whole no more successful, except insofar as it allows for the computation of the heat of adsorption of electropositive atoms such as the alkali metals.

Metals vary widely in their ability to chemisorb gases. We take as a rough criterion for chemisorption a value for the heat of adsorption >20 kJ mol^{-1} when dealing with relatively small molecules, and Table 1.4 summarizes a selection of the available data. These values for the heat of adsorption are differential enthalpy changes (as $-\Delta H$) at approximately zero coverage, and refer to polycrystalline or polyfaceted adsorbent surfaces. Thus, since the differential heat of adsorption either remains constant or becomes numerically smaller with increasing coverage, the listed values represent maximum numerical values.

The interpretation of the heat of adsorption in terms of the metal–adsorbate bond depends on knowing the stoichiometry of the adsorption process, and this depends on the adsorption conditions. With molecules such as hydrogen, oxygen, nitrogen and saturated hydrocarbons, dissociative chemisorption is the dominantly important mode on transition metals if the temperature is high enough. However, it is known that in the case of hydrogen and nitrogen at a low temperature and at a high coverage there is some weaker molecular chemisorption. Non-dissociative chemisorption is also important with molecules such as the alkenes and aromatics for which there is an interaction between the surface metal atoms and the π-electrons of the molecules.

The type of bonding by which non-metallic atoms such as H, O, S and N interact with the surface varies. In the case of H, S and N there can be little doubt that the bonding is essentially covalent in all cases. In the case of O there is evidence that in certain circumstances the oxygen carries a substantial negative charge. The problem is complicated by a lack, in most cases, of a precise knowledge of adsorbate–adsorbent stereochemistry. Nevertheless, the fact is that the change in the metal work-function caused by the adsorption of such species never exceeds about 2 eV. Thus, unless the adatom lies very close to the plane of the surface metal atoms so that the effective length of the surface dipole in a direction normal to the surface is very short, the work-function change cannot be consistent with the existence of adsorbed ions. However, there is at least one case, the adsorption of O on a nickel (100) surface, where the existence of something approaching an oxygen ion appears possible: rearrangement of the surface nickel atoms places the adatom nearly in the plane of the

STRUCTURE OF METALLIC CATALYSTS

TABLE 1.4 Adsorption of gases on metals

Gas	IA	IIA	IIIA	IVA	Period
	Li	Be	B	C	
H_2					
O_2	✓	✓		400	
N_2					
CO					2
CO_2					
CH_4, C_2H_6					
C_2H_4					
	Na	Mg	Al	Si	
H_2			†		
O_2	✓	✓	525	870	
N_2			†		
CO					3
CO_2					
CH_4, C_2H_6					
C_2H_4					
	K	Ca	Ga	Ge	
H_2	†	✓		✓	
O_2	✓	✓	✓	550	
N_2	†	✓			
CO	†	✓			4
CO_2					
CH_4, C_2H_6	(†)				
C_2H_4	†				
	Rb	Sr	In	Sn	
H_2		✓	†	†	
O_2	✓	✓	✓	✓	
N_2		✓	†	†	
CO		✓	†	†	5
CO_2			†	†	
CH_4, C_2H_6				(†)	
C_2H_4			†	†	
	Cs	Ba	Tl	Pb	
H_2		✓		†	
O_2	✓	✓	✓	✓	
N_2		✓		†	
CO		✓		†	6
CO_2				†	
CH_4, C_2H_6				†	
C_2H_4				†	

Gas	IIIB	IVB	VB	VIB	VIIB	VIII_1	VIII_2	VIII_3	IB	IIB	Period
	Sc	Ti	V	Cr	Mn	Fe	Co	Ni	Cu	Zn	
H_2		✓	✓	189	✓	142	101	155	34;†	†	
O_2	(✓)	990	✓	730	630	570	420	500	✓	✓	
N_2		✓	(✓)	✓	(✓)	168		*†	†	†	
CO		640	(✓)	326	260	192	197	176	39	†	4
CO_2		787	(✓)	✓	✓	280	152	222	†	†	
CH_4, C_2H_6		✓	✓	✓	✓	✓	✓	✓	†	†	
C_2H_4		✓	(✓)	✓	✓	285	✓	244	76	†	
	Y	Zr	Nb	Mo	Tc	Ru	Rh	Pd	Ag	Cd	
H_2		✓	✓	170		~118	117	117		†	
O_2	(✓)	✓	870	755		✓	503	294	✓	✓	
N_2		630	✓	272			†	†	†	†	
CO		620	553	310		✓	193	180	†	†	5
CO_2		(✓)	626	449		✓	†	†	†	†	
CH_4, C_2H_6		✓	(✓)	✓		✓	✓	✓	†	†	
C_2H_4		✓	(✓)	✓		✓	✓	✓	36	†	
	La	Hf	Ta	W	Re	Os	Ir	Pt	Au	Hg	
H_2	(✓)	(✓)	188	184	✓	✓	109	109			
O_2	✓	(✓)	890	490	✓	✓	✓	294	†	✓	
N_2	✓	(✓)	845	>210		(✓)	243	†	†		6
CO	✓	(✓)	560	420	✓	✓	✓	201	36		
CO_2		(✓)	750	504	(✓)	(✓)	(†)	†	(†)		
CH_4, C_2H_6		(✓)	✓	✓		(✓)	✓	✓			
C_2H_4		(✓)	580	420	✓	(✓)	✓	✓	87		

surface nickel atoms, and the orbital energies of the oxygen obtained by ion-neutralization spectroscopy indicate that the oxygen carries a substantial negative charge[18].

The question of the geometric relationship between an adatom and the surface metal atoms in most cases remains unresolved. We have already referred to this problem in discussing hydrogen chemisorption. Low energy electron diffraction data with a (usual) geometric interpretation give the geometry of the two-dimensional adatom array, but do not show the positions of the adatoms relative to the surface metal atoms (either laterally or vertically). Nevertheless the data due to Hagstrum and Becker[18], based on ion-neutralization spectroscopy, work-function measurement and low energy electron diffraction, show that for O, S or Se adsorbed on a nickel (100) surface the most probable adatom location is a surface site in which it will be co-ordinated to more than one surface metal atom.

The infrared adsorption spectra of chemisorbed carbon monoxide have been interpreted to indicate the existence of two forms (cf. refs. 20, 21), a bridged form involving two surface metal atoms and a linear form involving one, and the bonding in these adsorption modes may be represented on a dangling orbital model: for instance, the bridged form involving t_{2g} orbitals on next nearest neighbour surface atoms of f.c.c. (100), and the linear form involving a t_{2g} and a vertical e_g orbital on one surface atom in f.c.c. (100). Non-dissociatively adsorbed ethylene may also be represented in a manner that bears some similarity to that suggested for linear carbon monoxide: that is, on f.c.c. (100) it is bonded to a single metal atom with the vertical e_g surface orbital overlapping the bonding π-orbital of the olefin, and the two lobes of the t_{2g} surface orbital each overlapping a lobe of the antibonding π-orbital of the olefin. These models involving a single metal atom are in fact the same as have often been written for ligand bonding in mononuclear metal complexes. This is hardly surprising since the metal orbitals are postulated to be the same in the two cases.

Dissociative chemisorption of nitrogen is known on many transition metals although the adsorption process is activated in a number of cases; again the adatoms are probably located in surface interstitial positions.

Notes to Table 1.4
* Weak molecular chemisorption on extremely small nickel crystals.
Numbers indicate heats of adsorption at surface coverage tending towards zero and at about room temperature for clean polycrystalline surfaces, kJ (mol of gas) $^{-1}$.
Ticks indicate that chemisorption occurs.
† Means no chemisorption at 273 K.
Entries in parentheses indicate reasonably reliable estimate. Numerical data assembled from Cerny and Ponec[14]; Hayward[15]; Geus[16]; Ehrlich[17].

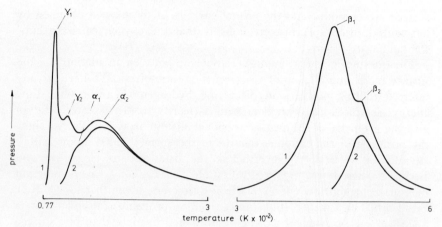

Fig. 1.10 Desorption spectra for gases adsorbed on polycrystalline tungsten.
a. Hydrogen. Left: curve 1, after adsorption at 77 K; curve 2, after adsorption at 158 K. Right: curve 1, after adsorption at 273 K; curve 2, after adsorption at 473 K. After Ricca F., Medana, R. and Saini, G. *Trans. Faraday Soc.* **61**, 1492 (1965).

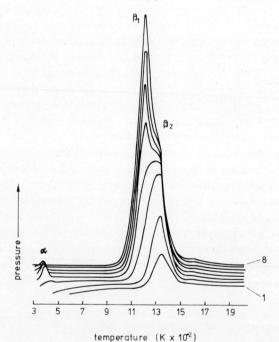

b. Nitrogen. Adsorption at about 300 K for increasing periods of time on going from curve 1 to 8: curve 8 corresponds to saturation adsorption at about 10^{-4} Pa (about 10^{-6} Torr). After Rigby, L. J. *Can. J. Phys.* **43**, 532 (1965).

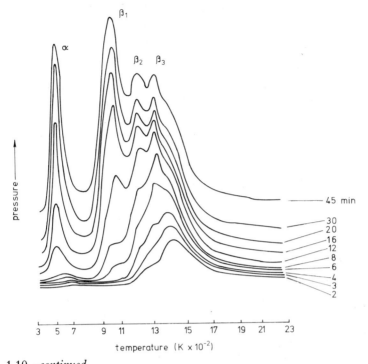

FIG. 1.10—*continued*

c. Carbon monoxide. Adsorption at about 300 K for various times (indicated). The zero levels have been shifted to prevent overlap. After Redhead, P. A. *Trans. Faraday Soc.* **57**, 641 (1961).

The adsorption behaviour of even the simple gases such as hydrogen, oxygen, nitrogen and carbon monoxide on transition metals is complicated by the existence of more than one binding state. These are clearly revealed in flash desorption spectra, examples of which are shown in Fig. 1.10 for hydrogen, nitrogen and carbon monoxide on polycrystalline tungsten. With hydrogen and nitrogen the weakly bound states are probably molecular in nature, adsorbed by charge-transfer interactions, in distinction to the several strongly bound states which are atomic. To some extent at least, the existence of more than one strongly bound state is a consequence of the crystallographic heterogeneity of a polycrystalline adsorbent surface. However, it cannot be entirely attributed to this cause since it is also found using single crystal adsorbents, and some examples are shown in Fig. 1.11. The detection of more than one strongly bound adsorption state on a single crystal metal surface should come as no surprise in view of the multiplicity of ordered adsorption states which have been found by LEED for many adsorbate–adsorbent systems. For instance, with hydrogen adsorption on tungsten (100), it is likely that the β_2 state (cf. Fig. 1.11) is associated with the known c(2 × 2) structure, and the β_1 state with a (1 × 1) structure.[22]

In addition to the effect of crystal face with massive specimens, there are some data for the effect of metal particle size on chemisorption behaviour. Thus, in the infrared absorption spectrum of carbon monoxide adsorbed on transition metals, using very small metal particles produces more intense carbonyl stretching bands at lower frequencies, and this has been interpreted[23] as due to adsorption at corner sites on these crystallites where it is postulated stronger metal–carbon bonding occurs, and this results in a weaker and more polarizable carbonyl group. Small particle effects have also been observed with adsorbed nitrogen[24]. Using silica- and alumina-supported nickel, palladium and platinum, it appears that

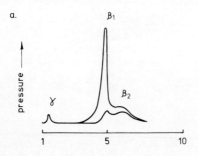

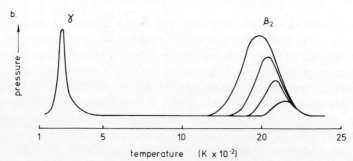

FIG. 1.11 Desorption spectra for gases adsorbed on (100) tungsten surface.
a. Hydrogen. Adsorption at 77 K. Two surface coverages are shown.
b. Nitrogen. Adsorption at 77 K. Several surface coverages are shown. The γ state can be resolved into two separate states, and there is also a β_1 state (not shown) which can be populated by adsorption with temperature cycling between 77 K and 300 K. After Han, H. R. and Schmidt, L. D. *J. Phys. Chem.* **75**, 227 (1971).

metal particles of an average diameter <7 nm are needed to generate adsorbed nitrogen which gives rise to infrared absorption (presumably molecularly adsorbed nitrogen).

A detailed account of the surface chemistry of alloys remains in a rudimentary state. We have already mentioned the inadequacy of a rigid band model and, furthermore, it seems illogical from a chemical point of view to ignore entirely the different identity of the two sorts of atoms in the surface of a binary alloy. Nevertheless, in a few cases ideas based on a rigid band model have had some success in accounting for alloy catalyst activity as a function of composition. The idea is, for instance, that as a group IB element is alloyed with a group VIII element, the valence s electrons from the former fill electron holes in the d band of the latter, and the d band

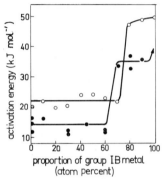

FIG. 1.12 Activation energy for ortho–parahydrogen conversion on Pd–Au alloy wires (–●–) and Ni–Cu alloy foils (–○–) as a function of the proportion of group IB metal. After Couper, A. and Eley, D. D. *Discussions Faraday Soc.* **8**, 172 (1950) and Rienacker, G. and Vormun, G. *Z. Anorg. Allg. Chem.* **283**, 287 (1956).

just becomes completely filled at a particular composition, at which point a substantial change in catalytic behaviour might be expected. Thus in the case of nickel, palladium or platinum with about 0·4–0·6 electron holes in the d band, on alloying with a group IB metal which contributes one s electron per atom, the d band should become filled at a composition of about 40–60 atom % group IB metal. Reactions such as H_2–D_2 exchange or ortho–parahydrogen conversion do show that the catalytic activity as indexed by the activation energy changes rapidly (an increase in activation energy) on increasing the proportion of group IB metal to the vicinity of about 60 atom %. This is illustrated in Fig. 1.12 However, with reactions of greater molecular complexity, although there is often a general decreasing trend in activity as the proportion of group IB metal increases, the identification of a sharp change at the region expected to correspond to filling of the *d* band becomes at best rather more indefinite than is the case with parahydrogen conversion (e.g. nitrous oxide decomposition, methane–deuterium exchange), or else there is no reasonable correlation to be found at all (e.g. ethylene hydrogenation, ethylene oxidation, carbon monoxide oxidation, formic acid decomposition). Rossington[25] and Eley[19] summarize some important examples. With the latter it is clear that the reaction is strongly influenced by the chemical identity of the atoms in the catalyst surface. In this situation one must ask: what is the average composition of the alloy surface relative to the overall composition of the bulk, how are the constituent metal atoms arranged in the surface region, and what are the relative propensities of surface metal atoms or groups of atoms for bonding with adsorbed species? The question of average surface composition of alloys is discussed at some length in later sections; for the present it suffices to say that relative enrichment with the component of lower surface energy is expected under equilibrium conditions, and it frequently is found in practice. In addition, chemisorption specificity with respect to one of the metallic constituents in an alloy can readily lead to an enrichment. The complexities that these factors can introduce in interpreting alloy catalyst activity will be obvious.

The idea that in most cases the catalytic activity of an alloy is dominated by the chemical properties of the surface atoms, is related directly to the question of the detailed arrangement of the constitutent atoms in the catalyst surface, since there is an implication that a particular catalytic or adsorption process may require a specific local arrangement of surface metal atoms for the purpose. A beginning in this direction has recently been made by Dowden[26] following earlier ideas of Kobozev.[44] In an alloy there will exist atom clusters of varying compositions and sizes, and there will be corresponding atom ensembles in the surface. If the alloy is assumed to behave as an ideal solution, the proportions of the various possible ensembles can be readily evaluated statistically. If n is the total number of atoms in a particular type of ensemble, the probability $p_{n,m}$ of finding m atoms of type A in an ensemble is given by

$$p_{n,m} = {}^nC_m \, x^m \, y^{n-m} \qquad (1.7)$$

where x and y are the atom fractions of the alloy components A and B. The probability $P_{n,m}$ of finding m *or more* atoms of A in an n-ensemble is given by summing $p_{n,m}$ over m from m to m = n–1.

The number of atoms in the ensemble can, in principle, take any value, but Dowden suggests the most reasonable values should relate to the surface geometry. Thus, on a f.c.c. (100) surface, the surface atoms lie on a square net which, together with the atoms in the layer immediately beneath, forms an array of square pyramidal interstices: the surface region can thus be considered to be made up from a repetition of five-atom ensembles. Furthermore, each surface atom possesses four nearest neighbours in the surface and four in the layer immediately below, so the surface region could also be described by the repetition of nine-atom ensembles. If the ensembles are limited to units which generate the surface structure by repetition, and are also limited to the pyramidal interstices or to aggregates of these, or to a surface atom and its shell of nearest, next-nearest neighbours, etc., n takes values 5, 9, 14. . . .

Any attempt to use ensemble theory for interpreting the adsorption or catalytic behaviour of alloy surfaces requires a combination of data concerning ensemble concentration (taken proportional to ensemble probability) together with information about the bonding properties of various sorts of ensembles. The latter are, however, largely unknown in any detailed way. Dowden[26] has attempted an interpretation of the dependence of the heat of adsorption of hydrogen on palladium–silver alloys in which the bonding energy of a hydrogen atom to various ensembles is parametized from a d band filling model. The net result is at best a rough semi-quantitative representation of the main features of the experimental data.

3. Metals as Catalysts

The transition metals catalyse a wide range of reactions and the best known of these are summarized in Table 1.5.

Catalysts act by chemisorbing one or more of the reactants, and it is the consequent perturbation of the electronic and geometric structure of the molecule which leads to the catalytic enhancement in reaction rate. In general, therefore, one would expect maximum catalytic activity to be achieved at some intermediate strength of chemisorption—very weak adsorption will imply too little modification to the reactant molecule for its reactivity to be greatly affected, while very strong adsorption will imply the formation of a stable "compound" which will cover most of the catalyst surface. Broadly speaking, the catalytic activity of the transition metals agrees with this idea. The chemical reactivity of the transition metals tends to decrease on moving from left to right across each of the periods; on the other hand, if one moves out of the transition metals into group IB, the ability of the metal for chemisorption by covalent binding falls dramatically because of the filling of the d band orbitals. A compromise in terms of activity is struck at the right hand side of the transition periods, that is with the group VIII metals, and it therefore comes as no surprise that this is where many of the active metal catalysts are found. Figure 1.13 illustrates this optimal activity for the metal-catalysed decomposition

TABLE 1.5 Some reactions catalysed by metals†

Reaction	Metals known to show catalytic activity†	Examples of metals with high activity
hydrogen–deuterium exchange	most transition metals: some non-transition metals >600 K	W, Pt
deuterium-saturated hydrocarbon exchange		
(a) involving σ-bonding to surface	most transition metals	W, Rh
(b) involving π-allyl bonding to surface	Pd, Pt, Ni, Rh, W	Pd
hydrogenation of alkenes	most transition metals, Cu	Rh, Ru, Pd, Pt, Ni
deuterium–alkene exchange	most transition metals, Cu, Au	W, Rh, Pd
double bond shift	most group VIII metals	Pd
hydrogenation of alkynes	most group VIII metals, Cu	Pd
hydrogenation of aromatics	most group VIII metals, W, Ag	Pt, Rh, Ru, W, Ni
deuterium–aromatic exchange	most group VIII metals, W, Cu, Ag	W, Pt
hydrogenolysis of C–C bonds	most transition metals	Os, Ru, Ni
skeletal isomerization of hydrocarbons	Pt, Ir, Pd, Au	Pt
deuterium exchange with NH_3, $-NH_2$, H_2O, $-OH$	Pt, Rh, Pd, Ni, W, Fe, Ag	Pt
hydrogenolysis of C–N bonds	most transition metals, Cu	Ni, Pt, Pd
hydrogenolysis of C–O bonds	most transition metals, Cu	Pt, Pd
hydrogenation of carbonyl group	Pt, Ni, Fe, W, Pd, Au	Pt
hydrogenation of carbon monoxide	most group VIII metals, Cu, Ag	Fe, Co, Ru-(Fischer–Tropsch) Ni-(methanation)

Reaction	Metals
hydrogenation of carbon dioxide	Co, Fe, Ni, Ru
hydrogenation of nitrogen oxides	most platinum group metals
hydrogenation of nitro group	most group VIII metals, Cu
hydrogenation of nitriles	Co, Ni
hydrogenation of nitrogen (ammonia synthesis)	Fe, Ru, Os, Re, Pt, Rh (Mo, W, U, probably as nitrides)
ammonia decomposition	most transition metals, group IB
dehydrogenation, cyclization, aromatization of hydrocarbons	most group VIII metals
decomposition of alcohols	most transition metals, group IB
decomposition of formic acid	most transition metals, group IB
oxidation of hydrogen	*platinum group metals, Au
oxidation of hydrocarbons	
(a) ethylene to ethylene oxide	Ag
(b) oxidation of other hydrocarbons	*platinum group metals, Ag
oxidation of carbon monoxide	*platinum group metals
oxidation of carbon monoxide with steam (water–gas shift)	Cu
oxidation of ammonia	*platinum group metals
oxidation of sulphur dioxide	*platinum group metals, Au
oxidation of alcohols, aldehydes	*platinum group metals, Ag, Au
oxidation of methane with steam	Ni, Co, platinum group metals

	Ru, Ni
	Ru, Pd, Pt
	Pt, Pd, Ni
	Co, Ni
	Fe
	Pt
	Pt
	Cu, Ni
	Pt, Ir
	Pt
	Ag
	Pd, Pt
	Pd, Pt
	Cu
	Pt
	Pt
	Ag, Pt
	Ni, Pt

* Activity of metals outside the platinum group and noble metals is known, but under reaction conditions the metal oxide is the active catalyst.
† The compilation is not exhaustive.

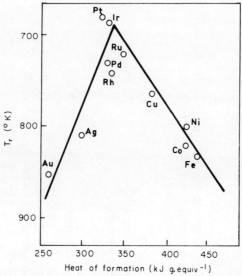

FIG. 1.13 Activity of various metals for the decomposition of formic acid. The activity is expressed by T_r, the temperature for the rate of reaction to equal 0·16 molecules site^{-1} s^{-1}, and this is related to the heat of formation of the corresponding metal formates. After Fahrenfort, J., van Reyen, L. L. and Sachtler, W. M. H. *In* "The Mechanism of Heterogeneous Catalysis" (J. H. deBoer, ed.) Elsevier, Amsterdam (1960), p. 23.

of formic acid. Again, in the hydrogenation of ethylene there is a general trend for catalytic activity to increase as the heat of adsorption of ethylene or hydrogen decreases (Fig. 1.14). However, it is important not to press this idea too far, since it has obvious limitations. In many cases the strength of adsorption of the catalytically important intermediate will bear little or no relation to a directly measured heat of adsorption because there may be a spectrum of adsorbed species of which the catalytically important one may be only a minor component. The general approach also ignores the stereochemical specificity often present in chemisorption and which is of particular importance to catalysis. As pointed out in an earlier paragraph, one is very often left with the construction of empirical correlations between catalytic activity and some metallic parameter. Illustrations of this abound in the catalytic literature: for instance, the correlation between activity for deuterium–ethane exchange and metal–metal bond strength (e.g. ref. 27), or between activity for ethane hydrogenolysis and percentage *d*-character of the metallic bond[28]. The danger with this sort of correlation is that it may tempt one to believe that one understands something about the factors which control catalytic activity at a fundamental level. In fact, it offers no insight of that sort at all, unless the reaction pathway itself is known in detail at the molecular level, and this is seldom the case.

There has long been an interest in the way that catalytic specificity towards some reactions may be dependent on particle size and on the surface structure of the metal; however, to some extent at least these factors are coming to assume the role of controllable variables as better methods become available for the study of the fine detail of catalyst structure.

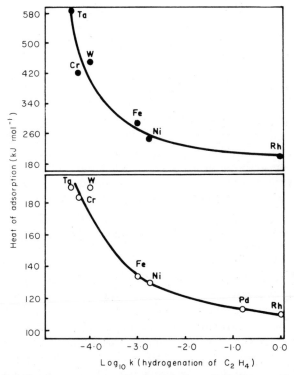

FIG. 1.14 Relation between rate of hydrogenation of ethylene and heat of adsorption of ethylene (●) and hydrogen (○). After Beeck, O. *Discussions Faraday Soc.* **8**, 118 (1950).

In 1925 H. S. Taylor[29] proposed that there are active sites on the catalyst surface, each consisting of some sort of small excrescence containing a relatively small number of metal atoms which have an abnormally low number of nearest neighbours. The other factor influencing the structure of the catalyst surface is the nature of the crystallographic planes exposed.

Despite considerable experimental work the situation still remains imperfectly understood, particularly in relation to Taylor's active sites and the role of surface imperfections. Some examples will illustrate the problems which are met. Shooter and Farnsworth[30] studied hydrogen/deuterium exchange on UHV-clean nickel crystals and found no difference in catalytic activity for ion-bombarded surfaces before and after thermal annealing. It seems safe to conclude with some certainty that this reaction system is uninfluenced by surface topography. On the other hand, Uhara et al.[31] found that on increasing the extent of annealing of cold-worked nickel, the catalytic activity for ortho–para hydrogen conversion decreased. The reason for the conflict between these two sets of results undoubtedly lies with the influence of surface contaminant which Uhara et al. made no serious effort to control.

A similar sort of interpretative dilemma due to contamination problems is presented by the conflicting results reported for the influence of surface defects on single crystal silver surfaces for the decomposition of formic acid. One set of results found a positive correlation[32] while another reported no correlation with surface defects, but a dependence on the extent to which a (111) surface was exposed.[33] The results of Robertson and his collaborators[34] with regard to changes in the catalytic activity of wires for formic acid decomposition as a result of rapid quenching from high temperatures (hopefully freezing in surface defects), or from cold-working, is another example where the effects observed were mostly dominated by adventitious contaminant. It is interesting to see again that provided the wire was UHV-clean, the degree of cold-working was without influence on its catalytic activity.

Although these examples have stressed the undesirability of adventitious surface contamination in experimental catalytic studies it is necessary to put this into perspective. In a laboratory study where the aim is to understand the reaction mechanism, anything less than a detailed knowledge of the composition and structure of the catalyst surface is an unsatisfactory state of affairs. However, practical metallic catalysts seldom have surfaces that are clean in a rigorous sense, and because of this and their configuration, a detailed surface characterization is usually extremely difficult. They are valuable catalysts none the less, but they need to be accepted on a more empirical level than catalysts for which a more detailed characterization is possible.

There are catalytic systems where the influence of surface structure and topography on catalyst activity has been fairly unequivocally demonstrated. We may note the following examples.

On UHV-prepared nickel film catalysts, the nature of the products from the exchange of ethane or propane with deuterium depends on the surface structure in a manner that has been interpreted to indicate that the formation of a carbon–metal double bond was favoured on low index surface planes.[35] In the hydrogenolysis and skeletal isomerization of alkanes on platinum catalysts there is more than one reaction pathway and, depending on the type of reaction, the importance of both low co-ordinated (e.g. corner) surface metal atoms (a concept reminiscent of Taylor's active sites) and low index surface planes could be seen.[36,37] The reaction of benzene with deuterium or hydrogen over nickel catalysts is sensitive to catalyst structure,[38–40] although the results from the various groups of workers have not yet been fully reconciled (cf. ref. 41). It is not our purpose here to explore the chemistry of such reactions in depth, but merely to draw attention to the need to carry out experimental work with catalysts which have been structurally characterized in as much detail as possible.

That some catalytic processes are sensitive to surface structure while others are not should come as no surprise. Reactions vary in the steric requirements which the adsorbed reaction intermediates make on the catalyst, and in general one would expect that the greater the degree of molecular complexity, the greater is likely to be the degree of surface specificity for catalysis.

Adsorbed species other than the specific reaction intermediate can have a pronounced effect on the performance of a catalyst. These adsorbed species can arise in various ways. They may be formed, for example, from the reactant itself. This is frequently found to occur, for instance, in hydrocarbon reactions over transition metals, particularly with those catalysts which are initially prepared in a state of high surface purity: these residues are strongly adsorbed and result from extensive dehydrogenation and fragmentation of the hydrocarbon molecules. Some of the important features have recently been reviewed.[27] Alternatively, material may be deposited on the metal surface deliberately during catalyst preparation, and the best known example is still probably the addition of potassium oxide as a chemical promotor to the surface of iron synthetic ammonia catalysts. Finally, strongly adsorbed material may occur in an adventitious or uncontrolled manner. This is undoubtedly a frequent occurrence with technical metallic catalysts where the method of catalyst preparation has been optimized with respect to performance in an empirical manner.

Although the presence of strongly adsorbed material will obviously reduce the area of virgin metal surface available for reactant adsorption, its prescence is by no means necessarily inimical to catalyst performance. Some of the factors which may operate are as follows. The adsorbed material may alter (usually reduce) the heat of adsorption of the reactant so as to increase its reactivity. This appears to be the case with potassium oxide promotor on iron synthetic ammonia catalysts. Furthermore, an adsorbed promotor may function to suppress self-poisoning by irreversibly adsorbed reactant, so that the concentration of kinetically significant adsorbed intermediate may be increased. Finally, the adsorbed material may function to generate reactant adsorption sites of specific configuration: thus if the reaction mode undergone by adsorbed reactant is sensitive to site geometry, the course of the overall reaction will be modified. There is more than one way in which this can happen. The catalytic reaction may be confined to small groups of surface metal atoms left bare of strongly adsorbed residues; alternatively, strongly held adsorbate *and* surface metal atoms may together function as the reactant adsorption sites. Again, these comments serve to emphasize the need for catalyst characterization if reaction mechanisms are to be understood.

Supports for metallic catalysts are not necessarily inert. Apart from the way in which the support may control the morphology of the supported metal, it may play a direct role in the catalytic reaction itself: that is, the total catalyst consisting of metal plus support may be bifunctional. Undoubtedly the best known examples are reforming catalysts such as platinum/alumina or platinum/zeolite in which the acidic support functions as the seat of carbonium ion isomerization activity, while the platinum provides a hydrogenation/dehydrogenation function. Because of these sorts of possibilities, one should always treat the support as an integral part of a supported metal catalyst. In other words, the characterization of such a catalyst requires the characterization of both the metallic component *and* the support.

References

1. Mott, N. F. and Jones, H. "The Theory of the Properties of Metals and Alloys", Oxford University Press, London (1936).
2. Raimes, S. "The Wave Mechanics of Electrons in Metals", North-Holland, Amsterdam (1961).
3. Wood, J. H. *Phys. Rev.* **117**, 714 (1960).
4. Mott, N. F. *Adv. Phys.* **13**, 325 (1964).
5. Trost, W. R. *Can. J. Chem.* **37**, 460 (1959).
6. Goodenough, J. B. "Magnetism and the Chemical Bond", Interscience, New York (1963).
7. Pauling, L. *Proc. Roy. Soc.* **A196**, 343 (1949).
8. Pauling, L. "The Nature of the Chemical Bond", Cornell University Press, Ithaca (1960).
9. Altmann, S. L., Coulson, C. A. and Hume-Rothery, W. *Proc. Roy. Soc.* **A240**, 145 (1957).
10. Grimley, T. B. *Proc. Phys. Soc.* **90**, 751 (1967).

11. Eley, D. D. *Discussions Faraday Soc.* **8**, 34 (1950).
12. Stevenson, D. P. *J. Chem. Phys.* **23**, 203 (1955).
13. Higuchi, I., Ree, T. and Eyring, H. *J. Amer. Chem. Soc.* **79**, 1330 (1957).
14. Cerny, S. and Ponec, V. *Catal. Rev.* **2**, 249 (1969).
15. Hayward, D. O. *In* "Chemisorption and Reactions on Metallic Films" (J. R. Anderson, ed.), Academic Press, London, (1971), p. 225.
16. Geus, J. W. *In* "Chemisorption and Reactions on Metallic Films" (J. R. Anderson, ed.), Academic Press, London, Vol. 1 (1971), p. 327.
17. Ehrlich, G. *In* "Transactions 8th National Vacuum Symposium and 2nd International Congress" (L. E. Preurs, ed.), Pergamon, Oxford (1961), p. 126.
18. Hagstrum, H. D. and Becker, G. E. *Phys. Rev. Letters* **22**, 1054 (1969).
19. Eley, D. D. *J. Res. Inst. Catal. Hokkaido Univ.* **16**, 101 (1968).
20. Little, L. H. "Infrared Spectra of Adsorbed Species", Academic Press, London (1966).
21. Little, L. H. *In* "Chemisorption and Reactions on Metallic Films" (J. R. Anderson, ed.), Academic Press, London, Vol. 1 (1971), p. 489.
22. Han, H. R. and Schmidt, L. D. *J. Phys. Chem.* **75**, 227 (1971).
23. Blyholder, G. *J. Phys. Chem.* **68**, 2772 (1964).
24. van Hardeveld, R. and van Montfoort, A. *Surface Sci.* **4**, 396 (1966).
25. Rossington, D. R. *In* "Chemisorption and Reactions on Metallic Films" (J. R. Anderson, ed.), Academic Press, London, Vol. 2 (1971), p. 211.
26. Dowden, D. A. *In* "Proceedings 5th International Congress on Catalysis" (J. W. Hightower, ed.), North-Holland, Amsterdam, (1973), p. 621.
27. Anderson, J. R. and Baker, B. G. *In* "Chemisorption and Reactions on Metallic Films" (J. R. Anderson, ed.), Academic Press, London, Vol. 2 (1971), p. 63.
28. Sinfelt, J. H. *Advances in Catalysis* **23**, 91 (1973).
29. Taylor, H. S. *Proc. Roy. Soc.* **A108**, 105 (1925).
30. Shooter, D. and Farnsworth, H. E. *J. Phys. Chem. Solids* **21**, 219 (1961).
31. Uhara, I., Hikino, T., Numata, Y., Hamada, H. and Kageyama, Y. *J. Phys. Chem.* **66**, 1374 (1962).
32. Sosnovsky, H. M. C., Ogilvie, G. J. and Gillman, E. *Nature* **182**, 523 (1958); Sosnovsky, H. M. C. *J. Phys. Chem. Solids* **10**, 304 (1959).
33. Bagg, J., Jaeger, H. and Sanders, J. V. *J. Catal.* **2**, 449 (1963); Jaeger, H. *J. Catal.* **9**, 237 (1967).
34. Duell, M. J. and Robertson, A. J. B. *Trans. Faraday Soc.* **57**, 1416 (1961); Willhoft, E. M. A. and Robertson, A. J. B. *J. Catal.* **9**, 348 (1967); Willhoft, E. M. A. *Chem. Comm.* **1968**, 146.
35. Anderson, J. R. and Macdonald, R. J. *J. Catal.* **13**, 345 (1969).
36. Anderson, J. R., Macdonald, R. J. and Shimoyama, Y. *J. Catal.* **20**, 147 (1971); Anderson, J. R. and Shimoyama, Y. *In* "Proceedings 5th International Congress on Catalysis" (J. W. Hightower, ed.), North-Holland, Amsterdam (1973), p. 695.
37. Anderson, J. R. *Advances in Catalysis* **23**, 1 (1973).

38. Crawford, E. and Kemball, C. *Trans. Faraday Soc.* **58**, 2452 (1963).
39. van Hardeveld, R. and Hartog, F. *In* "Fourth International Congress on Catalysis", Moscow, 1968, Paper No. 70.
40. Coenen, J. W. E., van Meerten, R. Z. C. and Rijnten, H. Th. *In* "Proceedings 5th International Congress on Catalysis" (J. W. Hightower, ed.), North-Holland, Amsterdam (1973), p. 671.
41. Moyes, R. B. and Wells, P. B. *Advances in Catalysis* **23**, 121 (1973).
42. Fassaert, D. J. M., Verbeek, H. and van der Avoird, A. *Surface Sci.* **29**, 501 (1972).
43. Bond, G. C. *Discussions Faraday Soc.* **41**, 200 (1966).
44. Kobozev, N. I. *Acta Physicochim. URSS* **9**, 805 (1938).
45. Horiuti, J. and Toya, T. *In* "Solid State Surface Science" (M. Green, ed.), Dekker, New York, Vol. 1 (1969), p. 1.

CHAPTER 2

Support Materials

	page
1. SILICA	39
High area silica	39
Kieselguhr	44
Low area silica	44
2. GLASSES	45
3. ALUMINA	46
Active alumina	46
Low area alumina	54
4. CHROMIA	54
5. TITANIA	56
6. ZIRCONIA	61
7. MAGNESIA	63
8. THORIA	65
9. METAL SULPHATES, PHOSPHATES, CHLORIDES AND CARBONATES	67
10. SILICA–ALUMINA AND OTHER MIXED OXIDES	68
11. ZEOLITES AND NATURAL CLAYS	74
12. CARBON	81
13. MISCELLANEOUS SUPPORTS	86
Silicon carbide, mullite, zircon and calcium aluminate	87
Asbestos	87
Monolithic porous support	87
Inorganic precipitates	89
Polymers and resins	89
14. SINGLE CRYSTAL SURFACES	89
Mica	90
Alkali halides	92
Miscellaneous	93

There is a very wide range of conditions under which supported metallic catalysts are used, and as a consequence there is a wide range of possible support materials. Many of these are particulate or granular, although fibrous materials are also used, and monolithic porous ceramic has recently become important. Depending on the nature of the material, the specific surface area (surface area per unit mass) and the pore structure can vary enormously.

Earlier general accounts of catalyst supports are given by Berkman et al.[101] and by Innes.[102]

Whatever the actual chemical composition of the support material, the fact is that the surface atoms must have a markedly different environment to those in the bulk. The termination of the three dimensional atomic arrangement at the surface is achieved at an energetic penalty, and this penalty can be minimized in various ways, including a modification or a rearrangement of the positions of the surface atoms, or the bonding of foreign atoms or groups at the surface. For substances which are predominantly ionic and which are of simple crystal structure such as the alkali halides or oxides of the alkaline earth metals, there is both experimental and theoretical evidence that, in the absence of adsorbed foreign material, the lattices terminate without major change of the ionic positions, at least if attention is confined to the low index surface planes. It seems clear that the only change of any significance is a small alteration in the spacing between the first and second ionic layers, and this does not exceed 3–4%. The data are summarized by Benson and Yun.[103] It also appears likely that with this type of substance there is no departure of the average lattice parameter from the value characteristic of the bulk material for particles down to a diameter of only a few nanometres, except for this quite small contribution from the outermost layer.

With substances of more complex crystal structure or of lower ionicity, ordered surfaces which depart from the ideal surface structure have been detected by low energy electron diffraction (LEED). Thus we may note the modification of an α-alumina hexagonal (001) surface consequent upon heating in vacuum above 1170 K,[104] and the extensively rearranged surfaces of covalent materials such as diamond, silicon and germanium,[105, 106] and the III–V semi-conductors.[107, 108] Substances of intermediate bonding type such as II–VI wide-band semiconductors vary in their behaviour: for instance the hexagonal (001) surfaces of zinc oxide and zinc sulphide are ideal but those of cadmium sulphide and cadmium selenide are rearranged.[109] The tendency is for such rearrangements to minimize the number of dangling orbitals, and probably also involves a rehybridization of the orbitals of the surface atoms. The basal plane of graphite is ideal, and this plane has no dangling orbitals except at the edges. In addition, it is possible for a rearranged ordered surface to be formed when the new surface structure corresponds to that for an ordered defect phase which is known in the bulk. An example is the electron beam conversion of a (010) surface of V_2O_5 to an oxygen deficient structure which corresponds to the surface of $V_{12}O_{26}$.[110]

However, if a surface is modified without the retention of long range two-dimensional order, structural information cannot be obtained by LEED, and indeed, experimental evidence is quite difficult to obtain: one has to rely on inferential chemical evidence together with information available via infrared spectrophotometry and to a more limited extent n.m.r., and e.s.r. if there are unpaired electrons present in surface traps or if paramagnetic molecular species are present. This is the situation to be faced with most materials we are interested in as catalyst supports which are not present in anything approaching single crystal form.

A pristine oxide surface is usually of relatively high chemical reactivity and, in particular, the dissociative adsorption of molecular water is virtually a universal process. The importance of this lies in the common occurrence of water in the environment in which the oxide is likely to be prepared and handled, as well as in the modified chemical properties which the adsorbed water confers. On an oxide surface for which an ionic model is adequate, this adsorption can be written

$$O^{2-}_{(s)} + H_2O \rightarrow 2OH^-_{(s)} \tag{2.1}$$

where we have at this stage avoided specifying stereochemical details of the surface species. If a covalent model is more appropriate the corresponding reaction is better written after the style

$$M_2O_{(s)} + H_2O \rightarrow 2M(OH)_{(s)} \tag{2.2}$$

Much of the surface chemistry of oxides is dominated by the extent to which reactions such as 2.1 or 2.2 occur and the associated stereochemical details, and the reactivity of the surface hydroxyl groups and aquo groups towards various reagents, including their Brønsted acid/base behaviour, together with the extent to which metal atoms are exposed at the surface so as to generate Lewis acid sites.

The chemistry of an oxide surface in contact with an aqueous environment is thus quite complex. First of all, one can reasonably expect hydroxyl and aquo groups at an oxide surface to undergo many of the reactions known for these groups in normal hydroxyaquo metal ions. The hydroxyaquo ion $[M(OH)_x(H_2O)_y]^{(n-x)+}_{(aq)}$ is, in principle, amphoteric and may function as a proton donor or a proton acceptor according to reactions of the type (n is the metal oxidation number),

$$[M(OH)_x(H_2O)_y]^{(n-x)+}_{(aq)} \rightleftharpoons [M(OH)_{x+1}(H_2O)_{y-1}]^{(n-x-1)+}_{(aq)} + H^+_{(aq)} \tag{2.3}$$

and similar behaviour is possible if the metal ion exists in an oxide surface.

However, there is also the possibility that a surface hydroxyl can itself function as a proton donor,

$$S^+OH^-_{(s)} \rightleftarrows S^+O^=_{(s)} + H^+_{(aq)} \qquad (2.4)$$

where S^+ represents a surface site carrying a net positive charge. The surface may also function as a donor of a hydroxyl,

$$S^+OH^-_{(s)} + H^+_{(aq)} \rightleftarrows S^+_{(s)} + H_2O \qquad (2.5)$$

where $S^+_{(s)}$ is probably hydrated rather than being co-ordinatively unsaturated, although this is not shown explicitly.

The hydrated surface of an oxide exhibits ion exchange properties[13, 14] and there is a correlation between ion exchange ability and the net surface charge carried by the oxide. The latter is pH dependent via the ability of the surface to function as a Brønsted acid (leaving a negative surface charge) or as a Brønsted base (leaving a positive surface charge) (cf. reactions 2.4 and 2.5). The oxides are cation exchangers in an environment which generates a negative surface charge, and are anion exchangers in an environment which generates a positive surface charge. As an index of the propensity of a surface to become either positively or negatively charged as a function of pH, one may note the pH required to give zero net surface charge, and values for this quantity for various materials are indicated in Table 2.1.

TABLE 2.1 Values for pH of zero net surface charge for some oxides in aqueous environment

Material	pH of zero net surface charge
silica	~2*
alumina	~9
chromia	~7
titania	~5
zirconia	4–7
magnesia	~12

* Charge density very low at pH < 6.

Taken as a whole, ion adsorption at an oxide surface is complicated because there are a variety of possible processes. We deal first with ion exchange. There are two ways in which cation exchange can occur. If the oxide surface is only protonated, that is if it contains no adsorbed heterometal ions, then adsorption may be brought about by treatment with a solution containing the metal ion in question at a sufficiently high pH so that the surface exchanges hydrogen ions for metal ions. In practice, this is a very important reaction in providing for metal ion adsorption, and an archetypal reaction may be written

$$nS^+OH^-_{(s)} + M^{n+}_{(aq)} + nOH^-_{(aq)} \rightleftharpoons [S^+O^=]_n M^{n+}_{(s)} + nH_2O \qquad (2.6)$$

The pH dependence of the position of equilibrium is obvious, and the reverse (desorption) reaction may be formally regarded as a hydrolysis reaction. However, raising the pH to drive reaction 2.6 to the right requires care that the pH and the metal ion concentration are not sufficiently high to precipitate metal hydroxide. With some metals this problem may be overcome by using aqueous ammonia to make the solution alkaline, in which case metal ammine ions are formed and it is these rather than aquo cations which are adsorbed. The second method exchanges the surface with (say) alkali metal or alkaline earth metal ions, perhaps at relatively high pH, then following washing to remove occluded solution, the alkali or alkaline earth ions are exchanged for another metal, and this may be done at, for instance, near neutral pH.

In addition to adsorption in the manner just described, a complex ion can become bound to a surface by a ligand exchange process in which, for example, a surface oxygen enters the co-ordination sphere of the metal by replacing a ligand originally present.

Some data obtained by Burwell *et al.*[175] for exchange of cobalt complexes onto a sodium-exchanged silica gel illustrate some of these features. It was shown that doubly charged ions were bound less strongly than triply charged ions ($[Co(en)_2(NH_3)Cl]^{2+}$ versus $[Co(en)_3]^{3+}$); an anion was not adsorbed at all ($[Co(NH_3)_2(NO_2)_4]^-$), and an uncharged complex was only very slightly adsorbed ($[Co(NH_3)_3(NO_2)_3]^0$). Adsorbed metal cations such as copper (II) or nickel (II) showed roughly normal reactivities such as the replacement of aquo groups by other ligands. Examples of ligand exchange involving a group on the silica gel surface were also observed. Thus it was found that a surface group, probably in an ionized form as a siloxy anion, could enter the co-ordination sphere of some cobalt complexes by replacement of an original aquo or chloro ligand to give strong ion adsorption. It is also likely that the siloxy anion may enter the co-ordination sphere of aquo cations such as copper (II) and nickel (II) by ligand replacement, but in these cases the interaction is weaker.

However, the simple features outlined above are not adequate as a complete description of ion adsorption. There are at least two other important factors. The first is that the adsorption is strongly influenced by hydrolytic equilibria involving the adsorbing cation, and the second is that the solubility of the solid oxide can often not be ignored.

If we refer to metal cation adsorption onto an oxide surface containing no heterometal ions, it is found that for the adsorption of simple aquated metal cations onto a given oxide surface, the more readily hydrolysed is the aquocation, that is the greater the tendency for the reaction 2.3 to lie to the right, the lower will be the pH at which this cation will be adsorbed. This trend is illustrated in Fig. 2.1 which shows the pH dependence of the adsorption of some metals from aqueous solution onto silica gel.[9] To quantify the relation between the pH required for cation adsorption and aquocation hydrolysis, we may characterize each adsorption curve such as appear in Fig. 2.1 by the pH for 50% of maximum adsorption (pH*), and use as an index of the ease of hydrolysis the equilbrium constant (K*) for the first hydrolysis step

$$[M(H_2O)_y]^{n+}_{(aq)} \rightleftarrows [M(OH)(H_2O)_{y-1}]^{(n-1)+}_{(aq)} + H^+_{(aq)} \qquad (2.7)$$

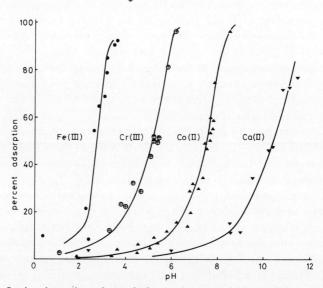

FIG. 2.1 Ionic adsorption of metals from aqueous solutions of the metal nitrates onto the surface of silica gel as a function of pH. Temperature 298 K. Aqueous solutions $1 - 2 \times 10^{-4}$ mol dm^{-3}. After James, R. O. Ph.D. Thesis, University of Melbourne, Australia (1971).

and define $pK^* = -\log_{10} K^*$. Figure 2.2 shows pH* as a function of pK* for various metals with a silica gel surface (data from a literature summary in reference 9). This sort of behaviour has been interpreted[9, 176] as being due to preferential adsorption of the partly hydrolysed cation, perhaps due to a reduction in the secondary solvation energy which facilitates close approach to the interface.

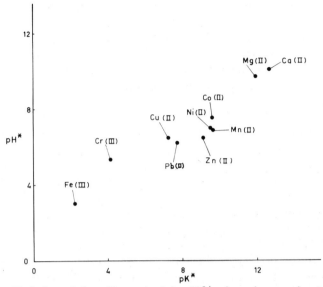

FIG. 2.2 Variation of the pH required for 50% of maximum adsorption (pH*) of metal ions from aqueous solution onto silica gel, as a function of the ease of hydrolysis of the aquometal cation, the latter being expressed by pK* in terms of the equilibrium constant (K*) for the first hydrolysis step. Temperature 298 K. Aqueous solutions $10^{-3} — 10^{-4}$ mol dm^{-3}. Literature data summarized by James, R. O. Ph.D. Thesis, University of Melbourne, Australia (1971).

The solubility of the solid oxide is usually appreciable. Thus at 298 K the solubility of silica in an aqueous medium at pH < 8 is about 2×10^{-3} mol dm^{-3} with the solute being monomeric silicic acid: above a pH value of 8 the solubility rises rapidly and polymeric solute species are formed, and at a pH of 11, for instance, the solubility is in the region of 5×10^{-2} mol dm^{-3}. With γ-alumina the solubility at a pH of 7 is low, about 10^{-12} mol dm^{-3}, but at pH of 4 the solubility is 10^{-3} mol dm^{-3}, while at pH of 11 it is 3×10^{-4} mol dm^{-3}. As Sacconi has pointed out,[176] this pH-dependent solubility of the oxide is a process which couples with the pH-dependent hydrolysis of the adsorbing ion. In the case of alumina, for instance, some aluminium ions are displaced into solution by the hydrogen ions which are generated by the hydrolysis reaction, and the aluminium ions are subsequently readsorbed together with the other adsorbing ions, the net effect being a sort of buffering action together with a mixed adsorption process. This mixed adsorption may well contribute towards the formation of some metal aluminate or metal silicate at the oxide surface.

These ion adsorption phenomena are important in understanding some of the processes which are used for the introduction of metal onto a support, and further reference will be made to this in Chapter 4.

Some recent evidence[152] indicates that, although single crystal oxide surfaces (Al_2O_3, MgO, TiO_2, ZnO) closely parallel their amorphous counterparts in terms of both the infrared spectra and heats of adsorption of adsorbed water, they appear to differ intrinsically with regard to the concentration of isolated surface hydroxyls, since this is considerably lower on single crystal surfaces.

A pristine carbon surface is also highly reactive, except for the basal plane of graphite, and the existence of chemically bound oxygen has been known for many years. In fact, a wide and complex spectrum of functional groups can exist on the surface of carbon, and all these groups serve to terminate the three-dimensional covalent structure. Certain of these groups also confer on the carbon surface Brønsted acidity and basicity when in contact with an aqueous environment. In general, the surface acidity falls and the basicity rises with increasing temperature of thermal treatment of the carbon in the range 670–1270 K. There are certainly several types of acidic and basic groups present. This complicated and rather confused situation has been reviewed by Garten and Weiss[174] and by Boehm.[10] It appears that Brønsted acidity is due to phenolic groups, lactone groups and carboxylic acid groups, while basicity is due to a chromene-like structure which functions as a base in the presence of oxygen (as is observed). These acidic and basic functions imply cationic and anionic exchange properties.

In discussing the surface chemistry of materials used as catalyst supports, we are not necessarily concerned with all the ramifications of the chemistry of all possible surface groups. Nevertheless, some of the main features are outlined in the subsequent sections devoted to individual materials.

Virtually any solid can, in principle, be used as a catalyst support for dispersed metal. However, in practice the range of materials in reasonably common use is restricted, and there are a few such as silica, alumina, silica-alumina, the zeolites and carbon which overshadow the rest. The latter are also the supports which have chiefly been used when detailed structural studies of supported catalysts have been undertaken. We have therefore been selective in this chapter, and most detail is given for those supports which are most commonly met with.

Except in special cases such as asbestos, most support materials can be obtained in physical forms ranging from powders, through granules to larger aggregates of irregular or regular shapes. The conversion of a substance to a particle size below that in which it was initially prepared is seldom a problem, and the usual methods of comminution and size grading are well known. However, the formation of a substance into a form in which the units are larger than those initially produced may be more difficult if the resulting aggregate is to have substantial mechanical integrity. Purely for laboratory use, it is possible to cold-press a finely divided material such as microspheroidal silica and then to crush the pressed cake to give lumps or grains which can be handled conveniently. However, these lumps or grains would not have sufficient mechanical strength to allow for their use in a technical or process situation. The latter will, in general, require that aggregation be done by sintering or by fusion. The use of a binder or a flux may help, but at the risk of alteration in properties if there is an appreciable change in chemical composition due to the added material. A common technique is to form the powder into a paste, using a liquid in which the powder material is slightly soluble. After shape-forming by, for instance extrusion or pellet formation, the dissolved material is deposited between the grains on drying and acts as a binder. An example is the formation of a paste from high area alumina powder with dilute aqueous acetic acid solution. On calcination of the extrudate or pellets, the binder is converted to alumina. Attack on the internal pore structure of the alumina by the acetic acid solution is minimized by first filling the pores with water, and by working as quickly as possible.

The process of particle aggregation may well generate an interparticular pore structure which will be additional to any intraparticular pore structure retained by the individual particles. These interparticular pores will mostly be considerably larger than any intraparticular pores. The pore structure is important because the pores offer voids in which most of the metal in a supported catalyst is deposited, and because they are required to provide for reactant and product transport during catalysis.

Within the scope of the present book, it is impossible to provide structural and specification details for the very large number of support materials which are commercially available. Details are readily available in the manufacturers' technical and catalogue literature which is available on request.

Some typical physical forms of various supports are illustrated in Fig. 2.3.

1. Silica*

High Area Silica

High area silica (e.g. silica gel) is amorphous. The polymerization of silicic acid, which proceeds by the condensation of silanol groups

---

* Well-known suppliers of various grades include: W. R. Grace, Davison Chemical Divn.; Johns-Manville, Celite Divn.; Girdler-Sudchemie; Air Products, Houdry Divn.; General Refractories, Dicalite Divn; Cabot Corpn.; Degussa.

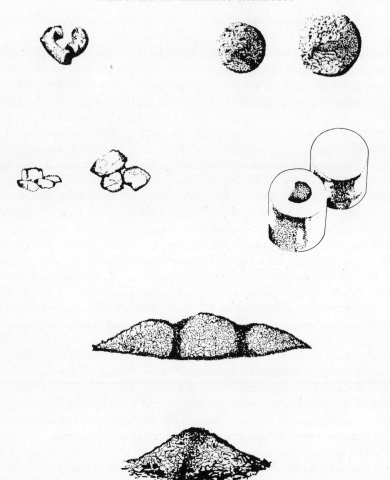

FIG. 2.3   Physical forms of some typical commercial catalyst supports, including extrudate, spheres, granules, cylinders, powder and flakes.

[–Si–OH] (some of which may be in the ionized form) to form siloxane groups (–Si–O–Si–) when carried far enough, yields a silica hydrosol. The sol particles are approximately spherical and Fig. 2.4 is a somewhat idealized representation. The particles consist mainly of a non-ordered arrangement of SiO_4 tetrahedra.

The general morphology (but not the detailed molecular architecture) of silica hydrosol and silica hydrogel is typical for many hydrous oxides which are precipitated in an amorphous condition (even though some subsequently recrystallize to a greater or lesser extent), and the details which are given for the silica system are intended to provide a general descriptive background for other systems described subsequently.

Primary sol particles may themselves aggregate into chains and networks, and the stability of these strongly suggests that siloxane linkages are formed between the particles. This process uses silanol groups at the surface of the particles. Again, interaction between particles no doubt also occurs

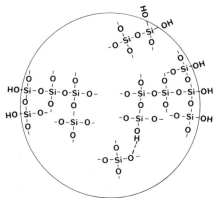

FIG. 2.4 Schematic representation of structure of particle in colloidal silica or silica gel fully hydrated at the surface.

via hydrogen bonds, but this is probably of lesser importance, except in the initial stages of aggregation. When this aggregation proceeds so that the network extends throughout the volume of the specimen, the result is a gel. Alternatively, if the sol is initially caused to flocculate by, for instance, neutralizing the surface charge on the sol particles, the flocculate consists of aggregates of sol particles between which linkages are then formed. The essential difference between these two products is that in a gel the network of linked particles is quasi-infinite in extent, while in a flocculate it is limited to (say) a few hundred nanometres: furthermore in a flocculate the particles are usually more closely packed and the pores correspondingly smaller (cf. Fig. 2.5).

The average diameter of the primary particles in silica gel has been estimated by various methods, including X-ray scattering, and a typical figure of 3–6 nm has been quoted.[1] However, there is no doubt that the average size of the primary particles can vary widely depending on conditions of preparation. The likely range 3–30 nm is probably conservative, but corresponds to a commonly observed range of specific surface areas of 100–1000 m^2g^{-1} (if each particle has the density of bulk vitreous silica and this is a reasonably accurate assumption).[2]

As indicated by Fig. 2.5 the gel has a highly porous structure. The pore

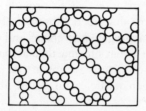 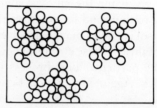

FIG. 2.5 Schematic representation of structure of silica gel (left) and silica colloid flocculate (right). Reproduced with permission from Iler, R. K. "The Colloidal Chemistry of Silica and Silicates" Cornell University Press, Ithaca (1955).

structure (e.g. the average pore diameter) is much dependent on the method of preparation and dehydration.* If the water is removed directly (after removing electrolyte) the product ("xerogel") always suffers from pore shrinkage in the process, with a consequent reduction in the internal surface. This is irreversible, and is due mainly to the mechanical force of retreating water menisci in the pores, and the effect is therefore minimized by either replacing the water with another liquid of lower surface tension before drying, or by removing water vapour at a temperature above its critical point ("aerogel").

The physical characteristics of the gel such as specific surface area and average pore diameter can be controlled by the detailed method of preparation, and Iler provides a useful summary.[2] However, to give a feeling for the magnitude of the variables, common xerogels have specific surface areas in the range 200–800 m^{2}g^{-1}, with average pore diameters in the range 7–2 nm. Common aerogels range from 500 to 800 m^{2}g^{-1}, with pores of 5–2 nm. Both types of gel are reasonably stable to heat up to 770 K, but some loss in internal surface area occurs above this. Typical figures are that on heating from 770 to 1270 K, a xerogel suffers an area reduction by a factor of about three, while an aerogel suffers an area reduction of about 20%.

In addition to these forms of porous silica, non-porous finely divided silica powder is also important as a catalyst support for experimental work. This material (e.g. "Cabosil", "Aerosil") is made by the flame hydrolysis of silicon tetrachloride, and has a particle size in the region of 40–5 nm diameter, with specific surface areas in the range 50–400 m^{2}g^{-1}. It has the advantage of high purity.

It is now reasonably well established[3,4] that by dehydration of a hydrated gel at about 390 K only physically adsorbed water is removed, while

* Where appropriate we shall use the classification of pores: macropore, width greater than about 50 nm; micropore, width not greater than about 2 nm; mesopore, intermediate widths.

heating above this temperature results in the progressive loss of chemically bound hydroxyl groups. The latter are present both on the surface and within the elementary particles of a freshly prepared gel. The relative proportion of surface hydroxyl increases with increasing dehydration temperature, but the internal component is still significant up to at least 870 K.[5]

A well hydrated surface of a structurally stable gel prepared by dehydration at 420–470 K has a concentration of chemically bound hydroxyl[6–8] of about 5 OH per nm². Nevertheless, there are also surface siloxane groups

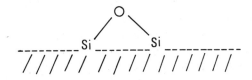

their concentration reaching[10] possibly about 2 per nm². Some of these siloxane groups are probably highly strained and can react with ammonia to yield adsorbed NH_2 and OH groups[94] by a heterolytic process which is quite analogous to that which is undergone by water.

A surface hydroxyl concentration of 5 OH per nm² agrees well with the value expected for the area per hydroxyl (0·218 nm²) assuming one hydroxyl per surface silicon atom at the octahedral surface of β-cristobalite or the basal and prism faces of β-tridymite. Nevertheless, hydration under vigorous conditions (e.g. hot aqueous media, particularly hydrothermal) readily leads to higher concentrations of bound hydroxyl. Much of this increase is probably due to the formation of internal hydroxyl groups. However, there is also the possibility of the formation of gem disilanol groups $\left[\begin{array}{c} \diagdown \quad \diagup OH \\ Si \\ \diagup \quad \diagdown OH \end{array} \right]$, a model particularly emphasized by Vleeskens,[11] and more recently by Peri and Hensley.[8]

Surface groups of the type ($>$ Si $=$ O) which might be formed by the elimination of water within a gem disilanol group, are improbable.[8] The result of dehydration is that after heating to 770 K the concentration of surface silanol groups is no more than about 20–30% of the saturation value, and after heating to 1270 K is 10–15%.[10, 12] This process is rapidly reversible by rehydration with water or water vapour, provided heating does not exceed about 770 K. After heating above 770 K, rehydration occurs only slowly in liquid water, but more rapidly in the presence of strong base, or under hydrothermal conditions. The residual OH groups on the surface of a dried silica are relatively immobile even at 1070 K, and there is evidence that they tend to occur in pairs as vicinal OH.[94]

The silanol groups at the surface of a silica particle are weakly acidic. No Lewis acid behaviour is displayed (provided the silica is pure). However, even quite low levels of impurity may be of considerable significance; witness for instance, the presence of Lewis sites on porous Vycor glass[15] which may well be associated with residual aluminium. Although its high surface area makes the gel valuable for use as a catalyst support, pure silica itself is relatively very inert as a catalyst. It does possess very weak activity for alcohol dehydration,[123] but for most purposes this is of negligible importance, and may be due to Al^{3+} impurity. Nevertheless, irradiation of silica either with γ-radiation alone or in a pile does confer upon it a degree of catalytic activity. The various centres produced by irradiation and their reactivity have been summarized by Taylor.[99] For the present purpose we may merely note that the colour centre, which is probably a positive hole trapped at an oxide ion vacancy adjacent to Al^{3+} (the latter present as impurity), is probably responsible for hydrogen chemisorption and for catalysis of H_2/D_2 exchange. Provided the silica is adequately outgassed, irradiation also generates acidic sites which can catalyse reactions such as double bond isomerization in olefins, and olefin polymerization.

Kieselguhr

Kieselguhr is a naturally occurring silica (70–90 % SiO_2, generally 80–90 %) which still finds extensive use as a support for metallic catalysts. Most grades are powders, but aggregated granules are also available. Various grades are available but specific surface areas and average pore diameters fall into the following ranges: natural kieselguhr: 15–40 m^2g^{-1}, 0·7–0·2 μm; calcined kieselguhr: 2–6 m^2g^{-1}, 5–2 μm; with the larger specific surface areas being associated with smaller average pore sizes. Natural kieselguhr also contains some pores of average diameter <10 nm, and these are eliminated on calcination. Kieselguhr's main attraction is that it offers a support with reasonable thermal and chemical stability which combines a moderate specific surface area with relatively large pores which facilitate reactant transport. Data on available kieselguhrs have been summarized.[145]

Low Area Silica

This is available either as powdered silica glass in which individual grains have no internal porosity at all, or alternatively, as highly sintered silica powder compacts in which each pellet retains some internal porosity. With the latter, the silica contents of commercial samples are typically around 95%, the most important other components being alumina and oxides of the metals magnesium, calcium and sodium. The specific surface areas lie in the range $0 \cdot 1$–$0 \cdot 6$ m^{2}g^{-1}, with average pore diameters in the range 40–2 μm.

2. Glasses

Silica glass and the various sorts of borosilicate, soda and lead glasses, are of no importance as supports for commercial catalysts. Nevertheless, these glasses are useful as catalyst supports in laboratory studies, either in a massive form as a substrate for evaporated metal film catalysts, or occasionally in the form of crushed powders as supports for dispersed metals.

A freshly fire-polished glass surface has a very high degree of smoothness on a molecular scale, but it is subject to corrosion by aqueous media or water vapour, and when this occurs the smoothness is destroyed and a porous surface is generated. This phenomenon has been known for many years.[140-142] If the corrosion is not too severe, as may occur with water or limited exposure to dilute alkali solution, the surface pores are of average diameter $>3 \cdot 5$ nm, and the increase in surface area is not greater than a factor of about 3–4. However, alkali solution can lead to more severe attack and to the generation of pores of average diameter $<3 \cdot 5$ nm. These porous surface structures are destroyed by heating close to the glass softening point (e.g. fire polishing), but they are not destroyed at the temperatures used for bake-out in normal vacuum technique (620–720 K).

The surface produced by fracture of a glass is not flat.[143,144] The surface consists of stippled areas together with regions of relative flatness, while on a more macroscopic scale the surface contains undulations typical of a concoidal fracture. The stippled regions are also associated with sub-surface cracks. This then is the structure expected for the surface of the particles in a crushed glass powder.

As would be expected, a glass surface becomes hydroxylated in contact with water. However, a pristine surface produced by fracture or crushing under UHV conditions cannot have surface hydroxyl groups to terminate the three-dimensional structure. In the case of silica, there are data[111] to indicate the presence of some dangling orbitals. These dangling orbitals belong to silicon atoms and are partly paramagnetic and neutral, and partly diamagnetic and charged. However, they amount in number to no more than about 1% of the maximum number which could exist in the surface, and presumably some surface reconstruction occurs to eliminate most of them.

3. Alumina*

Active Alumina

The structural chemistry of alumina is complicated by the existence of a number of different phases, and by the effects of various methods of preparation. When a precipitate is formed from an aluminium solution, the nature of the precipitate, that is whether it is an amorphous hydrogel or is crystalline, and if crystalline of what structure, is determined by the precipitation conditions such as speed, temperature and pH. However, even if the initial precipitate is a hydrogel, an amorphous structure is never retained (except transiently) on dehydration. Conversion to a crystalline form will also occur on standing in an aqueous environment. For instance, Lippens[16] describes a typical gradual conversion of an amorphous alumina hydrogel, via gelationous boehmite to bayerite, on standing at 300 K in contact with an aqueous ammoniacal solution at pH 9. Such ageing is also accompanied by a change in morphology in which the spherical particles (diameter typically 2–5 nm) are converted into fibrils of length about 10 nm.

In general, one is left with the use of essentially empirical recipes for the preparation of the various hydrated aluminas which themselves are used as precursors for dehydrated alumina. However, these processes are not easy to control, and details given in the literature are sometimes contradictory. If a hydrate of a specific structure is desired, its characterization after preparation is desirable. Some recipes are given by de Boer et al.[18] In particular this reference gives details of preparation of gibbsite and bayerite samples (both of composition $Al(OH)_3$) with varying dehydration reactivities. The literature contains numerous other recipes for preparing various dehydrated aluminas, for instance references 17, 19 and 20.

A variety of dehydration products have been characterized, depending on starting materials, reaction conditions and ultimate impurity contents (both foreign metal ions and residual water). The situation is complicated because a number of these products are not thermodynamically distinct and stable phases, and because many of the structures are closely related. Under all conditions, α-alumina (corundum) is the high temperature (1470 K) end product of dehydration. All of the others which result from dehydration at lower temperatures approximate to a greater or lesser extent to a model in which the oxygens are arranged as in spinel, but in

---

* Well-known suppliers of various grades include: Alcoa; Air Products, Houdry Divn.; Girdler-Sudchemie; Kaiser Chemicals; Pechiney; Harshaw; Degussa.

which not all the cation positions are occupied, and in which there are varying degrees of cation disorder. An order–disorder model was adopted by Ervin[21] in agreement with the conclusions of Leonard.[22] There are, however, two other factors which need recognition: one is distortion away from the cubic symmetry of the spinel structure; the second is stacking disorder in the arrangement of the close-packed oxygen layers. In general terms, the dehydration products fall into two main categories depending on the dehydration temperature, and the structure of the important products may be summarized as follows.[23] At low temperatures ($\not> 720$ K) the products are γ-, η- and χ-alumina. Both γ- and η- are tetragonally distorted from cubic symmetry to an extent that decreases with decreasing residual water content. Both have cation disorder, particularly on the tetrahedral sites, and in addition η- has stacking disorder. χ- is rhombohedrally distorted and with substantial stacking disorder. Both θ- and κ- are higher temperature dehydration products (1070–1170 K). θ- is monoclinically distorted with the cations mainly in tetrahedral positions, while κ- is rhombohedrally distorted, again with stacking disorder.

The dehydration sequences involving gibbsite, bayerite and boehmite (composition AlO(OH)) have been summarized by Lippens and Steggerda.[23]

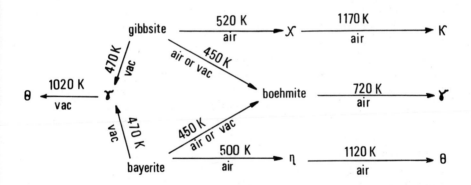

The extent to which reaction occurs via alternative pathways is controlled by other variables such as particle size and degree of crystallinity, as well as temperature. Thus, for instance, very finely divided gibbsite and bayerite yield no reaction to or via boehmite when heated in air, although the latter pathway can account for up to 25% of the total reaction at larger particle sizes. Moreover, because bayerite as prepared is much less well crystalline than gibbsite, the boehmite pathway is much more important for gibbsite. The significance of internal hydrothermal conditions in promoting the formation of boehmite was emphasized by de Boer et al.[17-19]

The work of de Boer et al. on gibbsite dehydration illustrates how the pore structure of the product develops. As measured by BET nitrogen adsorption, a maximum in the surface area is obtained for a dehydration temperature of about 570–620 K and, with the particular starting material used in this work, this maximum in the specific surface area was about 300 m^2g^{-1}; on further heating to about 1070 K this falls to a figure of something less than 100 m^2g^{-1}. This maximum in the specific surface is achieved at a water content of about 10% (calculated on anhydrous Al_2O_3): the water content falls to close to zero at 820–870 K, at which temperature the specific surface has fallen by a factor of about two from its maximum value. By heating gibbsite at 470–520 K water is formed by elimination between OH groups in the (001) planes so that rupture of the crystal occurs along these planes. At this stage this results in the formation of plate-like particles some 20 nm thick, separated by gaps of about 3 nm and the specific surface area is in the region of 60 m^2g^{-1}. With further heating these plates divide up into parallel rod-like particles separated by new gaps (micropores) about 1 nm wide, at this stage the specific surface area reaches its maximum value. This general structure of rod-like particles is retained at higher temperatures, although the micropores tend to be eliminated, thus accounting for the fall in the specific surface area. This presumably occurs by the collapse of some micropores and the simultaneous widening of others.

An empirical study has been made[24] of the dehydration in dry air of amorphous alumina hydrate and various gibbsite and bayerite samples. In all cases, the maximum specific surface area was achieved at 620–670 K. Although there was a range of values in the maxima of specific surface areas, the largest to the smallest spanned a change of only about 20%. On the other hand, dehydration of a well crystalline boehmite gave a low area product: thus the maximum specific surface area obtained (770 K) from this starting material was only about $\frac{1}{3}-\frac{1}{4}$ of that obtained from gibbsite or bayerite samples. Above about 820 K the micropores (again slit-shaped) are eliminated and the specific area falls by a factor of about five. If the starting boehmite is poorly crystalline (e.g. gelatinous boehmite) the specific surface of the dehyration product is much greater, and values up to 600 m^2g^{-1} have been reported after dehydration at only 420 K, while after heating to 770 K the value is still 400 m^2g^{-1}.

In practice there is a wide available range of specific surface areas and average pore sizes for dehydrated aluminas depending on the nature of the starting material, its particle size and the conditions of dehydration:

de Boer's data (*vide supra*) appear to be typical. Commerically available samples have specific surface areas up to about 400 m^2g^{-1}.

Except when the temperature is so high that α-alumina is formed, all aluminas prepared by dehydration contain residual water in the range a few tenths to some 5 wt. %, the amount decreasing with increasing dehydration temperature. Only physically adsorbed water is removed up to 390 K; however, heating somewhat above 390 K also causes some physically adsorbed water to react before desorption to yield chemically bound hydroxyl groups. Above about 570 K surface hydroxyl groups are gradually removed by the elimination of water between adjacent pairs, and after dehydration in the region of 720–870 K the surface hydroxyl concentration is about 8–12 OH per nm^2. By 1070–1270 K the residual water content is quite low, being no more than a few tenths of a per cent.

Peri[25, 26] has pointed out that, if during dehydration adjacent pairs of hydroxyl groups are eliminated at randon, only about $\frac{2}{3}$ of the total number present in an ordered array can be removed if hydroxyl or hydrogen mobility is absent. Thus, at the drying temperature needed to reduce the surface hydroxyl coverage to below about 10% (>920 K), surface mobility must exist, and this is in agreement with infrared and other evidence which also indicate some hydrogen mobility down to 670 K. Residual hydroxyl groups can occur with varying numbers of oxide and hydroxyl nearest neighbours, and Fig. 2.6 due to Peri[26] indicates possible arrangements around isolated hydroxyl groups, based on the idealized model of a (100) plane of a cubic close-packed structure: a surface hydroxyl coverage of about 10% is, on this model, the limit below which only isolated hydroxyl groups are present. Peri correlates these differing hydroxyl group environments with the range of O-H stretching frequencies observed in the infrared. Peri has also pointed out that with limited surface mobility, dehydration (dehydroxylation) must generate an oxide surface in which the oxide ions occur in domains, and defects occur at the domain boundaries. These defects are closely related to the various types of hydroxyl groups pictured in Fig. 2.6, while they also allow varying degrees of exposure for aluminium ions situated in the layer below, a situation which is relevant to the possible structure of Lewis acid sites in a dehydroxylated surface. The defects referred to here are defined with reference to a hypothetical fully packed layer of site density equal to that of the aluminium ions or to that of the hydroxyl ions in the hypothetical fully hydroxylated surface. Inasmuch as dehydration leaves behind one oxide ion for every two hydroxyls removed, it is obvious that a dehydrated surface contains, on this definition, a large number of single vacancy defects within the oxide domains, but triple and double vacancy defects occur only at domain boundaries.

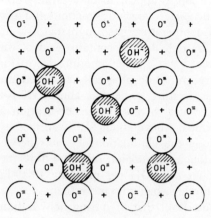

FIG. 2.6 Types of isolated hydroxyl groups formed during dehydration of alumina. The model is an idealized one based on a (100) plane of a cubic close-packed structure. + denotes Al^{3+} in a lower layer. Reproduced with permission from Peri, J. B. *J. Phys. Chem.* **69**, 220 (1965). Copyright: the American Chemical Society.

These surface hydroxyl groups have but very weak Brønsted acid strength, weaker than those on silica[30] in an aqueous environment. Virtually no NH_4^+ is formed by the adsorption of ammonia on γ-alumina predried at 1070 K.[26] However, they are probably involved in the dehydration of tertiary alcohols[27] in which some rearrangement of the carbon skeleton can occur, probably via a carbonium ion. Whether these Brønsted acid sites which result from residual surface hydration can also act as proton donors in carbonium ion formation in hydrocarbon isomerization reactions, is not yet finally settled. Finch and Clark[29] have recently concluded that the possibility cannot be entirely ruled out. Certainly one concludes from the results of MacIver *et al.*[30] that reactions such as olefin isomerization and hydrocarbon cracking can occur at both Brønsted and at Lewis sites: these workers concluded that Lewis sites became of increasing importance as the temperature of alumina dehydration increased above 770 K. The concentration of hydroxyl groups may be particularly low when the halogen content is appreciable: thus Finch and Clark[29] report a surface hydroxyl concentration of only about 1 OH per nm^2 for alumina containing 3–7 wt. %F which had been dehydrated at 820 K. This is about a factor of ten lower than the hydroxyl concentration in the absence of halogen. However, Finch and Clark[29] also now have evidence to support the view that protonic sites can become populated via a dehydrogenative hydrocarbon polymerization reaction on the alumina surface, and that these sites are active for carbonium ion formation. This process appears to be more important at higher fluorine contents of the alumina. In the protonated state these sites are almost certainly not hydroxyl, but probably consist of a protonated Lewis site of the type described below.

Although both the strength and concentration of Brønsted acid sites are extremely low on a dehydrated alumina surface, the strength and concentration of Lewis acid sites are relatively high. The evidence has been summarized by Boehm[10] and Tanabe.[63]

An indication of the acidity of alumina as measured by the amine titration method is contained in Figs 2.7 and 2.8, in terms of the Hammett acidity function, H_0.

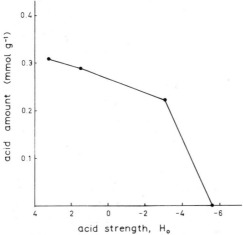

FIG. 2.7 Amount of surface acidity for a commercial activated alumina after heating in air at 770 K for 3 hours. The plot is cumulative so that each point indicates the amount of surface acidity having H_0 values equal to or less than the indicated H_0. Measurements made by amine titration using colour indicators. Data from Tanabe, K., "Solid Acids and Bases" Kodansha, Tokyo, and Academic Press, New York (1970).

Both Boehm and Peri have suggested a model for a Lewis acid site the main feature of which is that dehydration of an alumina surface leads to the generation of a vacancy in the co-ordination sphere of an aluminium ion. The precise site structure is still open to speculation, but it is very probable that a variety of different structures exists with a variety of different acid strengths. These would be expected to arise from the exposure of different crystallographic planes in the alumina surface, as well as from the various defects which dehydration may generate in a given surface. For instance, with reference to the idealized model based on a

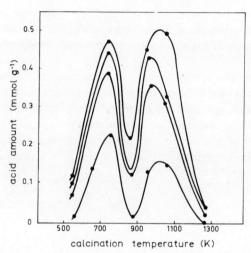

FIG. 2.8 Amount of surface acidity of alumina as a function of calcination temperature at varying H_0 values. Curve a, $H_0 < +3\cdot3$; b, $H_0 < +1\cdot5$; c, $H_0 < -3\cdot0$; d, $H_0 < -5\cdot6$. Structures: 720–870 K, η-Al_2O_3; 1070 K, η-Al_2O_3 plus θ-Al_2O_3; 1270 K, α-Al_2O_3. Measurements made by amine titration using colour indicators. Reproduced with permission from Tanabe, K. "Solid Acids and Bases" Kodansha, Tokyo, and Academic Press, New York (1970).

(100) plane of a cubic close-packed structure, Peri suggests that a triple vacancy in the surface oxide layer provides an unusually high degree of exposure of an aluminium ion in the layer beneath, and that this may constitute a strong Lewis acid site, particularly when this vacancy is neighbouring to an abnormally high oxide ion concentration as may occur with a triplet of adjacent oxide ions. On the other hand, a single vacancy may constitute only a very weak Lewis acid site; in any case, the concentration of single vacancy defects would certainly be very much higher than the concentration of Lewis acid sites as determined by techniques such as base adsorption combined with spectrophotometry.

On γ-alumina predried at 1070 K some 90% of the total ammonia adsorption occurs in a molecular form, and there can be no doubt that this is adsorbed on Lewis acid sites. However, there is also evidence for dissociative ammonia adsorption in which the nitrogen atom of the adsorbing molecule becomes located at a vacant co-ordination position of a surface aluminium atom in the form of adsorbed NH_2, while the dissociated proton becomes attached to an adjacent oxide ion which is converted to hydroxyl. Dissociative adsorption thus occurs at an acid–base pair site, and the process is quite analogous to that of rehydroxylation of a dehydrated surface. Peri suggests that dissociative ammonia adsorption occurs only at the strongest Lewis acid sites, and there is also evidence that these sites are among those reponsible for the adsorption and double bond isomerization of olefins. Pines and his collaborators[17] have suggested that acid–base pairs are involved in dehydration reactions of primary and secondary alcohols (cf. Eucken and Wicke[28]). In fact, alumina has a long history as a catalyst for dehydration and hydration reactions, and a summary has been given by Winfield.[123] There is no doubt that reaction with alcohol can generate alkoxyl groups on an alumina surface, but whether these are involved in alcohol dehydration catalysis remains a matter for dispute. In view of the wide range of surface properties which alumina may possess, it is possible that dehydration via alkoxyl or allied surface intermediates and via acidic and basic surface sites may be alternative reaction pathways, depending on the history and surface structure on the alumina. The Lewis acid strength of alumina (but not the Brønsted acid strength[10, 31]) is enhanced by replacement of surface hydroxyl with halogen such as fluorine or chlorine. Presumably, if a halogen atom is adjacent to a Lewis site, the acid strength is enhanced by the higher electronegativity of the halogen compared to hydroxyl, the effect being essentially electrostatic. As judged from a variety of evidence,[10] γ- and η-alumina have about the same concentration of surface Lewis acid sites, but the acid strength of the sites on η-alumina is the stronger. A dehydrated alumina surface also has electron donating properties (Lewis base): BF_3 is adsorbed at least as strongly as is NH_3 after dehydration at 770 K. Presumably a suitably located oxide ion may act as a Lewis base.[23] Nevertheless, Lewis acid sites appear to be of greater significance.

The problem of whether reactions such as hydrocarbon isomerization and cracking involve mainly Lewis or Brønsted sites still remains a matter for dispute. For the present purpose it suffices to say that we incline to the view that Brønsted hydroxyl acid sites are of little importance, except possibly for fresh alumina catalysts dehydrated at <770 K, where the catalytic activity is in any case relatively poor. For the rest, Lewis sites and protonated Lewis sites are of greater importance. Certainly the adsorption of water, presumably by co-ordination at the Lewis sites, results in catalyst poisoning.

The effect of γ-radiation alone or irradiation in a pile on the catalytic activity of alumina has been summarized by Taylor.[99] The main effect is an increase in activity for reactions such as double bond isomerization in olefins due to the removal of surface hydroxyl groups and adsorbed water. The net result is the same as that due to dehydration. On a thoroughly dehydrated specimen, irradiation is without significant effect on this activity. Irradiation also generates centres which can specifically catalyse ortho–para hydrogen conversion by the magnetic mechanism, and these centres are probably trapped unpaired electrons; they are destroyed, at least at the surface, by exposure to air. There is a general trend towards a reduction in surface area on irradiation, although the effect is usually not very large.

Low Area Alumina

Calcining active alumina at >1470 K results in the formation of α-alumina and, of course, also results in a drastic loss of surface area. There is some LEED evidence that if α-alumina is heated above about 1170 K, the hexagonal (001) surface becomes oxygen deficient and the aluminium atoms are relocated in tetrahedral sites.

Low area alumina (α-alumina) is available as sintered granules or pellets. The specific surface area is typically in the range $0 \cdot 1$–5 m^{2}g^{-1}, with the average pore diameter in the region $2 \cdot 0$–$0 \cdot 5$ μm. The composition is generally $>99 \cdot 5\%$ Al$_2$O$_3$. With some formulations, a small proportion of silica (1–4%) is incorporated, and this increases the ease of sintering and reduces the porosity. If much larger amounts of silica are added (10–20%), the product may be prepared by fusion: again granules or pellets are available with specific areas in the region $0 \cdot 02$–$0 \cdot 3$ m^{2}g^{-1} and average pore diameters 60–5 μm.

4. Chromia

Chromia gel is generally prepared via hydrosol and hydrogel by the addition of ammonia to an aqueous solution of a chromium (III) salt such as the nitrate. Chromia has, of course, a very considerable catalytic activity in its own right, and both the physical characteristics of the gel and its catalytic activity are very sensitive to the conditions of preparation, storage and thermal treatment. Slow addition of ammonia is advantageous, and the slow generation of ammonia by the hydrolysis of urea in a boiling solution[130] is a very convenient technique.

The amorphous hydrogel is formed as a result of polycondensation reactions of a type reminiscent of the genesis of silica gel, and the general morphology of chromia hydrogel closely resembles that of silica hydrogel. The uncalcined chromia gels are always either amorphous or, at best, very poorly ordered.

After drying at 390 K chromia gel still contains about 3·5 mol of water per mol of Cr$_2$O$_3$, and after drying at 570 K the figure is still about 0·5 mol of water. After drying at 720 K a substantial concentration of chemically bound surface hydroxyl groups remains, and these are not completely removed until much higher temperatures (about 1170 K). In the process of dehydration, the hydrogel may undergo partial hydrothermal conversion to the orthorhombic CrO(OH) at about 520 K.[129] The extent to which this occurs is dependent on the pore structure of the gel, and the process is similar to the formation of AlO(OH), boehmite, in the corresponding aluminium system.

Recrystallization of chromia gel may occur during heating, but the extent to which this occurs is strongly dependent on the conditions, particularly if oxidation or reduction of the chromium occurs.[178] Burwell and his co-workers[132] showed that dehydration by heating in an inert atmosphere to 670 K or so leads to a stable gel of amorphous structure. On the other hand, if the initial thermal treatment consists of heating to 670 K in hydrogen, the result is the formation of microcrystalline α-Cr_2O_3. If the thermal treatment is carried out in air, the product depends on the way the treatment is carried out. Rapid heating may result in a strongly exothermal recrystallization to α-Cr_2O_3 in the region of 620–670 K: however, if the dehydration is carried out slowly it is possible for an amorphous structure to be retained. The progressive increase in conversion to crystalline α-Cr_2O_3 at progressively higher temperatures, from 620 K where the first signs become apparent, to 970 K where the fully resolved α-Cr_2O_3 X-ray diffraction pattern is obtained, has been described by Deren et al.[133] If in the initial dehydration of the gel, 670 K is reached without recrystallization, the gel is relatively stable at least to 820 K, although a slow recrystallization will occur at temperatures >770 K, and this is assisted by thermal and/or oxidation–reduction cycling.

Oxidation of some of the chromium in the surface layers of the gel occurs to an appreciable extent by heating in air or oxygen at temperatures in the range 370–870 K. Oxidation at 370 K results in the generation of Cr^{4+} and Cr^{5+}, but not much Cr^{6+}. However, the proportion of oxidized chromium present as Cr^{6+} increases rapidly with increasing temperature being 67% at 470 K and 95% at 620 K, and the maximum extent of the oxidation occurs also at about 620 K.[129, 133]

The surface area of chromia gel decreases with increasing temperature of thermal treatment. The initial surface after, say, dehydration at 370 K, is highly variable depending on the preparative conditions, but values in the range 80–300 m^2g^{-1} are typical. The data of Deren et al.[133] and Carruthers et al.[129] indicate that for gels dehydrated by heating in air, the specific surface area is relatively temperature independent above about 770 K, being about 10–30 m^2g^{-1}, but for a gel dehydrated in an inert gas atmosphere this does not occur until about 970 K. Chromia with a high specific surface area, >200 m^2g^{-1}, is certainly microporous, with pores of an equivalent diameter <2 nm: when recrystallization to α-Cr_2O_3 occurs, this micropore structure is destroyed. Evidence has recently been presented[134] to show that when chromia gel is prepared by the urea hydrolysis technique and outgassed below 470 K the micropores are of great uniformity and of very small diameter, so that molecular sieve properties result.

Ignoring for a moment the occurrence of chromium valence variability, a hydroxylated chromia surface resembles, at least in a formal sense, that which we have already discussed in a previous section for alumina. The process of dehydroxylation and hydroxylation, and the generation and destruction of various sorts of acidic and basic surface sites are analogous for the two systems.

The various types of chemisorptive processes occurring on chromia have been discussed by Burwell *et al.*[131, 177] and these include both dissociative and nondissociative processes such as we have also discussed for alumina. The main qualitative distinction between chromia and alumina in terms of chemisorptive behaviour lies in the influence of the variable valency of chromium. One result of this is the ready ability of chromia to adsorb oxygen with the generation of surface chromium ions of charge greater than three. On the other hand, chromia also dissociatively chemisorbs molecules such as hydrogen or the alkanes more readily than alumina, and this arises from the ability of a surface Cr^{3+} to form an essentially covalent bond to a hydrogen atom or to an alkyl group.

Chromia has catalytic activity for a wide range of processes, the more important being olefin hydrogenation (e.g. refs 131, 177), alkane dehydrogenation, dehydrocyclization and aromatization.[137] Chromia also has some catalytic activity for alcohol dehydration,[123] and for the oxidation of substances such as hydrocarbons or carbon monoxide,[135] but the level of activity for these functions is relatively low.

5. Titania*

Synthetic titania is prepared by hydrolytic methods including hydrolysis in aqueous media and oxidative hydrolysis in a flame, as well as by direct oxidation of titanium tetrachloride. Although hydrogels of titanic acid are readily obtained as a primary hydrolytic product from aqueous media, an amorphous structure is never completely retained on dehydration, and the degree of crystallinity tends to increase with increasing temperature of dehydration or calcination, and is also subject to control during the precipitation stage, particularly by the speed of precipitation. The calcination of an initially poorly crystalline specimen is also accompanied by a very substantial reduction in specific surface area, and a reduction by a

---

* Well-known suppliers of various grades include: British Titan Products; Degussa; Laport Industries; National Lead.

factor of three on increasing the calcination temperature from 470 to 770 K is typical (e.g. ref. 77).

Both anatase and rutile are tetragonal and differ in their axial ratio, and the former transforms into the latter at about 1188 K, so high temperature calcining can be used to acquire the rutile modification, although the required temperature will generally result in some loss of surface area by particle agglomeration or growth. The transformation is, however, sluggish and accelerating additives are employed in commercial production. Nevertheless, complete purity of a given modification is difficult to achieve, and in practice, samples usually contain a small proportion of the other modification. However, even assuming that the material is not calcined above the transformation temperature, the modification obtained is still very dependent on the preparative details. The primary product from flame hydrolysis of titanium tetrachloride is mostly anatase (cf. refs. 77, 79), and a specific surface area of 40–80 m^2g^{-1} is typical for this product. On the other hand, rutile has been prepared by the flame hydrolysis of titanium tetraisopropoxide.[78] Hydrolytic preparations in aqueous media using either titanium tetrachloride or titanium sulphate solutions can yield either anatase or rutile depending on the conditions. The presence of sulphate ions tends to favour the formation of anatase at the expense of rutile, while the nature of the alkali used for pH control is also important. Seeding the precipitate with rutile particles is also employed in order to encourage this modification when desired. Rutile or anatase powders are available with specific surface areas up to about 200 m^2g^{-1}, but specimens calcined at 1070 K or above can generally be expected to result in an area of less than about 10 m^2g^{-1}, and the more commonly available specimens have areas in the range 3–80 m^2g^{-1}.

Equilibration of a rutile or anatase surface with water or water vapour leads to the presence of both chemisorbed hydroxyl groups and to adsorbed molecular water. Compared to silica, the metal–oxygen interaction on titania is more polar, and chemical evidence such as reactivity with thionyl chloride suggests that hydroxyl groups on a titania surface are more ionic in character than on silica.

The nature of the adsorbed hydroxyl groups has been extensively studied both by infrared spectrophotometry[77-85] and by chemical reactivity.[86-92] There is good evidence that there are two types of chemisorbed hydroxyl groups on the surface of rutile powder, and the nature of the hydrated surface has been rationalized in terms of the surface structures.[10, 83, 85, 93] Rutile is known to crystallize so that most of the crystal surface is composed of planes of three sorts; (110), (100) and (101). Of these, the (110) plane has the greatest preponderence (~60%). This surface structure also probably applies to fine rutile powders provided they are well crystalline. In bulk rutile, titanium ions are six-fold co-ordinated by oxide ions. In the (100) and (101) surface planes, the titanium ions are five-fold co-ordinated by oxide ions, while in the (110) surface plane (cf. Fig. 2.9) there are equal numbers of surface cations which are five-fold (A) and four-fold (B) co-ordinated. On a (110) plane it has thus been suggested that the two types of chemisorbed hydroxyl groups arise from the dissociative chemisorption of water, much after the style of the process already considered for silica and alumina. The oxygen atom of the incoming water molecule becomes located at the vacant co-ordination position of a five-fold co-ordinated surface titanium ion ($OH_{(a)}$), and the dissociated proton becomes attached to an adjacent superficial oxide ion (represented by an open circle in Fig. 2.9) which is converted into $OH_{(b)}$. On this basis, $OH_{(a)}$ has one Ti^{4+} adjacent to it, while $OH_{(b)}$ has two. It has been pointed out that there is no real evidence

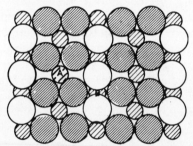

FIG. 2.9 Rutile (110) surface. The titanium ions (small shaded circles) and the oxygen ions (large shaded circles) are in the same plane. The oxygen ions shown as large open circles lie above the plane containing the titanium ions. The surface titanium ions of type A are five-fold co-ordinated by oxide ions, those of type B are four-fold co-ordinated.

that the $OH_{(b)}$ retain this structure, and they may move to become singly co-ordinated like $OH_{(a)}$: however, general electrostatic arguments would suggest that the former configuration should be more likely. An analogous sort of dissociative adsorption process has been suggested for the (100) and (101) rutile planes, but in these cases if the proton is accepted by a superficial oxide ion, only a single type of hydroxyl group will be generated. With this model it has been argued[93] that a fully hydroxylated (100) or (101) surface is unlikely because it would require two hydroxyl groups co-ordinated to each surface titanium, and this is sterically difficult; as an alternative it was suggested that (100) and (101) surfaces can only accommodate undissociated water ligands which occupy the vacant positions on the five-fold co-ordinated titanium ions. This steric argument is much less compelling for only partly hydroxylated surfaces since the need to accommodate two hydroxyls per surface titanium ion is not then required.

This model is of course based upon that of an ideal regular surface structure. Departures from this regularity, that is the presence of surface defects after the style of those discussed by Peri for γ-alumina, have not yet been studied. However, the structure of the (110) titania surface is much more open than that discussed previously for γ-alumina and vacancy defects are likely to be less important in any case.

There is infrared evidence for the dissociative adsorption of ammonia with the formation of adsorbed OH and NH_2,[88] and a process similar to that described above for water is very probable.

The maximum formation of chemisorbed hydroxyl groups on a rutile surface occurs to the extent of 4–7 OH per nm^2 but the figure found most commonly is about 5 OH per nm^2: for anatase, figures of 4–5 OH per nm^2 have been reported. These data correspond roughly to the dissociative adsorption of one water molecule for every two five-fold co-ordinated surface titanium ions (with a heat of adsorption of about 107 kJ mol^{-1}).[85] The remaining five-fold co-ordinated surface titanium ions then appear to function as sites for the adsorption of molecular water as a co-ordinating ligand (with a heat of adsorption of 75–50 kJ mol^{-1}). This molecular adsorption is irreversible at room temperature and should be distinguished from further weak molecular adsorption, which is reversible at room temperature, and which is probably due to adsorption of water molecules at the surface oxide ions. A fully hydrated surface at about room temperature has about 11 OH per nm^2 (with the OH existing in the various forms indicated above). Assignments made from thermal desorption data[85] indicate that the strongly bound molecular water is mostly desorbed at about 520 K and desorption is complete by 620 K, while the removal of surface hydroxyl groups occurs at about 640 K although it is not complete until rather higher temperatures. There is evidence that rutile samples prepared by hydrolysis of $TiCl_4$ may contain residual chlorine which facilitates the removal of surface hydroxyl groups by desorption of water. Equilibrium with respect to the adsorption and desorption of molecular water is established rapidly, but the position with respect to hydroxylation and dehydroxylation is more equivocal. The ease of rehydroxylation depends on the degree of crystallinity of the specimen. A very poorly crystalline specimen was reported as resistant to rehydroxylation even at a water vapour pressure of $2 \cdot 7 \times 10^3$ Pa* (about 20 Torr) at 670 K,[77] but with a well crystalline specimen rehydroxylation readily occurred even at room temperature. However, the behaviour was also dependent to some

---

* 1 Pa (pascal) = 1 N m^{-2} = $7 \cdot 50 \times 10^{-3}$ Torr.

extent on the number of previous hydroxylation cycles. On balance, the evidence suggests that with well crystalline specimens, equilibrium with respect to hydroxylation is established reasonably easily at room temperature.

Both Brønsted and Lewis acidity of titania has been investigated using diagnostic adsorbates combined with infrared spectrophotometry. No matter what the state of surface hydration or hydroxylation, titania shows no Brønsted acidity towards ammonia or pyridine;[87-89] however, some hydroxyl groups on anatase have a sufficient acidic character to protonate trimethylamine.[87] The formation of surface bicarbonate[87] from carbon dioxide adsorption indicates the presence of some basic hydroxyl groups. However, this basic character is very weak since the bicarbonate species is destroyed by outgassing the specimen at 300 K. The conversion of a surface oxide ion into hydroxyl during dissociative water adsorption also amounts to Brønsted basicity. Nevertheless, a titania surface in contact with an aqueous environment can show both Brønsted acidity (adsorption of NaOH) and basicity (adsorption of H_3PO_4), and it has been suggested that these two functions are associated with the two types of hydroxyl groups thought to be present.[86]

Two types of Lewis acid sites have been identified on titania[87] and the nature of these will also be apparent from the previous discussion concerning the modes of water adsorption. The stronger of the two Lewis sites consists of a five-fold co-ordinated surface titanium to which, for instance, becomes attached the hydroxyl group from an incoming water molecule during dissociative adsorption. The weaker site is the one on which water adsorbs molecularly. Thus, on a previously hydrated surface, only the weak sites will exist on a specimen dehydrated at 520 K, and both types are present after dehydration at 670 K. The strong sites appear to be comparable in strength to those on γ-alumina. This model clearly involves the existence of an acid–base surface pair which is utilized on the dissociative adsorption of molecules such as water or ammonia. Tanabe[63] quotes some evidence that some laboratory-prepared titania specimens which were generated from chloride by treatment with ammonia solution followed by heat treatment at 670–770 K, had higher acidity than commercially prepared specimens. The difference is probably due to residual surface impurity (possibly chloride) on the laboratory prepared specimens.

Surface Ti^{4+} ions are readily reduced to Ti^{3+} by heating titania in the presence of reducing agents such as hydrogen or carbon monoxide, typically at 470–570 K, and some Ti^{3+} ions may be generated by prolonged heating in vacuum at similar temperatures, or more readily with u.v. illumination at room temperature either in vacuum or in the presence of hydrocarbons.[79,10] Data from e.s.r. show that the Ti^{3+} ions are of two types: those present in either normal surface lattice positions or in interstitial positions, and those associated in the surface with one or two oxygen ion vacancies. Although there are claims to the contrary,[10] more recent evidence suggests[79] that the presence of Ti^{3+} has no significant effect on acid–base interactions at a titania surface. This is an important question which still awaits further clarification.

Titania generally has fairly low catalytic activity. It possesses some activity for alcohol dehydration and dehydrogenation although, of the two, the former function is the more important.[123] It also has some activity as an oxidation catalyst for substances such as hydrocarbons and hydrogen, but again the activity is relatively very low.[125] However, its oxidation activity is augmented by u.v. irradiation and the photocatalytic activity of titania has been the subject of a good deal of study (cf. ref. 79); it is associated with centres which are deficient in oxygen.

6. Zirconia*

Hydrolytic precipitation is usually carried out from an aqueous zirconium solution, but the hydrolysis of zirconium tetra-alkoxide has also been described.[136] Aqueous zirconium chemistry, including hydrolytic polymerization has recently been summarized by Clearfield[116] and by Rijnten.[112] In a structural sense, Clearfield distinguishes two stages in the formation of a hydrogel from zirconium halide. The first is the formation of a tetramer in which four zirconium atoms are located at the corners of a slightly distorted square and which are linked by bridging oxygens; the second stage consists of linking up these tetrameric units again by bridging oxygens. The degree of crystallinity depends on the degree of regularity with which the tetrameric units are arranged. The primary particles of which the gel is composed are typically 3-6 nm diameter, which is much the same as with silica. Drying the precipitate at 390 K leads to a material for which the weight loss on ignition at 1370 K is about 10–11 wt. % and which is usually amorphous. However, some degree of crystallinity may be induced by protracted washing or standing in an aqueous environment. Materials of this sort typically have specific surface areas in the range 150–350 m²g⁻¹. Some typical preparative details have been given by Rijnten.[112] Even so, it has recently been concluded by the use of X-ray and neutron diffraction[113] that apparently amorphous zirconia contains very small ordered regions of diameter in the vicinity of 2 nm and one unit cell in thickness, which take the form of platelets based on a structure very similar to that in the region of a (111) plane of the tetragonal modification.

* Well-known suppliers of various grades include: Zircoa Corpn.; Norton; Harshaw; Carborundum Refractories.

The structure of this platelet is also very similar to that proposed by Clearfield as resulting from an ordered arrangement of tetrameric units.

Heating above 390 K leads to progressive further dehydration and to an increasing degree of crystallinity. There is an extremely rapid and exothermic recrystallization which occurs at 683–703 K. Below 1370 K the stable crystallographic modification is monoclinic, but the modification which is first formed from the amorphous material may be either tetragonal or monoclinic depending on the preparative conditions: the tetragonal form is the more usual both when the zirconia is prepared by precipitation from alkaline aqeuous solution and when it is prepared by calcining a salt such as the nitrate at low temperatures. However, maintaining the tetragonal modification for prolonged periods in the region of about 870 to <1370 K always results in at least partial conversion to the monoclinic modification and this is a conversion which starts at the particle surface and progresses inwards towards the centre of the crystal. There appears to be a critical size of about 30 nm diameter above which zirconia particles cannot exist in the tetragonal modification at room temperature, and Garvie[114] has suggested that the occurrence of the tetragonal modification under apparent metastable conditions below 1370 K is due to a difference in the surface energies of the two forms.

There is no doubt that chemically bound hydroxyl groups are present on a zirconia surface after equilibration with water or water vapour.[112, 113, 115] Some of these hydroxyls are removed as low as 650 K,[113] but complete dehydration is not achieved until about 1170 K.[115] The stereochemistry and detailed reactivity of hydroxylated zirconia surfaces have not yet been explored. Nevertheless, some insight into the possibilities may be gained by reference to the structure of the tetragonal modification shown in Fig. 2.10. One would expect the low index (111) plane to be

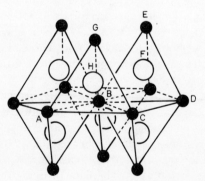

FIG. 2.10 Structure of zirconia in the tetragonal modification. Filled circles zirconium, open circles oxygen. AB, 0·35; CE, 0·50; CF, 0·42; AD. 0·61; BA, 0·21 nm. After Teufer, G. *Acta Cryst.* **15**, 1187 (1962).

important in an actual surface (the zirconium plane in Fig. 2.10 containing atoms A, B, C), and an exposed surface based upon this plane would presumably carry a superficial layer containing oxygen ions, hydroxyl groups and vacancies; and one would expect the general chemical principles governing the behaviour of this sort of surface to be much the same as we have already described for other systems. Surface acidity has been studied[128] by a combination of pyridine adsorption and infrared spectrophotometry. This demonstrated the existence of Lewis acid sites, although these appear to be of lower strength than those present on alumina or titania.

Pure zirconia possesses some activity for alcohol dehydration and dehydrogenation, and these are of comparable importance.[123] Nevertheless, its level of activity in this direction is very low. It also has a very low activity as an oxidation catalyst.

7. Magnesia

Details of magnesia formation by the dehydration of the hydroxide have been discussed by Anderson and co-workers,[71-74] who showed that decomposition *in vacuo* at 570 K gave an oxide of area about 220 m^2g^{-1} and average crystallite size about 7·5 nm: at this temperature the surface retains hydroxyl groups to a concentration of about 8 OH per nm^2. These crystallites exist in aggregates which are relics of the original hydroxide crystals. Heating in vacuum to 970 K reduces the surface hydroxyl concentration to about 0·5 OH per nm^2, and the concentration falls below the limit of detection by heating to 1170 K. Increasing dehydration temperatures result in a decrease in the specific surface area and an increase in the average crystallite size, but the magnitude of these changes is much dependent on the dehydration conditions, in particular on the water vapour pressure existing over the specimen. It appears that adsorbed hydroxyl greatly increases the surface mobility of atoms in the oxide surface. For instance, by sintering at 1320 K in water vapour at a pressure of about $6·7 \times 10^2$ Pa (about 5 Torr) the surface area and average crystallite size reached ultimate (time independent values) of about 30 m^2g^{-1} and 26 nm in about 2 h, while at the same temperature in vacuum these changes had only proceeded to $<50\%$ of this extent in periods >100 h. In these cases, most of the surface area is attributable to the presence of micropores (probably slit-shaped) of average size <2 nm for specimens prepared at 570 K, and about 2·5 nm for specimens sintered in vacuum at 1320 K.

The rehydration of a hydroxide-free magnesia surface is an activated process, the maximum coverage by chemisorbed hydroxyl being in the vicinity of 11 OH per nm^2.

de Vleesschauwer has given details for the generation of magnesia from magnesite ($MgCO_3$) and nesquehonite ($MgCO_3.3H_2O$).[75] From nesquehonite, complete conversion (>99%) to magnesia was obtained at >770 K in 24 h, while complete conversion of magnesite required >920 K. The pores formed in the calcining process are of two types: those formed by the interstices between the magnesia crystals (macropores), and those within individual crystals (mesopores). After calcining at 870–1070 K, the macropores were in the region of 100 nm average size. The mesopores were more strongly dependent on calcining conditions: thus, increasing the calcining temperature from 870 to 1070 K resulted in an increase in the average mesopore size from about 15 to 32 nm, with a reduction in the surface area from about 60 to 30 m^{2}g^{-1}, and an increase in the average magnesia crystallite diameter from about 15 to 55 nm.

Mikhail et al. have studied the preparation of magnesia by the decomposition of magnesium oxalate in the range 670–870 K.[95] The only residue present was water. For temperatures 670 and 700 K, a dual pore distribution was obtained, with most probable pore diameters of about 2 nm and 4 nm respectively. Above about 730 K these consolidated into a distribution with a single maximum, and at 830 K the most probable width was about 10–12 nm.

Hydroxyl groups are generated on the surface of magnesium oxide by the dissociative adsorption of water. As with other oxides, one expects two types of hydroxyl groups, one situated above a magnesium ion and containing the oxygen from the water molecule, the other being formed from an adjacent oxide ion by the addition of a proton. In this way, complete hydroxylation of (100), (110) and (111) surfaces would lead to about 11, 8 and 6·5 OH per nm^2 respectively. The experimental figure is about 11 OH per nm^2, which suggests that the crystallites in magnesia powder expose (100) faces predominantly, and this is reasonable as (100) is a low energy face on a substance with a rocksalt structure. However, the nature of the dominant crystal face exposed is almost certainly dependent on the thermal history of the specimen and samples prepared by hydroxide dehydration under conditions mild enough for the morphology to show a relic structure of the original hydroxide crystals, probably have (111) faces dominant, since this relates to the (001) plane in the hexagonal hydroxide crystal structure. On the other hand, Ramsey[96] has reported that magnesia prepared by vapour condensation is much more resistant to the formation of surface hydroxyl groups by water adsorption than are specimens prepared by conventional methods. It is hard to escape the conclusion that vapour condensation leads to a different surface structure, but the nature of this difference is not clear. Moreover, a (111) face is probably not the one of lowest surface energy, and recrystallization may well occur by heating to high temperatures.

Surface hydroxyl groups can also be generated by γ or neutron irradiation in gaseous hydrogen.[97] The hydroxyl is formed in the vicinity of a surface F-centre which is created at the same time. If irradiation is carried out in vacuum, surface F-centres alone are created, and these may also be destroyed by exposure to oxygen which results in immediate adsorption as O_2^-.

The acid strength of a magnesia surface increases with increasing temperature of dehydration in the range 870–1270 K.[67]

Magnesia is relatively inert catalytically. Hydrocarbons appear to undergo no reaction, but there is some activity as a catalyst for alcohol dehydration and dehydrogenation.[75] Of these, the former is only retained if the calcination temperature does not exceed about 820 K. The specific activity for alcohol dehydrogenation is, however, many orders of magnitude lower than that of (say) copper at the same temperature. There is some infrared spectrophotometric evidence for the formation of adsorbed methoxyl groups resulting from the adsorption of methyl alcohol on magnesia.[76]

8. Thoria

Mihail and Fahim[117,118] have studied the dehydration of thoria hydrogel and established that the material remains amorphous at all dehydration temperatures below about 570 K, but heating above this temperature results in increased crystallinity and a general trend towards decreased specific surface area. A typical gel had a specific area of about 80 m^2g^{-1} after dehydration at 420 K: after heating to 770 and 1270 K the specific surface areas were 21 and 1·5 m^2g^{-1} respectively. Dehydration at 670 K removes all but about 10 wt. % of the adsorbed water, but ignition at >1270 K is required to reduce the residual water content to <1 wt. % of its saturation value. These thoria gels were found to have an equivalent average pore diameter of about 1–2 nm.

Thoria also results from the decomposition of the oxalate, and the properties of material of this origin have been thoroughly studied by Holmes and collaborators.[119-122] The thoria appears as square prismatic relic crystals, each of which may be several μm across, but which is made up from an agglomerate of smaller particles each several tens nm across. Specific surface areas in the range 5–15 m^2g^{-1} are typical, depending on the preparative details and the dehydration temperature (380–620 K). Dehydration at 1270 K results in a surface area of typically 1–2 m^2g^{-1}.

There is ample evidence for the presence of chemically bound surface hydroxyl groups. After equilibration with water or water vapour at about room temperature the concentration of these is in the vicinity of 7–8 OH per nm^2 and dehydration at 700 K reduces the figure by a factor of about two.

Not much evidence is available concerning the possible acidic and basic properties of the surface of thoria. In an aqueous environment it has a weak acidic function shown by adsorption of NaOH, while there is also evidence for basic behaviour of surface hydroxyl groups in the sense that they react with formic acid.[127] The basic function appears to predominate. Thoria has a substantial history as a dehydration catalyst, and Winfield provides a useful summary.[123] Thoria is also known to function as a catalyst for some oxidation reactions, and the oxidation of carbon monoxide is an example.[124] However, it is a relatively unimportant oxidation catalyst.

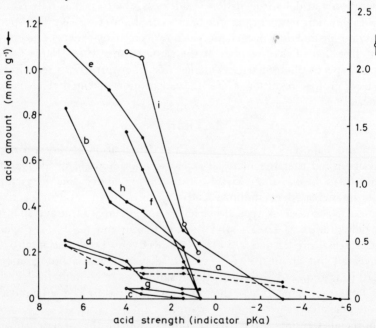

FIG. 2.11 Amount of surface acidity of some metal sulphates. Curve a, $CuSO_4.5H_2O$ after heating to 573 K; b, $MgSO_4.7H_2O$, 523 K; c, $CaSO_4.2H_2O$, 503 K; d, $ZnSO_4.7H_2O$, 448 K; e, $Al_2(SO_4)_3.18H_2O$, 623 K; f, $Cr_2(SO_4)_3.18H_2O$, 373 K; g, $MnSO_4.7H_2O$, 503 K; h, $FeSO_4.7H_2O$, 473 K; i, $Fe_2(SO_4)_3.xH_2O$, 373 K; j, $NiSO_4.7H_2O$, 623 K. The plot is cumulative so that each point indicates the amount of surface acidity having pK_a values equal to or less than the indicated pK_a. Measurements by adsorption of colour indicator. Data from Tanabe, K. "Solid Acids and Bases" Kodansha, Tokyo and Academic Press, New York (1970).

9. Metal Sulphates, Phosphates, Chlorides and Carbonates

The acidic properties of these substances have recently been summarized by Tanabe.[63] As judged for instance by amine titration, the acidity of metal sulphates is zero for completely anhydrous material, is relatively low for fully hydrated material, and is a maximum for heavily but incompletely dehydrated sulphate. Figure 2.11 summarizes the behaviour of a range of metal sulphates. Tanabe makes the reasonable proposal that partial dehydration leaves a surface metal ion incompletely co-ordinated, thus generating a Lewis site, while a proton of a residual water molecule of hydration provides for Brønsted acidity. The acid strengths of metal sulphates will be seen to be generally very low to medium, being considerably lower than, for instance, silica–alumina.

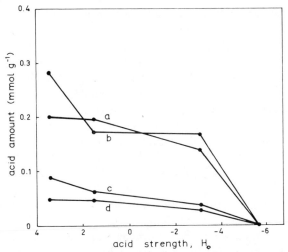

FIG. 2.12 Amount of surface acidity of some metal phosphates. Curve a, $AlPO_4$; b, $FePO_4$; c, $Cu_3(PO_4)_2$; d, $Ni_3(PO_4)_2$. The plot is cumulative so that each point indicates the amount of surface acidity having H_0 values equal to or less than the indicated H_0. Measurements made by amine titration using colour indicators. Reproduced with permission from Tanabe, K. "Solid Acids and Bases" Kodansha, Tokyo, and Academic Press, New York (1970).

Metal phosphates also have surface acidic properties with values for the Hammett function, H_0, falling roughly in the same range as for the sulphates: some data quoted by Tanabe[63] are given in Fig. 2.12.

A wide range of metal chlorides show surface acidic reactions, and these include chlorides of Al, Sb, Sn, Fe, Zn, Ca, Hg, Cr, Cu and Pb. Again it seems likely that the acidity is a maximum when a small amount of water is present.

The carbonates of the alkali metals and of the alkaline earth metals are weakly basic as judged from the colour of adsorbed indicators such as methyl red (in non-aqueous solvent).[63, 64]

10. Silica–Alumina and other mixed Oxides*

A silica–alumina hydrogel is usually prepared by co-precipitation of both components, or by precipitation of an alumina hydrogel in the presence of a freshly prepared silica hydrogel. Preparative methods based on ester hydrolysis have also been used for laboratory specimens. Ryland et al.[32] give some typical details together with references to the literature. The general morphology of silica–alumina gels is quite similar to that of silica gel, and the material remains non-crystalline to X-rays provided it is not heated above about 1070 K. Typically, the primary spherical particles are of average diameter 3–5 nm and this is also the size range for the average pore diameter. Surface areas are usually found in the range 200–700 m^2g^{-1}. Practical catalysts have an alumina content in the range 10–30 mol %. The aluminium is distributed in the gel structure with varying degrees of uniformity depending on the preparative details: in a number of cases (e.g. refs 32–34) an essentially uniform distribution has been reported, and this is probably the most common situation. On the whole, co-precipitation can be expected to give a more uniform distribution than does precipitation of an alumina hydrogel on to a silica hydrogel. For alumina contents <30 ml %, the possibility of discrete alumina aggregates can be neglected.

Both the general structure of silica–alumina as well as the specific nature of the catalytically active sites should be directly relatable to the known chemistry and crystallography of similar but better defined materials. This point of view is clearly emphasized by the rather similar catalytic activity of a number of naturally occurring crystalline aluminosilicates and particularly the activity of the synthetic crystalline aluminosilicates (zeolites). We first consider the structure within the bulk since it is only in terms of this that the surface structure can be described. If we consider the dehydrated gel, then in quasi-crystallographic terms the structure should be related to the ways in which SiO_4 tetrahedra and AlO_x (x = 4 or 6) tetrahedra or octahedra can be linked together by corner, edge or face sharing, and what structural defects are used to accommodate the

* Well-known suppliers of various grades include: Harshaw; W. R. Grace, Davison Divn.; Air Products, Houdry Divn.; American Cyanamid.

differing valence requirements of aluminium and silicon. The only additional restriction is that silica–alumina gel, like silica gel, does not have long range order. This model is directly derivable from the point of view adopted by Thomas.[35] In its simplest (and original) form it proposes a structure based entirely on corner sharing of SiO_4 and AlO_4 tetrahedra, the latter being randomly distributed. If full tetrahedral oxygen coordination is retained around the aluminium, charge neutrality may be achieved with an associated positively charged ion (e.g. Fig. 2.13a). Alternatively, ordinary valence requirements can be satisfied if the aluminium has only three-fold co-ordination, together with an adjacent silanol group (e.g. Fig. 2.13b).

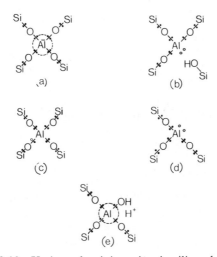

FIG 2.13 Various aluminium sites in silica–alumina.

If the centres in Fig. 2.13a and b were accessible to external reagents, they would obviously function as Brønsted and Lewis acid centres respectively. In fact, for the latter to occur, the centres would have to reside in the gel surface and in this circumstance their structural description becomes a good deal more uncertain. In addition to these proposals, Weisz[36] has suggested a model which retains full tetrahedral co-ordination of the aluminium, but with an electron hole trapped at the aluminium (Fig. 2.13c). When present in the surface, sites of the type in Fig. 2.13c would provide for the dissociative chemisorption of hydrogen (by which process they would be converted to a or b).

There is obviously a basic distinction between models a and b on the one hand, and c on the other, in so far as the former require the presence of an extra atom. The X-ray fluorescence results of Leonard et al.[37]

strongly suggest that at alumina contents <30% most aluminium is present in tetrahedral co-ordination, although at higher alumina contents the dehydrated gel contains increasing amounts of a structural unit consisting of two (or more) AlO_4 tetrahedra joined by edge sharing. This structure can, in fact be considered as an adjacent pair of Lewis sites.[38] In fact Leonard et al.[98] have recently detected phases similar to mullite and η-alumina in silica–alumina containing as little as 21 mol % alumina. For practical purposes there seems no doubt that the simple model we have already described involving only corner-shared tetrahedra is adequate at alumina contents <20 mol %. On the other hand, at >30 mol % alumina there is also no doubt that these other more complex structures are important.

One needs to consider the relation between structures such as shown in Fig. 2.13a and b which mainly differ by the location of the hydrogen. Weisz[36] has suggested an intermediate structure which (were it accessible to reagent attack) would manifest either Brønsted or Lewis acidity depending on the nature of the attacking reagent. On the other hand, Uytterhoeven et al.[46] have suggested (with zeolites) a tautomeric equilibrium which can be displaced by reagent attack; infrared evidence suggests that the latter proposal is correct, and that the equilibrium is heavily in the direction of structure b.

It is obvious that if an aluminium unit exists in the surface of a gel particle it cannot retain the same structure as in the bulk, since a surface aluminium must be surrounded by less than four –O–Si groups. Moreover, with primary gel particles as small as a few nm, a substantial proportion of the atoms must reside in the surface.

The voluminous literature on the surface structure and surface properties of silica–alumina has been summarized on a number of occasions.[10, 32, 39] For the present purpose we only need to note the salient points. There is much evidence that, in response to appropriate experimental conditions, silica–alumina shows Brønsted and Lewis acid behaviour by virtue of its surface structure. Infrared evidence shows the presence of both the protonated and unprotonated forms of adsorbed ammonia or pyridine.[34, 40-42] The amount of the Brønsted form is reduced by exchange with alkali metal ion and is increased by increased hydration, and the catalytic properties of the gel for hydrocarbon cracking follow a roughly similar trend. On various gels dehydrated at 770 K, the ratio of unprotonated to protonated adsorbed pyridine is in the region 1–6. The degree of hydration over the range 370–770 K is intermediate between that of alumina and silica gel, but for dehydration at 670–770 K, which is the range of practical catalytic interest, it is only slightly in excess of that for silica gel.[43]

We consider first the situation where the alumina content is <20 mol %. Since alumina is the minor component and is normally well distributed, much of the surface resembles that of silica gel. By analogy with the structures of Fig. 2.13a and b, two types of surface structures are possible for a surface aluminium (e.g. Fig. 2.13d and e). These are obviously related by the addition or removal of a water molecule. Infrared and n.m.r. studies[44, 45] have indicated that the proportion of surface aluminium carrying a hydroxyl group (i.e. e) does not exceed 10%. Basila et al.[34] have attempted to reconcile this result with the quite appreciable adsorption of ammonia and pyridine in the protonated form by suggesting that the latter may be formed using a Lewis site together with a hydroxyl carried on an adjacent part of the surface. In the light of the known facts, this model is probably the best yet available, although the situation could hardly be described as completely satisfactory. Other structures in which three-fold co-ordinated silicon functions as a Lewis site have been proposed.[38, 46] Although they would presumably be of very high acid strength, they would also be of high energy and hence of relatively sparse occurrence.

At higher alumina contents, that is possibly above 20 mol % and more probably above 30 mol %, Lewis acid sites are also likely to occur on the η-alumina phase, and these would be generally similar in nature to those described in a previous section dealing with alumina itself. A mullite-like phase would probably also generate Lewis sites.

Recently, Panchenkov and Kolesnikov[61] have argued strongly in favour of a model in which the catalytic properties of silica–alumina are due entirely to Lewis acid sites. The evidence, *inter alia* that hydrocarbon conversion continues to take place long after exchange has ceased between hydrogen in the hydrocarbon and deuterium previously introduced into the catalyst, certainly strongly points to Lewis site activity. Nevertheless, it is still difficult to exclude Brønsted site activity altogether, and at the moment it seems likely that depending on conditions, both sites can have catalytic activity.

Silica–alumina is relatively strongly acidic, and has a maximum acid strength as measured by amine titration with colour indicators, of at least $H_0 = -8·2$ after heating at 770 K. The amount of acidity varies with composition, but for alumina contents in the region of 20–40 mol %, a value of $0·2$–$0·4$ mol g^{-1} is fairly typical.

The results of γ-radiation alone or irradiation in a pile on the catalytic properties of silica–alumina, are varied: both increases and decreases in activity have been reported.[99] In any case, the changes in activity for reactions such as double bond isomerization in olefins and cumene cracking are relatively small and it seems clear that irradiation is at best of but marginal significance as a technique for the practical modification of catalyst performance. There is some evidence that the effect of irradiation is strongly dependent on the impurity level; thus it is reported[100] that irradiation of a highly pure silica–alumina leads to a small decrease in activity for cumene cracking, but with a less pure specimen the reverse trend was found. Heavy metal impurities are apparently involved.

A number of other oxide mixtures have been found to have acidic and/or basic properties, and these are therefore of interest for use as supports for dispersed metals in dual function catalysts. Some of the more important examples are summarized in Table 2.2. The properties of a number of other oxide mixtures, mainly those containing transition metal oxides, have been examined by Shibata et al,[148] while some comparative data are also given, *inter alia* by Nagarajan and Kuloor,[149] and by Zdzislaw.[150] Little is known about the detailed morphological structure, or of the nature of the acidic sites in these other mixed oxide systems. Often, but not always, these products are X-ray amorphous, and it is then at least reasonable to suppose that the morphology follows the general lines outlined previously for amorphous gels. The problem of suggesting likely structures for acidic sites is most acute for materials such as silica/zirconia where the nominal valency of each metal is the same, and where the detailed model outlined previously for silica/alumina cannot be directly applicable. One can readily propose structures for Brønsted acid

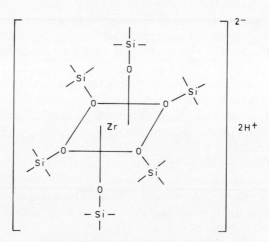

Fig. 2.14 Proposed Brønsted acid site in silica–zirconia.

sites based on the idea of an atom existing with a co-ordination number in excess of the value required to satisfy normal valence requirements, since electrical neutrality can be obtained by associating hydrogen ions with the centre. As an example, we may note the structure proposed by Hansford[156] following Thomas[35] for silica–zirconia, which is elaborated in Fig. 2.14. The proposal made by Plank,[66] originally for the Brønsted acidity in silica–alumina, is related to this. Plank proposed that there are aluminium atoms on the surface which are not substituted for silicon atoms into the silica structure, but which exist more or less isolated on the surface and which bind extra water molecules to become six-fold co-ordinated, as in Fig. 2.15. It would obviously be possible for analogous sites to those shown in Figs 2.14 and 2.15 to exist in other systems.

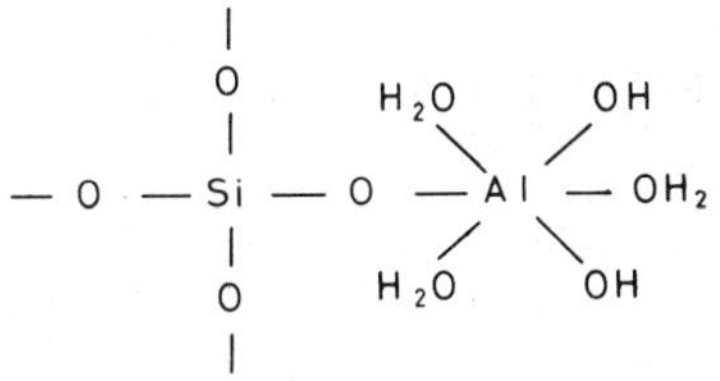

$$-O-\underset{\underset{O}{|}}{\overset{\overset{O}{|}}{Si}}-O-\underset{H_2O}{\overset{H_2O}{Al}}\underset{OH}{\overset{OH}{\diagup}}-OH_2$$

FIG. 2.15 Proposed Brønsted acid site in silica–alumina. After Plank, C. J. *J. Coll. Sci.* **2**, 413 (1947).

Lewis acidity can be pictured as any situation in which there is a vacancy in the co-ordination of a metal so it is possible for a molecule to be bound by electron pair donation, and the relation of this to the increase in the co-ordination in Plank's model for Brønsted sites will be apparent.

The idea suggested by Stone[151] to account for Lewis acidity in the α-alumina–chromia system should have applicability to other systems. α-Al_2O_3 is made up from linked AlO_6 octahedra, and an aluminium in the surface is likely to have an oxygen co-ordination of five rather than six. If a chromium replaces an aluminium isomorphously in the surface, the crystal field stabilization energy will tend to preserve six-fold co-ordination of the chromium, and this could occur by the transfer of an oxygen from an adjacent aluminium, thus generating an aluminium with an abnormally low four-fold co-ordination which should act as a strong Lewis acid site.

TABLE 2.2 Surface acidity of various mixed oxides

	maximum acid strength	
SiO_2/ZrO_2(12 mol % SiO_2)	$H_0^* \leqslant -8\cdot 2$	(a), (d)
SiO_2/Ga_2O_3(7·5 mol % SiO_2)	$\leqslant -8\cdot 2$	(a)
SiO_2/MgO(30 mol % SiO_2)	$\leqslant -6\cdot 4$(a); -3(d), (e)	
SiO_2/BeO(15 mol % SiO_2)	$\leqslant -6\cdot 4$	(a)
SiO_2/Y_2O_3(7·5 mol % SiO_2)	$\leqslant -5\cdot 6$	(a)
SiO_2/La_2O_3(7·5 mol % SiO_2)	$\leqslant -5\cdot 6$	(a)
Al_2O_3/B_2O_3(15 mol % B_2O_3)	$\leqslant -8\cdot 2$	(a), (c)
Al_2O_3/Cr_2O_3(17·5 mol % Cr_2O_3)	$\leqslant -8\cdot 2$	(a)
TiO_2/ZrO_2(50 mol % TiO_2)	$\leqslant -8\cdot 2$	(b)

* H_0, the Hammett acidity function.
(a) data from Tanabe, K., "Solids Acids and Bases", Kodansha, Tokyo and Academic Press, New York (1970).
(b) Shibata, K. and Kiyoura, T. *J. Res. Inst. Catal. Hokkaido Univ.* **19**, 35 (1971).
(c) Sato, M., Aonuma, T. and Shiba, T. *In* "Proceedings 3rd International Congress on Catalysis" (W. M. H. Sachtler, G. C. A. Schuit and P. Zwietering, eds) North-Holland, Amsterdam (1965), p. 396.
(d) Dzisko, V. A. ibid., p. 422.
(e) Benesi, H. A. *J. Phys. Chem.* **61**, 970 (1967), *J. Amer. Chem. Soc.* **78**, 5490 (1956).

11. Zeolites and Natural Clays*

Zeolites are crystalline aluminosilicates which are composed of ordered arrangements of SiO_4 and AlO_4 tetrahedra, again linked by corner sharing. The structures of catalytically important synthetic zeolites have been reviewed by Venuto and Landis[47] and by Barrer.[48] The commonly available synthetic zeolites can be thought of as consisting of interconnected cavities enclosed by an aluminosilicate skeleton. Thus, the basic

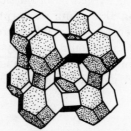

FIG. 2.16 Main structure of zeolite A.

* Well-known suppliers of various grades include: Union Carbide; Air Products, Houndry Divn.; W. R. Grace, Davison Chemical Divn.; Harshaw.

framework of zeolite A is shown in Fig. 2.16 and that of zeolite X and Y in Fig. 2.17. Zeolites in the chabazite group contain elongated-type cavities such as shown in outline in Fig. 2.18, while mordenite also contains tubular cavities. Also included in the chabazite group are offretite, gmelinite, erionite, levynite, Linde L and omega zeolite. Some characteristic parameters of the most important zeolites are given in Table 2.3, while Table 2.4 incorporates the specifications given by Meir[157] and Breck [158] for most of the known zeolites.

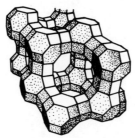

FIG. 2.17   Main structure of zeolites X, Y.

As with amorphous silica–alumina, electrical neutrality requires that for each AlO_4 unit there be one unit of positive charge provided by a cation. However, the cavity dimensions are dependent on the type of cation, and to a lesser extent on the Si/Al ratio, and these factors may be used to adjust the sieve properties of the zeolite. The lower limit of 1 on the Si/Al ratio is given by the empirical rule that AlO_4 tetrahedra can only be linked to SiO_4: to date there are no exceptions. The upper limit to this ratio generally lies in the region 4·5–5·0, although there are some exceptions such as a ferrierite where a ratio of 7 has been reported,[159] and a Linde L with a ratio of 6.4.

Zeolite Linde A provides an example of the effect of cation charge. In

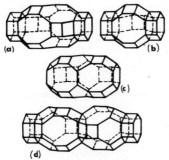

FIG. 2.18   Tubular-type cavities in zeolites of the chabazite group (a, chabazite; b, gmelinite; c, erionite; d, levynite).

TABLE 2.3 Parameters for some common zeolites

Type	Si/Al ratio	Approximate free dimensions of main features	Approximate number of guest molecules accommodated per cavity*
X, Y	~1·25(X) >2·4(Y)	"diameter" of main cavity, ~1·2 nm "diameter" of main cavity windows, ~0·8–1·0 nm†	32 H_2O; 17–19 Ar, N_2, O_2; 5·6 cyclopentane; 5·4 benzene; 4·6 toluene; 4·5 n-C_5H_{12}; 4·1 cyclohexane, 3·5 n-C_7H_{16}; 2·8 isoctane
A	1	"diameter" of main cavity (α-cage), 1·1–1·2 nm "diameter" of main cavity windows, 0·4–0·5 nm†	29 H_2O; 19–20 NH_3; 14–16 Ar, N_2, O_2; 15 H_2S; 12 CH_3OH; 10 SO_2; 9 CO_2; 5·4 n-C_3H_7OH; 4 n-C_4H_{10}
mordenite	~5	main tubular channels, roughly eliptical cross-section, 0·67 × 0·70 nm, 0·29 × 0·57 nm	
chabazite	2–3	windows (linking tubular channels), roughly eliptical cross-section, 0·37 × 0·42 nm	12–14 H_2O; 7·7 NH_3; 6–7 Ar, N_2, O_2; 4·9 CH_3NH_2; 4·3 CH_3Cl; 3·1 CH_2Cl_2

* From Barrer, R. M. *Endeavour*, **23**, 122 (1964).
† Depending on cation type.

TABLE 2.4 Zeolite parameters

Type	Diameter of main cavity windows (nm)*	Void fraction
analcime	0·26	0·18
laumontite	0·40 × 0·56	—
phillipsite	0·42 × 0·44; 0·28 × 0·48	0·31
gismondite	0·28 × 0·49; 0·31 × 0·44	—
yugawaralite	0·28 × 0·36; 0·32 × 0·43	0·27
Barrer's P1	0·35	0·41
Linde A	0·41	0·47
paulingite	0·39	0·49
faujasite	0·74	0·47
ZK–5	0·39	0·44
chabazite	0·37 × 0·42	0·47
gmelinite	0·69	0·44
Linde L	0·75	0·32
cancrinite	0·62	—
sodalite	0·26	—
levynite	0·33 × 0·51	—
erionite	0·36 × 0·48	0·35
offretite	0·36 × 0·48; 0·63	0·40
omega	0·75	0·38
brewsterite	0·23 × 0·50; 0·27 × 0·41	—
heulandite	0·24 × 0·61; 0·32 × 0·78	0·39
stilbite	0·41 × 0·62; 0·27 × 0·57	0·39
mordenite	0·67 × 0·70; 0·29 × 0·57	0·28
dachiardite	0·37 × 0·67; 0·36 × 0·48	0·32
epistilbite	0·32 × 0·53; 0·37 × 0·44	0·25
ferrierite	0·43 × 0·55; 0·34 × 0·48	—
bikitaite	0·32 × 0·49	—
natrolite	0·26 × 0·39	0·23
thomsonite	0·26 × 0·39	0·32
edingtonite	0·35 × 0·39	0·36

* Singly charged cations are assumed.

this case there is an increase in the window diameter when sodium ions are replaced by calcium ions ($2Na^+ \rightarrow Ca^{2+}$). The unit cell compositions are

Linde 4A $Na_{12}Al_{12}Si_{12}O_{48}$, $28H_2O$
Linde 5A $Na_6Ca_3Al_{12}Si_{12}O_{48}$, $30H_2O$

Table 2.5 shows the types of molecules capable of being adsorbed into some of the common zeolites (Mainwaring[179]).

78 STRUCTURE OF METALLIC CATALYSTS

TABLE 2.5 Typical molecular species adsorbed by some common zeolites

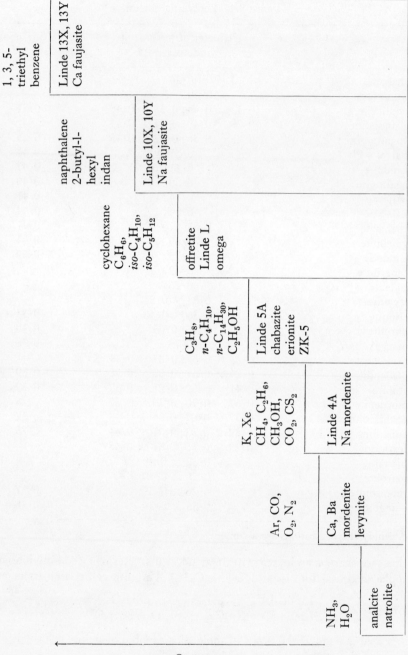

In Linde 4A, eight of the sodium ions lie at the centres of six-membered rings (Fig. 2.16), while the remaining four are adjacent to the eight-membered rings.

The cation sites in the faujasite (Linde X, Y) zeolites are shown in Fig. 2.19.[47] Site S_I is located at the centre of the hexagonal prism linking the two sodalite cages: it has 6-fold oxygen co-ordination. This site is well removed from the large channel system and is favoured by Ca^{2+} ions. Site S_{II} lies slightly forward of the free hexagonal face of the sodalite cage and has 3-fold oxygen co-ordination. S_{II} (and/or S_{III}) are favoured by singly charged cations. However, the precise site location and the associated electric field are dependent on the Si/Al ratio. This is particularly so for S_I sites because in this case the site is co-ordinated with oxygens from more than one tetrahedral unit. The fields in the zeolite cavities due to the presence of cationic charges have been calculated for some typical situations by Pickert *et al.*[49] and by Rabo *et al.*[50] The fact that these fields depend on cation type and Si/Al ratio has been used as a basis for discussing catalytic site specificity.[49, 50]

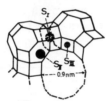

FIG. 2.19 Exchangeable cation positions in zeolites X, Y.

The total surface area as measured by gas adsorption is obviously dependent on the molecular size of the adsorbate: however, using nitrogen adsorption, 500–700 m^2g^{-1} is typical.

Hydrogen-exchanged zeolites are usually prepared by decomposition of the ammonium-exchanged material (NH_3 liberated) or by dehydration of the hydroxonium-exchanged material. There is a general tendency for a loss of crystallinity in the hydrogen-exchanged material but to varying extents: for instance, virtually complete with zeolite X,[51] but minimal with mordenite and chabazite.[52, 53]

The structural problem introduced by the presence of a surface exists just as much for the zeolites as for silica–alumina gel, and the general comments made previously concerning the latter also apply to the former (e.g. the structure of Brønsted and Lewis acid sites). It has been shown[46] that the sodium forms of zeolites X and Y contain enough hydroxyl groups to terminate the lattices at the surfaces. As with silica–alumina gel, the presence of a controlled amount of adsorbed water is important to the

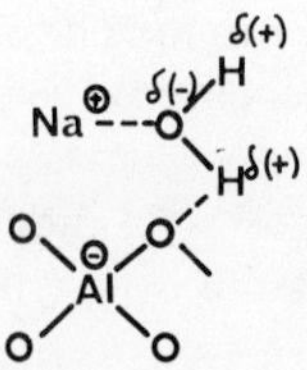

FIG. 2.20 Water adsorbed on Na-zeolite X to generate a Brønsted acid site. After Venuto, P. B. and Landis, P. S. *Advances in Catalysis*, **18**, 259 (1968).

catalytic properties of metal-exchanged zeolites, and this may also be discussed in terms similar to those used previously. However, there is a practical distinction between the use of silica–alumina gel and the zeolites in catalysts. The zeolites tend to lack crystallographic integrity in the hydrogen-exchanged form and they are usually used in a metal exchanged form. On the other hand, silica–alumina gel, being in any case non-crystalline, has been generally used in a hydrogen-exchanged form. Thus, in addition to the type of surface sites discussed previously, it has been suggested[47] that Brønsted acidity may be generated by a water molecule adsorbed in the vicinity of a cation, as in Fig. 2.20.

Zeolites which show an exceptionally high thermal stability have been prepared by the extraction of aluminium from dectionized Y zeolite,

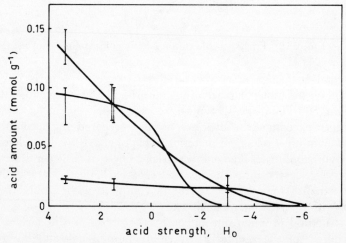

FIG. 2.21 Amount of surface acidity of some natural clays. Curve a, attapulgite; b, montmorillonite; c, kaolinite. The plot is cumulative so that each point indicates the amount of surface acidity having H_0 values equal to or less than the indicated H_0. Measurements made by amine titration using colour indicators. Reproduced with permission from Benesi, H. H. *J. Phys. Chem.* **61**, 970 (1957). Copyright: the American Chemical Society.

using EDTA.[153-155] Although the grosser features of the faujasite morphology are preserved, there are significant changes in the framework structure. The lost aluminium is, at least in part, replaced by silicon as a result of recrystallization of the framework, and the lost aluminium is also to be found in cation sites outside the framework. The concentration of weak Lewis acid sites is lower in the aluminium-deficient zeolite than in Y zeolite, but the concentration of strong Lewis or Brønsted acid sites is higher.[126]

Some other natural clays which have been used as catalyst supports are kaolinite, montmorillonite, and attapulgite. The acid strength of these materials has been measured[63] and the results are shown in Fig. 2.21. Clearly, in the original (unexchanged) form their acidity is relatively weak, but it can be significantly increased by hydrogen exchange.

12. Carbon*

The types of carbon which have been used as supports for metal catalysts are: charcoal, carbon black, graphite and synthetic molecular sieve carbon. Catalysts using these supports tend to be used mainly for laboratory and experimental purposes, although this is not exclusively so. The structure and chemistry of charcoal, carbon black and graphite have been reviewed previously on a number of occasions.[10, 54-59, 60]

Charcoal is prepared by the pyrolysis of natural or synthetic organic polymer, generally at <970 K. Subsequent activation consists of controlled oxidation, typically at about 1170 K, which removes residual pyrolysis products from the charcoal surface, as well as increasing the accessible surface area by removing some carbon so as to open blocked pores and by increasing the roughness of the internal surface. In the activated state, charcoal contains residual hydrogen (1–3 wt.%), oxygen (2–20 wt.%), sulphur (0–0·1 wt.%), nitrogen (0–0·2 wt.%) and inorganic residue. Most of the oxygen is adsorbed and is residual from the activation process. The nature of the inorganic residue is much dependent on the origin of the starting material, but for a good quality charcoal a total of 0·3–3 wt.% is typical, the residue containing compounds of the alkali metals, alkaline earths, iron, aluminium and silica.

The apparent density of charcoal lies in the region of 0·6–1·2 kg dm^{-3}, compared with graphite 2·27 kg dm^{-3} at room temperature. This indicates that the internal pore structure is very extensive, and specific surface areas up to the order of 1000 m^{2}g^{-1} may be readily obtained. In most cases the pores are far from uniform ("Saran" charcol from the pyrolysis

* Well-known suppliers of various grades include: Cabot Corpn.; Girdler-Sudchemie; Union Carbide.

of polyvinylidene chloride is an exception) and one can expect pores to occur over a very wide range of diameters, and the pores themselves to be of grossly non-uniform diameter along their lengths. Both open and closed pores occur. Typical pore structures have been summarized by Dubinin,[55] who reports three peaks in the pore diameter frequency distribution: micropores with an average diameter <2 nm and contributing >50% of the total pore volume; a restricted range of mesopores 10–20 nm, <15%, and large macropores >500 nm, <35%. As normally prepared, much of the charcoal structure is non-crystalline. Nevertheless, some degree of graphitization is present, and this consists of the presence of small, graphite-like crystallites within the amorphous matrix. The extent of graphitization increases if the charcoal is heated (in the absence of oxygen) above about 1270 K, but this would normally be avoided since it would be accompanied by a reduction in the pore volume and in the specific surface area. For typical charcoals the extent of graphitization would not exceed about 25%.

Carbon black is made by the controlled heating or burning of hydrocarbons with a limited supply of oxygen (air), and as usually produced contains 0–1 wt.% hydrogen, 0·05–10 wt.% oxygen, 0–0·5 wt.% sulphur, 0–0·05 wt.% nitrogen and 0·05–1·5 wt.% inorganic residue (the major inorganic constituents being the same as for charcoal). The primary carbon black particle is approximately spherical with a diameter in the range 3–500 nm. The majority of blacks have particles in the range 10–50 nm, and for any given black the particle size distribution is quite broad: for instance, a typical distribution for a black with a median diameter of 20 nm has an effective distribution stretching from about 4 to >50 nm. Although the primary particles of carbon black are very much less porous than charcoal, some porosity may still occur. When present, these are micropores of <2 nm average diameter. In practice, thermal treatment to increase the degree of graphitization involves heating to temperatures in the range 1270–3270 K in the absence of oxygen after initial preparation. The degree of graphitization for carbon blacks generally lies in the range 60–95%. As typical examples we may cite "Carbolac 1" which is a relatively porous black with a total specific surface of about 950 m^2g^{-1} of which about 40% is due to pores; the average pore diameter is about 1·7 nm and the average particle diameter is about 7 nm. At the other extreme, "Spheron MT" after graphitization at 3370 K has a specific surface of about 6·3 m^2g^{-1} and essentially zero internal surface and an average particle diameter of 560 nm.

High temperature treatment, in addition to increasing the degree of graphitization, also increases the homogeneity of the exposed surface, since there is an increasing tendency for this to consist of basal graphite planes. This is illustrated in Fig. 2.22 which shows the distribution of adsorption energies for argon adsorption on a series of blacks graphitized at temperatures in the range 1270–3370 K. Graphitization to a high degree, such as occurs at 3370 K, is also accompanied by a conversion of the particles from spherical to polyhedral shape; the model for this proposed by Kuroda and Akamatu[62] is illustrated in Fig. 2.23 in which each facet of the polyhedron consists of basal graphite planes due to the alignment of the graphite crystallites. Nevertheless, some carbons are resistant to extensive graphitization, even at 3370 K, and this appears to be due to cross-linking between the relatively disordered graphite crystallites which inhibits their reorientation and growth.

The basic particles of carbon black often aggregate into pellets of

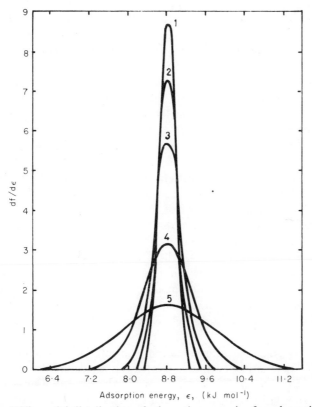

FIG. 2.22 Differential distribution of adsorption energies for adsorption of argon on carbon blacks thermally conditioned at various temperatures: 1, 3373 K; 2, 2973 K; 3, 2273 K; 4, 1773 K; 5, 1273 K. df is the fraction of surface having adsorption energy between ε and $\varepsilon + d\varepsilon$. After Ross, S. and Olivier, J. P. "On Physical Adsorption", Interscience, New York, p. 199 (1964).

macroscopic dimensions (in the range 0·1–1 mm diameter). This aggregation is a convenience for handling, but the particle packing is so loose that it has but a negligible influence on the properties of the black as a catalyst support or as an adsorbant.

FIG. 2.23 Schematic representation of carbon black particle before (a) and after (b and c) graphitization. After Kuroda, H. and Akamatu, H. *Bull. Chem. Soc. Japan*, **32**, 142 (1959).

Graphite may be either natural or synthetic. However, substantial quantities of high purity material are best obtained from synthetic sources, and it is this form which is the more attractive as a catalyst support. For use as a support this graphite has to be ground to a convenient particle size (e.g. in the range 20–100 mesh). Synthetic graphite is made by high temperature treatment of carbon compacted with a bituminous binder, and successive impregnations may be used to reduce the porosity. The final graphitizing temperature usually lies in the range 2870–3270 K. The degree of graphitization in a high quality synthetic graphite is certainly very high, but it remains a matter of dispute if it is in fact 100% or whether some non-crystalline material remains. The crystallites of the graphite in such a compact are variable in size depending on the conditions of graphitization, but 50–100 nm is a typical range. The density of synthetic graphite is appreciably less than that of pure crystalline graphite, 80% of the ideal density being not uncommon for a well graphitized sample. An alternative source is pyrolytic graphite formed by the decomposition of hydrocarbon gas at a hot surface at around 2270 K. Under suitable conditions, for instance simultaneous hot pressing with a shearing stress, extensive crystallite alignment can occur, and thus quite thick pieces of graphite can be produced whose properties very approximately match those of ideal graphite single crystals: the density can exceed 95% of the ideal crystal value.

In the early stages of graphitization, the interlayer spacing within the graphite-like crystallites is in excess of the ideal value of 0·335 nm; this spacing decreases towards the ideal value as graphitization proceeds. None the less, it should be remembered that diffusion of foreign atoms into the interplanar regions (intercalation) is possible even for an ideal graphite crystal and this process is presumably more facile for incompletely graphitized specimens. A non-ideal interplanar spacing will contribute towards a specimen density lower than for ideal graphite.

Despite the density of synthetic graphite being appreciably lower than the ideal value, the porosity is low in a well graphitized specimen, and is extremely low in nuclear grade material. Thus, one may consider graphite catalyst supports as essentially non-porous (except of course for the process of intercalation). Such pores as may be present are apparently completely closed off.

Powdered synthetic graphite would be expected to expose a large proportion of basal planes in the surface, but unless reheated to graphitizing temperatures after powdering, the surface is probably less uniform than that of a well graphitized carbon black. For a particle size in the region 0·2–1 mm the exposed surface area is only a few m^2g^{-1}.

As described above many activated carbons (particularly charcoals) have an extensive pore structure and have a substantial surface area. However, they have no molecular sieve properties because the average pore diameter is too large and the distribution of pore diameters is too wide. Trimm and Cooper[65] have described a series of molecular sieve carbons with a uniform pore structure of average diameter 0·4–0·6 nm. These were prepared by carbonizing (970–1070 K) various thermosetting organic polymers either directly or after coating the polymer into an activated carbon. These workers concluded that the pores are slit-shaped, and the best performance appeared to result from the use of polyfurfuryl alcohol. Heating above 1070 K results in a gross loss of available surface area due to pore closure. Unlike zeolite molecular sieves, carbon sieves are of course stable to acids and do not have strongly polar (hydrophilic) surfaces.

The result of surface oxidation of carbon by processes such as oxygen adsorption, or reactions with oxygen, or by reaction with other oxidizing agents such as permanganate, chromate, nitric acid, etc., leads to a wide range of oxygen-containing surface groups. The nature of these surface groups has been reviewed in detail by Boehm.[10] The oxygen in these surface groups is bound very strongly at the surface and removal is, in most cases, virtually impossible without simultaneous removal of some surface carbon atoms. Thus, outgassing at 1170 K is required to remove most of the surface oxygen, although removal first commences at temperatures as low as 370 K. Much, but not all, of this strongly bound surface oxygen is present in the form of various types of conventional functional groups, and these arise from attack on carbon rings which are situated at the periphery of the graphite sheets in the graphite-like crystallites. In those forms of carbon which are only poorly graphitized, oxygen is also no doubt bound at the surface of the non-crystalline component, but very little is known about the way in which this occurs. It will, in any case, be obvious that peripheral carbon rings in a graphite sheet must carry other atoms or groups of some sort on the non-fused carbon atoms, in order to satisfy ordinary valence requirements. It is really the nature of these latter atoms or groups which is at issue. In an idealized, oxygen-free graphite sheet, the peripheral rings would carry hydrogen atoms and the rings would remain aromatic in character. No doubt this description is appropriate to some parts of the carbon structure. However, oxygen is never totally absent, and its concentration increases as oxidation becomes more severe.

Oxygen-containing groups which have been identified with reasonable certainty are hydroxyl as phenolic groups, carbonyl in quinonoid structures, and carboxyl. Lactone and lactol groups are also likely but are less certain, and other types as yet unidentified are probable.[10] Phenolic and quinonoid groups can be formed by direct substitution into peripheral rings; however, the other groups require oxidative degradation with ring opening for their formation. In this connection it should be noted that maximum carbon acidity results from the most vigorous oxidizing conditions which no doubt produces carboxyl groups. However, the data summarized by Boehm[10] make it clear that a variety of surface groups is always present after oxidation, but their concentrations depend on the nature of the treatment. The detailed stereochemistry of these surface groups remains largely unknown, but will in any case, be heavily influenced by the detailed structure of the periphery of the graphite sheet. For instance, it is easy to see that only for a quite specific arrangement of peripheral rings is total conversion to a quinonoid structure possible. Finally we note that unpaired electrons are often detected (e.g. by e.s.r.) in carbons, and may well be associated with quinonoid structures.

Carbon is also able strongly to bind sulphur and chlorine but the way in which this occurs remains largely unknown, and one is reduced to inferential speculation based on comparison with the behaviour of oxygen.

The conclusion that oxidation of carbon leaves the surface of the basal graphite planes essentially free of oxygen (that is, that the oxygen is only bound at the periphery of these sheets) is important.[68] It implies that in a well graphitized specimen, much of the exposed surface will remain hydrophobic even after oxidation. It is known that oxidation does increase the proportion of surface which is hydrophilic[69, 70] but this is confined to that part of the exposed surface in which the edges of the graphite sheets occur.

13. Miscellaneous Supports

We indicate below some more important of the long list of various substances which at times have been used as "catalyst supports", but which have not been covered in the previous sections. We restrict ourselves to substances which are offered commercially for this purpose.

Silicon Carbide, Mullite, Zircon and Calcium Aluminate*

These are usually offered as materials of low specific surface area, and are prepared as sintered compacts in geometric shapes such as spheres, cylinders, rings, etc. Specific surface areas typically lie in the region $0 \cdot 1$–$0 \cdot 3$ m^{2}g^{-1}, and average pore diameters in the region 10–90 μm. It is the residual intergranular space in the compact which provides the porosity. Silicon carbide has a high thermal conductivity which can be valuable in highly exothermic reactions.

Asbestos

Several modifications are known, but the most important is chrysotile which is a magnesium silicate. Chrysotile has very high fibre flexibility so that it may be woven. The threads are made up from smaller fibrils which are individually about 15 nm diameter, and the specific surface area is typically about 10–20 m^{2}g^{-1}.

Monolithic Porous Support†

This type of support is usually made from a ceramic of the cordierite, mullite or alumina type, but metals have also been used to a limited extent. The basic form consists of a block of material through which pass holes of macroscopic dimensions, ranging in diameter from 1 to 20 mm. The holes may be parallel to or at an angle to the general direction of gas flow. Some examples are shown in Fig. 2.24. This type of support is intended to provide for rapid gas flow with a relatively low pressure drop. For instance, with a monolithic support with 3 mm diameter holes, the pressure drop for a given gas flow rate has been quoted as 5 % of that for a bed of the same thickness packed with 3 mm diameter spherical pellets. The geometric surface area exposed lies in the range 200–2000 m^{2}m^{-3}. In order to accommodate a reasonably large concentration of highly dispersed metal catalyst at the support surface, the monolithic ceramic may be coated with a thin (about 30 μm) layer of a porous material, e.g. γ-alumina or porous silica. This is readily done by treatment with a colloidal suspension of γ-alumina or silica (which are commercially available). With a monolithic metal support this is not possible because the porous layer lacks adhesion.

* Various suppliers include: Carborundum Refractories; Norton; Johns-Manville, Celite Divn.
† Various suppliers include: American Lava–3 M Co.; Corning; Dupont; Pilbrico.

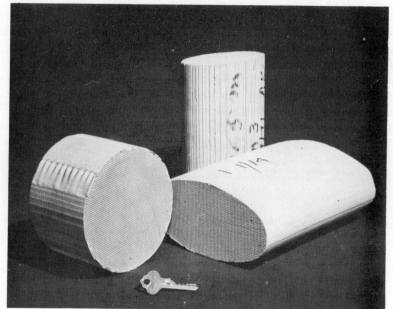

a

b

FIG. 2.24 Examples of monolithic porous ceramic catalyst supports, a. (upper); Corning W-1 supports, major component cordierite. b. (lower); DuPont Torvex supports, major component alumina or mullite. Reproduced with permission of Corning Inc. New York, and DuPont, Inc. Delaware.

2. SUPPORT MATERIALS

Inorganic Precipitates

Typical examples are barium sulphate, barium carbonate and calcium carbonate. Since these substances are generated without water of crystallization, dehydration leads to very little porosity, and the specific surface area is mainly dictated by the geometric particle size. Although this can vary a good deal, values of $0.5-5$ m^{2}g^{-1} are common.

Polymers and Resins

In a few cases, polymers such as nylon,[160-164] silk fibroin, polyterephthalate ester,[162,165] polyacrylonitrile[162,166] and polyvinylalcohol[162,167-170] have been used as supports for easily reducible metals such as platinum and palladium. These materials are usually used in granular or powder form. Their porosity is low: typical data are quoted by Bernard[161] for nylon 66 grains ($0.1-0.2$ mm diameter), when the BET surface area was 0.13 m^{2}g^{-1}, compared with a geometric surface area estimated for smooth spherical grains of about 0.05 m^{2}g^{-1}. Microporous plastic sheets are readily available commercially, and are often used as separators in lead–acid storage batteries. This material has a thickness of about 0.75 mm and the pores are very uniform with a diameter of about 5 μm. This material has found application as a support for metallic electrocatalysts in fuel cells.

Ion exchange resins have also been used as supports for platinum group metals.[171,172] These are available both in cationic and anionic exchangeable forms. Wolf[173] provides a useful summary, and reference also may be made to the manufacturers' technical literature. These resins are available with a wide range of specific surface areas and average pore diameters. For instance, Amberlite IR-120 which is a strong acid type cation exchange resin of the polystyrene-SO$_3$H type, is nonporous and has a specific surface area of <0.1 m^{2}g^{-1}. However, resins are also available with well developed porosity (macroreticular resins) and for these, specific surface areas in the range 50–500 m^{2}g^{-1} and average pore diameters in the range 10–100 nm are common.

All of these organic polymers and resins suffer to a greater or lesser extent from thermal instability by comparison with inorganic supports, and 420 K would be the maximum permissible temperature in most cases.

14. Single Crystal Surfaces

Single crystals can be cleaved or cut to expose surface planes of definite orientation. The extent of the surface obtained in this way obviously depends on the size of the single crystal available but, except in the case of mica which is exceptional, the upper limit is perhaps in the region of 10^3 mm². Single crystal surfaces of this sort are of no significance as supports for commercial catalysts, but they are useful as substrates for the preparation of evaporated metal film catalysts, particularly for the preparation of epitaxed films either as thick continuous, or as ultra-thin films. The latter consist of discrete very small metal particles, and can be used as laboratory models of technical catalysts.

Mica

Mica can, of course, be cleaved with great ease, so thin flexible sheets of relatively large dimensions (up to perhaps 0.02 m²) can be readily obtained. The preferred choice of material lies with natural muscovite mica, and only the highest quality should be considered so that reproducible behaviour from specimen to specimen can be established.

Mica cleaves on a plane which contains potassium ions, and these are approximately equally distributed between the two cleavage surfaces. Cleavage under UHV conditions produces a surface which is of ideal

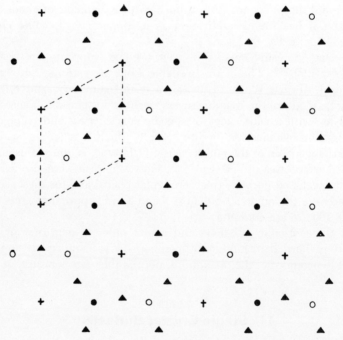

FIG. 2.25 Structure of mica. View of a mica surface. $+$, K$^+$ ($+$——$+$ = 0.53 nm), first layer; ▲, oxygen, second layer; ●, silicon, ○, ½ silicon, ½ aluminium, third layer. The surface unit mesh is indicated.

structure except that the potassium ions are disordered.[138] The structure of a muscovite mica surface is shown in Fig. 2.25.

Some depletion of the surface potassium concentration is caused by heating at 670–770 K in vacuum.[139] Poppa and Elliot[139] have examined the surface cleanliness of mica prepared in various ways. Specimens cleaved in air and subsequently examined by Auger electron spectroscopy showed the presence of appreciable carbon as a surface contaminant, and this was increased if, after cleavage, the surface was washed with an organic solvent. This carbon was not removed by UHV baking to 670 K for 6 h but removal may be effected by heating for 4 h in oxygen at about 10^{-3} Pa (about 10^{-5} Torr) at 720–770 K.

The exposure of mica to water or water vapour hydroxylates the surface,[146] and complete dehydroxylation requires heating to 870 K.[147] Mica has reasonable thermal stability up to 970 K, and excellent stability to 770 K.

A cleaved mica surface is of a very high degree of planarity on a molecular scale over quite big areas, and decoration reveals a very low concentration of topographic features.

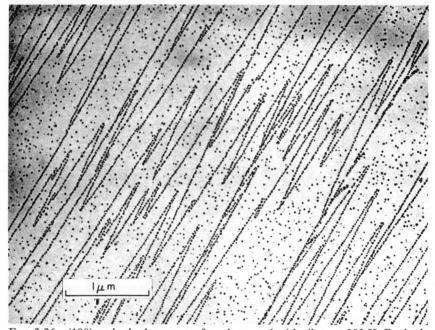

FIG. 2.26 (100) rocksalt cleavage surface decorated with silver at 293 K. Rocksalt crystal vacuum cleaved at 293 K. Reproduced with permission from Sella, C. and Trillat, J. J. *In* "Single Crystal Films" (M. H. Francombe and H. Sato, eds) Pergamon Press, London (1964), p. 201.

Alkali Halides

Although a number of the alkali halides have been used, rocksalt (sodium chloride) has been used far more extensively than any other. Surfaces have been prepared both by cleavage and by cutting and polishing. Crystals with the rocksalt structure always cleave to expose the (100) surface, so other orientations must be obtained by cutting and polishing. Solvent sawing is the preferred technique for cutting.

A rocksalt cleavage face is far from absolutely smooth, and Fig. 2.26 shows the large number of cleavage steps revealed by silver decoration of a vacuum cleaved rocksalt (100) surface. Moreover, a rocksalt surface is

FIG. 2.27 Electron micrograph of shadowed replica of (100) rocksalt surface after thermal etching for 1 hour at 673 K in vacuum. Reproduced with permission from Sella, C. and Trillat, J. J. *In* "Single Crystal Films" (M. H. Francombe and H. Sato, eds) Pergamon Press, London (1964) p. 201.

subject to thermal etching if heated to a sufficiently high temperature in vacuum, and if this occurs there is a marked increase in surface roughness. Figure 2.27 illustrates this for a (100) rocksalt face which has been heated for about 1 h at 670 K under vacuum.

Some use has been made of evaporated rocksalt as a substrate for subsequent film deposition (cf. ref. 20). Such an evaporated rocksalt layer is of course microcrystalline, but consists largely of single crystals each exposing (100) faces. For comparison with Fig. 2.26, Fig. 2.28 shows a micrograph of an evaporated rocksalt layer after decoration with gold: the presence of growth steps on the (100) surfaces is clearly revealed, and a high concentration of gold crystallites in some areas indicates the presence of part of the rocksalt surface consisting of high index planes or of highly faceted regions.

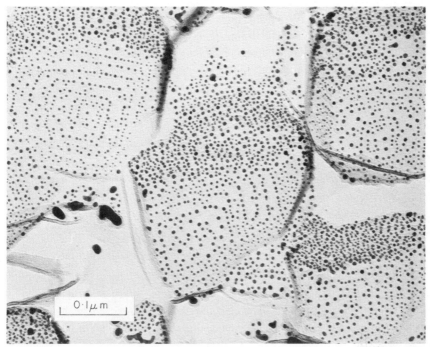

FIG 2.28 Evaporated rocksalt decorated with gold at 358 K. Reproduced with permission from Baker, B. G. and Bruce, L. A. *Trans. Faraday Soc.* **64**, 2533, (1968).

Miscellaneous

A very large number of substances are, of course, available as single crystal specimens of substantial size. However, only a few have been studied as evaporated film substrates. Single crystal magnesia has sometimes been used as an alternative to rocksalt. It can be cleaved on a (100) plane and has the advantage of greater thermal stability than rocksalt. Cleavage, however, is more difficult. No other substance matches mica in terms of ease of cleavage, size of cleaved surface or thinness and flexibility of specimen. However, there are some other substances which offer somewhat similar properties to mica as substrates for metal film epitaxy, for instance by favouring (111) orientation with f.c.c. metals. These are the hexagonal (001) surface planes of graphite, molybdenum disulphide and α-alumina. For both graphite and molybdenum disulphide, this is a cleavage plane. However, α-alumina cannot be cleaved, and the crystal must be cut and polished. The as-polished surface of α-alumina is highly disordered, and flaws show up on etching. To obtain a good LEED pattern requires removal of the disordered material by ion bombardment and annealing. This situation can also be expected with other surfaces prepared by cutting and polishing.

References

1. Alkins, P. B., Shull, C. G. and Roess, L. C. *Ind. Eng. Chem.* **37**, 327 (1945).
2. Iler, R. K. "The Colloid Chemistry of Silica and Silicates", Cornell University Press, Ithaca (1955).
3. de Boer, J. H. and Vleeskens, J. M. *Proc. Konin. Nederlandse Akad.* **61B**, 85 (1958).
4. Wirzing, G. *Naturwiss.* **30**, 13, 466 (1963).
5. Fripiat, J. J. and Uytterhoeven, J. *J. Phys. Chem.* **66**, 800 (1962).
6. de Boer, J. H. and Vleeskens, J. M. *Proc. Konin. Nederlandse Akad.* **61B**, 2 (1958).
7. Davydov, V. Y., Kiselev, A. V. and Zhuravlev, L. T. *Trans. Faraday Soc.* **60**, 2254 (1964).
8. Peri, J. B. and Hensley, A. L. *J. Phys. Chem.* **72**, 2926 (1968).
9. James, R. O. Ph.D. Thesis, Univeristy of Melbourne, Australia (1971).
10. Boehm, H. P. *Advances in Catalysis* **16**, 179 (1966).
11. Vleeskens, J. M. Thesis, Delft University of Technology, The Netherlands, (1959).
12. Haldeman, R. G. and Emmett, P. H. *J. Amer. Chem. Soc.* **78**, 2917 (1956).
13. Vesely, V. and Pekarek, V. *Tantala* **19**, 220 (1972).
14. Parks, G. A. *In* "Equilibrium Concepts in Natural Water Systems", *Advances in Chemistry Series* **67**, 121 (1967).
15. Cant, N. W. and Little, L. H. *Can. J. Chem.* **42**, 802 (1964).
16. Lippens, B. C. *Chem. Weekbl.* **62**, 336 (1966).
17. de Boer, J. H., Fortuin, J. M. H. and Steggerda, J. J. *Proc. Konin. Nederlandse Akad.* **57B**, 170 (1954)
18. de Boer, J. H., Fortuin, J. M. H. and Steggerda, J. J. *Proc. Konin. Nederlandse Akad.* **57B**, 434 (1954).
19. de Boer, J. H., Steggerda, J. J. and Zwietering, P. *Proc. Konin. Nederlandse Akad.* **59B**, 435 (1956).
20. Stumpf, H. C., Russell, A. S., Newsome, J. W. and Tucker, C. M. *Ind. Eng. Chem.* **42**, 1398 (1950).
21. Ervin, G. *Acta Cryst.* **5**, 103 (1952).

2. SUPPORT MATERIALS

22. Leonard, A. J., van Cauwelaert, F. and Fripiat, J. J. *J. Phys. Chem.* **71**, 695 (1967).
23. Lippens, B. C. and Steggerda, J. J. *In* "Physical and Chemical Aspects of Adsorbents and Catalysts" (B. G. Linsen, ed.), Academic Press, London (1970), p. 171.
24. Russell, A. S. and Cochran, C. N. *Ind. Eng. Chem.* **42**, 1336 (1950).
25. Peri, J. B. and Hannan, R. B. *J. Phys. Chem.* **64**, 1526 (1960).
26. Peri, J. B. *J. Phys. Chem.* **69**, 211, 220, 231 (1965).
27. Pines, H. and Manassen, J. *Advances in Catalysis* **16**, 49 (1966).
28. Eucken, A. and Wicke, E. *Naturwiss.* **32**, 161 (1944).
29. Finch, J. N. and Clark, A. *J. Catal.* **19**, 292 (1970).
30. Benesi, H. A., Curtis, R. M. and Studer, H. P. *J. Catal.* **10**, 328 (1968).
31. Webb, A. N. *Ind. Eng. Chem.* **49**, 261 (1957).
32. Ryland, L. B., Tamele, M. W. and Wilson, J. N. *In* "Catalysis" (P. H. Emmett, ed.), Vol. 7, Reinhold, New York (1960), p. 1.
33. Hall, W. K., Leftin, H. P., Cheselske, F. J. and O'Reilly, D. E. *J. Catal.* **2**, 506 (1963).
34. Basila, M. R., Kantner, T. R. and Rhee, K. H. *J. Phys. Chem.* **68**, 3197 (1964).
35. Thomas, C. L. *Ind. Eng. Chem.* **41**, 2564 (1949).
36. Weisz, P. B. *Ann. Rev. Phys. Chem.* **21**, 175 (1970).
37. Leonard, A., Suzuki, S., Fripiat, J. J. and de Kimpe, C. *J. Phys. Chem.* **68**, 2608 (1964).
38. Fripiat, J. J., Leonard, A. and Uytterhoeven, J. B. *J. Phys. Chem.* **69**, 3274 (1965).
39. Oblad, A. G., Milliken, T. H. and Mills, G. A. *Advances in Catalysis* **3**, 199 (1951).
40. Parry, E. P. *J. Catal.* **2**, 371 (1963).
41. Mapes, J. E. and Eischens, R. P. *J. Phys. Chem.* **58**, 1059 (1954).
42. Basila, M. R. and Kantner, T. R. *J. Phys. Chem.* **70**, 168 (1966).
43. Haldeman, R. G. and Emmett, P. H. *J. Amer. Chem. Soc.*, **78**, 2917 (1956).
44. Hall, W. K., Larson, J. G. and Gerberich, H. R. *J. Amer. Chem. Soc.* **85**, 3711 (1963).
45. Basila, M. R. *J. Phys. Chem.* **66**, 2223 (1962).
46. Uytterhoeven, J. B., Christner, L. G. and Hall, W. K. *J. Phys. Chem.* **69**, 2117 (1965).
47. Venuto, P. B. and Landis, P. S. *Advances in Catalysis* **18**, 259 (1968).
48. Barrer, R. M. *Endeavour* **23**, 122 (1964).
49. Pickert, P. E., Rabo, J. A., Dempsey, E. and Shomaker, V. *In* "Proceedings 3rd International Congress on Catalysis" (W. M. H. Sachtler, G. C. A. Schuit and P. Zwietering, eds), North-Holland, Amsterdam (1965), p. 714.
50. Rabo, J. A., Angell, C. L., Kasai, P. H. and Shomaker, V. *Discussions Faraday Soc.* **41**, 328 (1966).
51. Rabo, J. A., Pickert, P. E., Stamires, D. N. and Boyle, J. E. *In* "Actes du Deuxieme Congres de Catalyse", Editions Technip, Paris (1961), p. 2055.
52. Barrer, R. M. *Nature* **164**, 112 (1949).

53. Keough, A. H. and Sand, L. B. *J. Amer. Chem. Soc.* **83**, 3536 (1961).
54. Kipling, J. J. *Quart. Rev.* **10**, 1 (1956).
55. Dubinin, M. M. *Quart. Rev.* **9**, 101 (1955).
56. Courty, C. "Charbons Active", Gauthier-Villars, Paris (1952).
57. Mantell, C. L. "Industrial Carbon", Van Nostrand, New York (1946).
58. Hassler, J. W. "Active Carbon", Chemical Publishing Co., New York (1951).
59. Gibson, J. and Riley, H. L. *Fuel* **21**, 36 (1942).
60. van de Plas Th. *In* "Physical and Chemical Aspects of Adsorbents and Catalysts" (B. G. Linsen, ed.), Academic Press, London (1970), p. 425.
61. Panchenkov, G. M. and Kolesnikov, I. M. *Russ. J. Phys. Chem.* **44**, 500 (1970).
62. Kuroda, H. and Akamatu, H. *Bull. Chem. Soc. Japan* **32**, 142 (1959).
63. Tanabe, K. "Solid Acids and Bases", Kodansha, Tokyo and Academic Press, New York (1970).
64. Tanabe, K. and Katayama, M. *J. Res. Inst. Catal. Hokkaido Univ.* **7**, 106 (1959).
65. Trimm, D. L. and Cooper, B. J. *Chem. Comm.* **1970**, 477.
66. Plank, J. C. *J. Coll. Sci.* **2**, 413 (1947).
67. Malinowski, S., Szczepanska, S., Bielanski, A. and Sloczynski, J. *J. Catal.* **4**, 324 (1965).
68. Henning, G. R. *In* "Proc. Conf. Carbon 5th", Pergamon, London (1962), p. 143.
69. Healey, F. H., Yu, Y. F. and Chessick, J. J. *J. Phys. Chem.* **59**, 399 (1955).
70. Kraus, G., *J. Phys. Chem.* **59**, 343 (1955).
71. Anderson, P. J. and Horlock, R. F. *Trans. Faraday Soc.* **58**, 1993 (1962).
72. Horlock, R. F., Morgan, P. L. and Anderson, P. J. *Trans. Faraday Soc.* **59**, 721 (1963).
73. Anderson, P. J. and Morgan, P. L. *Trans. Faraday Soc.* **60**, 930 (1964).
74. Anderson, P. J., Horlock, R. F. and Oliver, J. F. *Trans. Faraday Soc.* **61**, 2754 (1965).
75. de Vleesschauwer, W. F. N. M. *In* "Physical and Chemical Aspects of Adsorbents and Catalysts" (B. G. Linsen, ed.), Academic Press, London (1970), p. 265.
76. Tench, R. J., Giles, D. and Kibblewhite, J. F. J. *Trans. Faraday Soc.* **67**, 854 (1971).
77. Primet, M., Pichat, P. and Mathieu, M.-V. *J. Phys. Chem.* **75**, 1216 (1971).
78. Jones, P. and Hockey, J. A. *Trans. Faraday Soc.* **67**, 2669 (1971).
79. Gravelle, P. C., Juillet, F., Merieudeau, P. and Teichner, S. J. *Discussions Faraday Soc.* **52**, 140 (1971).
80. Yates, D. J. C. *J. Phys. Chem.* **65**, 746 (1961).
81. Kiselev, A. V. and Uvarou, A. V. *Surface Sci.* **6**, 399 (1967).
82. Lewis, K. E. and Parfitt, G. D. *Trans. Faraday Soc.* **62**, 204, (1966).
83. Jackson, P. and Parfitt, G. D. *Trans. Faraday Soc.* **67**, 2469 (1971).
84. Herrmann, M. and Boehm, H. P. *Z. Anorg. Chem.* **368**, 73 (1969).

85. Munvera, G. and Stone, F. S. *Discussions Faraday Soc.* **52**, 205 (1971).
86. Boehm, H. P. *Discussions Faraday Soc.* **52**, 264 (1971).
87. Primet, M., Pichat, P. and Mathieu, M.-V. *J. Phys. Chem.* **75**, 1221 (1971).
88. Parfitt, G. D., Ramsbotham, J. and Rochester, C. H. *Trans. Faraday Soc.* **67**, 841 (1971).
89. Parfitt, G. D., Ramsbotham, J. and Rochester, C. H. *Trans. Faraday Soc.* **67**, 1500 (1971).
90. Lieflander, M. and Stober, W. *Z. Naturforsch.* **156**, 411 (1960).
91. Hollabaugh, C. M. and Chessick, J. J. *J. Phys. Chem.* **65**, 109 (1961).
92. Day, R. E., Parfitt, G. D. and Peacock, J. *Discussions Faraday Soc.* **52**, 215 (1971).
93. Jones, P. and Hockey, J. A. *Trans. Faraday Soc.* **67**, 2679 (1971).
94. Peri, J. B. *J. Phys. Chem.* **70**, 2937 (1966).
95. Mikhail, R. Sh., Nashed, Sh. and Khalil, A. M. *Discussions Faraday Soc.* **52**, 187 (1971).
96. Ramsay, J. D. F. *Discussions Faraday Soc.* **52**, 49 (1971).
97. Nelson, R. L., and Hale, J. W. *Discussions Faraday Soc.* **52**, 77 (1971).
98. Leonard, A. J., Ratnasamy, P., Declerck, F. D. and Fripiat, J. J. *Discussions Faraday Soc.* **52**, 98 (1971).
99. Taylor, E. H. *Advances in Catalysis* **18**, 111 (1968).
100. Trayward, P. and Orsini, L. *Comptes Rendus* **252**, 873 (1961).
101. Berkman, S., Morrell, J. C. and Egloff, G. "Catalysis", Reinhold, New York (1940).
102. Innes, W. B. *In* "Catalysis" (P. H. Emmett, ed.), Vol. 1, Reinhold, New York (1954), p. 245.
103. Benson, G. C. and Yun, K. S. *In* "The Solid–Gas Interface" (E. A. Flood, ed.), Vol. 1, Dekker, New York (1967), p. 203.
104. Charig, J. M. and Skinner, D. K. *In* "The Structure and Chemistry of Solid Surfaces" (G. A. Somorjai, ed.), Wiley, New York (1969), p. 34–1.
105. Lander, J. J. *In* "Progress in Solid State Chemistry" (H. Reiss, ed.), Vol. 2, Pergamon, New York, 1965, p. 26.
106. Marsh, J. B. and Farnsworth, H. E. *Surface Sci.* **1**, 3 (1964).
107. MacRae, A. U. and Gobeli, G. W. *J. Appl. Phys.* **35**, 1629 (1964).
108. Jona, F. *IBM J. Res. Develop.* **9**, 375 (1965).
109. Campbell, B. D., Haque, C. A. and Farnsworth, H. E. *In* "The Structure and Chemistry of Solid Surfaces" (G. A. Somorjai, ed.), Wiley, New York (1969), p. 33–1.
110. Fiermans, L. and Vennik, J. *Surface Sci.* **18**, 317 (1969).
111. Hochstrasser, G. and Antonini, J. F. *In* "The Structure and Chemistry of Solid Surfaces" (G. A. Somorjai, ed.), Wiley, New York, 1969, p. 36–1.
112. Rijnten, H. Th. *In* "Physical and Chemical Aspects of Adsorbents and Catalysts" (B. G. Linsen, ed.), Academic Press, London (1970), p. 315.
113. Livage, J., Vivien, D. and Mazieres, C. *In* "Reactivity of Solids: Proceedings 6th International Symposium on the Reactivity of Solids" (J. W. Mitchell,

R. C. DeVries, R. W. Roberts and P. Cannon, eds), Wiley-Interscience, New York (1969), p. 271.
114. Garvie, R. C. *J. Phys. Chem.* **69**, 1238 (1965).
115. Cypres, R., Wollast, R. and Raucq. J. *Ber. Deut. Keram. Ges.* **40**, 527 (1963).
116. Clearfield, A. *Rev. Pure Appl. Chem.* **14**, 91 (1964).
117. Mikhail, R. Sh. and Fahim, R. B. *J. Appl. Chem.* **17**, 147 (1967).
118. Fahim, R. B., Gabr, R. M. and Mikhail, R.Sh. *J. Appl. Chem.* **20**, 216 (1970).
119. Holmes, H. F., Fuller, E. L. and Secoy, C. H. *J. Phys. Chem.* **72**, 2293 (1968).
120. Holmes, H. F., Fuller, E. L., Gammage, R. B. and Secoy, C. H. *J. Coll. and Interface Sci.* **28**, 421 (1968).
121. Fuller, E. L., Holmes, H. F. and Secoy, C. H. *J. Phys. Chem.* **70**, 1633 (1966).
122. Holmes, H. F. and Secoy, C. H. *J. Phys. Chem.* **69**, 151 (1965).
123. Winfield, M. E. *In* "Catalysis" (P. H. Emmett, ed.), Vol. 7, Reinhold, New York (1960), p. 93.
124. Breysse, M., Claudel, B., Prettre, M. and Veron, J. *J. Catal.* **24**, 106 (1972).
125. Boreskov, G. K. *Discussions Faraday Soc.* **41**, 263 (1966).
126. Beaumont, R., Pichat, P., Barthomeuf, D. and Trambouze, Y. *In* "Proceedings 5th International Congress on Catalysis" (J. W. Hightower, ed.), North-Holland, Amsterdam (1973), p. 343.
127. Knozinger, H. *Discussions Faraday Soc.* **52**, 278 (1971).
128. Tret'yakov, N. E., Pozdnyakov, D. V., Oranskaya, O. M. and Filimonov, V. N. *Russ. J. Phys. Chem.* **44**, 596 (1970).
129. Carruthers, J. D., Fenerty, J. and Sing, K. S. W. *In* "Reactivity of Solids: Proceedings 6th International Symposium on the Reactivity of Solids" (J. W. Mitchell, R. C. de Vries, R. W. Roberts and P. Cannon, eds) Wiley-Interscience, New York (1969), p. 127.
130. Burwell, R. L. and Taylor, H. S. *J. Amer. Chem. Soc.* **58**, 697 (1936).
131. Burwell, R. L., Read, J. F., Taylor, K. C. and Haller, G. L. *Z. Phys. Chem.* (N. F.) **64**, 18 (1969).
132. Burwell, R. L., Taylor, K. C. and Haller, G. L. *J. Phys. Chem.* **71**, 4580 (1967).
133. Deren, J., Haber, J., Podgorecka, A. and Burzyk, J. *J. Catal.* **2**, 161 (1963).
134. Baker, F. S., Carruthers, J. D., Day, R. E., Sing, K. S. W. and Stryker, L. J. *Discussions Faraday Soc.* **52**, 173 (1971).
135. Dixon, J. K. and Longfield, J. E. *In* "Catalysis" (P. H. Emmett, ed.), Vol. 7, Reinhold, New York (1960), p. 281.
136. Mazdiyasni, K. S., Lynch, C. T. and Smith, T. S. *J. Amer. Ceramic Soc.* **50**, 532 (1967).
137. Steiner, H. *In* "Catalysis" (P. H. Emmett, ed.), Vol. 4, Reinhold, New York (1956), p. 529.
138. Muller, K. and Chang, C. C. *Surface Sci.* **14**, 39 (1969).

139. Poppa, H. and Elliot, A. C. *Surface Sci.* **24**, 149 (1971).
140. Frazer, J. C. W., Patrick, W. A. and Smith, H. E. *J. Phys. Chem.* **31**, 897 (1927).
141. Sewell, P. A. *Nature* **217**, 441 (1968).
142. Sewell, P. A. and Morgan, A. M. *Nature* **215**, 325 (1967).
143. Skatulla, W. *Silikattechnik* **13**, 19 (1962).
144. Johnson, J. W. and Holloway, D. C. *Phil. Mag.* **17**, 899 (1968).
145. Anderson, R. B., McCartney, J. T., Hall, W. K. and Hofer, L. J. E. *Ind. Eng. Chem.* **39**, 1618 (1947).
146. Shishelova, T. I., Metsik, M. S. and Baikovskaya, E. S. *Zh. Priklad. Spektrosk.* **11**, 921 (1969).
147. Brauer, K. H. and Simon, D. *Acta Cryst.* **21**, A192 (1966).
148. Shibata, K., Kiyoura, T. and Tanabe, K. *J. Res. Inst. Catal. Hokkaido Univ.* **18**, 189 (1970).
149. Nagarajan, V. and Kuloor, N. R. *Indian J. Tech.* **4**, 46 (1966).
150. Zdzislaw, C. *Zesz. Nauk. Politech. Slask. Chem.* **1968**, 3.
151. Pepe, F. and Stone, F. S. *In* "Proceedings 5th International Congress on Catalysis" (J. W. Hightower, ed.), North-Holland, Amsterdam (1973), p. 137.
152. Rice, R. W. and Haller, G. L. *In* "Proceedings 5th International Congress on Catalysis" (J. W. Hightower, ed.), North-Holland, Amsterdam (1973), p. 317.
153. Kerr, G. T. *J. Phys. Chem.* **72**, 2594 (1968).
154. McDaniel, C. V. and Maher, P. K. "Molecular Sieves", Soc. Chem. Ind., London (1968), p. 186.
155. Peri, J. B. *In* "Proceedings 5th International Congress on Catalysis" (J. W. Hightower, ed.), North-Holland, Amsterdam (1973), p. 329.
156. Hansford, R. C. *Advances in Catalysis* **4**, 17 (1952).
157. Meier, W. M. "Molecular Sieves", Soc. Chem. Ind., London (1968), p. 10.
158. Breck, D. W. "Molecular Sieves", Soc. Chem. Ind., London (1968), p. 47.
159. Wise, W. S., Nokleberg, W. J. and Kokinos, M. *Amer. Mineral.* **54**, 887 (1969).
160. Harrison, D. P. and Rase, H. F. *Ind. Eng. Chem. Fundamentals* **6**, 161 (1967).
161. Bernard, J. R. Ph.D. Thesis, University of Lyon, France, 1972.
162. Tyurenkova, O. A. *Russ. J. Phys. Chem.* **43**, 1167 (1969).
163. Tyurenkova, O. A. and Volkova, V. E. *Russ. J. Phys. Chem.* **43**, 1587 (1969).
164. Tyurenkova, O. A. and Zhakin, V. P. *Russ. J. Phys. Chem.* **44**, 210 (1970).
165. Tyurenkova, O. A. and Lankin, S. F. *Russ. J. Phys. Chem.* **43**, 76 (1969).
166. Tyurenkova, O. A. and Chimarova, L. A. *Russ. J. Phys. Chem.* **44**, 1289 (1970).
167. Tyurenkova, O. A. and Chimarova, L. A. *Russ. J. Phys. Chem.* **44**, 208 (1970).
168. Tyurenkova, O. A. *Russ. J. Phys. Chem.* **43**, 69 (1969).
169. Tyurenkova, O. A., Sokol'skii, D. V. and Chimarova, L. A. *Russ. J. Phys. Chem.* **43**, 650 (1969).

170. Tyurenkova, O. A. and Chimarova, L. A. *Russ. J. Phys. Chem.* **44**, 48 (1970).
171. Lazcano, L. R. Ph.D. Thesis, University of Lyon, France, 1971.
172. French Patents, Nos. 1583594, 1583593, 1583037.
173. Wolf, F. J. "Separation Methods in Organic Chemistry and Biochemistry", Academic Press, New York (1969).
174. Garten, V. A. and Weiss, D. E. *Rev. Pure and Appl. Chem.* **7**, 69 (1957).
175. Burwell, R. L., Pearson, R. G., Haller, G. L., Tjok, P. B. and Chock, S. P. *Inorg. Chem.* **4**, 1123 (1965).
176. Sacconi, L. *Discussions Faraday Soc.* **7**, 173 (1949).
177. Burwell, R. L., Haller, G. L., Taylor, K. C. and Read, J. F. *Advances in Catalysis* **20**, 1 (1969).
178. Dyne, S. R., Butt, J. B. and Haller, G. L. *J. Catal.* **25**, 378 (1972).
179. Mainwaring, D. E. *Proc. Roy. Soc. Austral. Chem. Inst.* **40**, 293 (1973).

CHAPTER 3

Massive Metal Catalysts

	page
1. IDEAL SURFACES	103
2. UNSUPPORTED BULK METAL	113
3. EVAPORATED METAL FILMS	130
4. ALLOY CATALYSTS	143

Catalysts which consist of metal in a massive state, that is not in a finely divided or dispersed form, are mainly of importance for laboratory studies where the aim is one of correlating catalytic activity with the chemical identity of the metal or with the nature of the metallic surface exposed to the gas. Compared with conventional supported catalysts, massive metal specimens allow much better opportunity for control of variables such as surface structure and surface contamination. Nevertheless, it must always be remembered that, to the extent that catalytic processes may be intrinsically dependent on metal particle size, massive metal catalysts may be intrinsically different from highly dispersed catalysts. Furthermore, clean catalysts differ from practical catalysts. However, it is by making comparisons of this sort that light can be shed on some catalytic reaction mechanisms, and the need to make such a comparison is one of the justifications for working with massive metal specimens. Although there are many catalytic data recorded in the literature which have been obtained with massive metal catalysts where neither surface contamination nor surface structure has been adequately controlled, with the techniques now available there is little excuse for a continuation of this state of affairs. In one way or another surface contamination can now usually be eliminated as a significant experimental variable, and with an appreciable number of metals it is possible to control both this and the surface structure.

One further point must also be made: no matter how much care is taken in specimen and surface preparation, the cleanliness and structure cannot simply be *assumed*. Efforts should always be made to assess surface cleanliness, and the available techniques include the measurement of workfunction, Auger electron spectroscopy and LEED. Surface structure needs to be studied directly: various electron microscopic techniques are available including replication, decoration and scanning microscopy. Obviously, electron microscopy cannot be done on a specimen before catalytic use without introducing surface contamination. Specimens can be examined after catalytic use, but with a distinct possibility that the reaction itself may have modified the surface structure; for the structure before reaction one is forced to use electron microscopy on duplicate specimens.

Examination by LEED of carefully prepared and well annealed clean metal single crystal surfaces of low indices has shown that in almost all cases the actual surface structure is such that the atoms occupy positions quite close to those expected from the known structure of the bulk crystal. This is one reason why it is worthwhile discussing ideal surface structures in some detail.

Having fewer neighbours than in the bulk, one would expect a surface atom to be bound rather less strongly, and there is evidence from the temperature dependence of LEED reflections and from Mössbauer spectroscopy with highly dispersed metal that this is so. For various (100), (110) and (111) surfaces of f.c.c. and b.c.c. metals (nickel, palladium, platinum, silver, copper, iridium, lead, tungsten, molybdenum, chromium, niobium), the ratio of Debye temperatures for surface atoms and bulk atoms is in the range $0 \cdot 4$–$0 \cdot 85$,[65,103] and the RMS vibrational amplitudes normal to the surface are in the range $1 \cdot 2$–$2 \cdot 5$ times greater than those for atoms in the bulk. Due to anharmonicity in the atomic vibrations, an increase in the RMS vibrational amplitude leads to a dilation of the surface layers of the metal in a direction normal to the surface. The extent to which this occurs is, however, relatively small, not exceeding 5% overall[63] and more probably being in the region of 1–2%.[64] On the other hand, with the (110) surface of aluminium (a non-transition element) there appears to be a normal surface contraction of 10–15%.[145] The reason for this exceptional behaviour is not known at present. With three metals, gold, platinum and iridium,[66,134] examination by LEED has revealed the presence of rearranged surface layers which are stable at room temperature and which exist under conditions apparently approaching an impurity-free situation. After ion-bombardment cleaning and annealing, the (100) face of these metals gave LEED patterns which could be indexed on the basis that the rearrangement is confined to an outermost layer of metal atoms. With Pt(100) both (1×2) and (1×5) patterns were observed from the overlayer, while with Au (100) and Ir (100), (1×5) patterns were observed. The (1×5) pattern undoubtedly arises from a coincidence lattice between the (100) substrate plane and the overlayer, the latter being a somewhat compressed $C(1 \times 2)$ structure. Both of these structures could be observed when no impurity could be detected by Auger electron spectroscopy. Moreover, the Au (100) (1 × 5) surface structure has been observed on gold grown epitaxially on silver where the impurity level would be expected to be quite low. The Auger electron emission technique had a lower detection limit of about 0·1 monolayer equivalent of impurity, so there still remains the possibility that these structures are impurity stabilized. However, it is clear the the necessary level of impurity must be very low. Model calculations on the energy of a (100) argon crystal surface have shown that rearrangement of the outermost layer to a C (1 × 2) structure can occur with only modest energetic penalty[69] and that the rearranged surface can be stabilized by surface impurity atoms. Irrespective of whether the rearranged structures of Pt(100), Au(100) and Ir(100) are intrinsic features of the clean surfaces, or whether impurity is required for their stability, it is clear that the rearrangements occur with unusual facility, presumably because bonding in these metals is somewhat less directionally dependent than with other metals.

1. Ideal Surfaces

It is useful to start with the model of the ideal flat crystal surface, formed by dividing an ideal crystal along a specified plane. Inasmuch as the simplest representation of the molecular structure of a crystal consists of portraying each atom as a spherical unit, the structure of an ideal surface can also be represented by an array of balls. An extensive atlas is now available[62] which gives pictorial representations of the more important ideal surface planes out to the eighth order for b.c.c., f.c.c., h.c.p. (as well as diamond and rocksalt) crystal structures. With b.c.c. and f.c.c. crystal structures, the structure of a surface plane is uniquely defined by the (hkl) indices of the dividing plane. However, in the h.c.p., diamond and rocksalt structures this is not necessarily so. Thus in a h.c.p. metal (and in a diamond structure) although all the atoms are chemically similar, these atoms can be divided into two classes depending on their environment: two surfaces are possible for each (hkl) if with a h.c.p. metal (2h + 4k + 3l) is not a multiple of 6.

For many purposes planes up to the second order are the most important, and Figs 3.1–3.11 represent the structures of ideal surface planes of metals of b.c.c., f.c.c. and h.c.p. crystal structures, out to the second order. Table A.1, Appendix 1, lists the crystal structures, lattice constants and nearest-neighbour distances for the most important metallic elements.

It will be clear from an inspection of Figs 3.1–3.11 that, in general, atoms with different numbers of nearest neighbours exist on a surface. An atom in the bulk of a crystal has a fixed number of nearest neighbours, so atoms in the surface have some nearest neighbour interactions missing,

FIGS 3.1–3.11 Plan views of primitive surface unit cells of various surface planes, viewed normally, showing surface atoms and some subsurface atoms. The tables list the positions of the various atoms with reference to the centre of the origin-atom labelled, O, and in the cartesian directions OX, OY (indicated), and OZ outward normal to the surface. For b.c.c. and f.c.c. structures the distances are measured in units of a/2 where a is the lattice parameter of the normal cubic unit cell. In the h.c.p. structures, it is assumed that the axial ratio, $r = c/a$ takes the ideal value $\sqrt{8/3}$, and the distances are measured in units of a. When two numbers are associated with one atom position, this implies that one atom lies directly below the other. The figures and tabulations have been abstracted from Nicholas, J. F. "An Atlas of Models of Crystal Surfaces", Gordon and Breach, New York (1965), and are reproduced with permission.

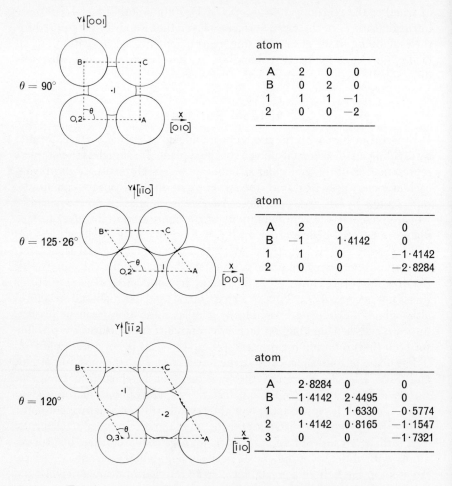

FIG. 3.1 b.c.c. surfaces. top (100); centre (110); lower (111).

3. MASSIVE METAL CATALYSTS   105

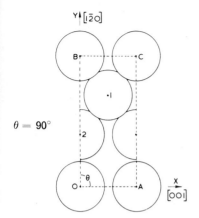

atom			
A	2	0	0
B	0	4·4721	0
1	1	3·1305	−0·4472
2	0	1·7889	−0·8944

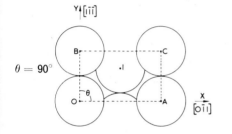

atom			
A	2·8284	0	0
B	0	1·7321	0
1	1·4142	1·1547	−0·8165

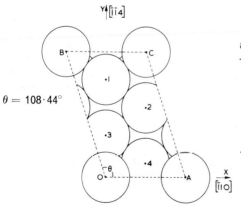

atom			
A	2·8284	0	0
B	−1·4142	4·2426	0
1	0	3·2998	−0·3333
2	1·4142	2·3570	−0·6667
3	0	1·4142	−1
4	1·4142	0·4714	−1·3333

FIG. 3.2 b.c.c. surfaces. top (210); centre (211); lower (221).

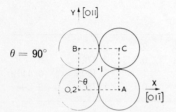

atom			
A	1·4142	0	0
B	0	1·4142	0
1	0·7071	0·7071	−1
2	0	0	−2

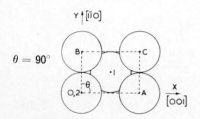

atom			
A	2	0	0
B	0	1·4142	0
1	1	0·7071	−0·7071
2	0	0	−1·4142

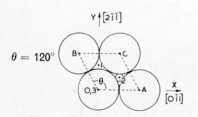

atom			
A	1·4142	0	0
B	−0·7071	1·2247	0
1	0	0·8165	−1·1547
2	0·7071	0·4082	−2·3094
3	0	0	−3·4641

FIG. 3.3 f.c.c. surfaces. top (100); centre (110); lower (111).

3. MASSIVE METAL CATALYSTS

$\theta = 114 \cdot 09°$

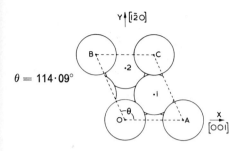

atom			
A	2	0·	0
B	−1	2·2361	0
1	1	0·8944	−0·4472
2	0	1·7889	−0·8944

$\theta = 90°$

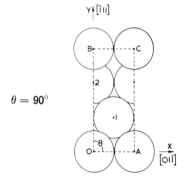

atom			
A	1·4142	0	0
B	0	3·4641	0
1	0·7071	1·1547	−0·4082
2	0	2·3094	−0·8165

$\theta = 90°$

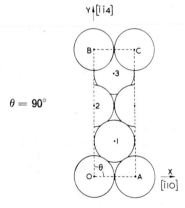

atom			
A	1·4142	0	0
B	0	4·2426	0
1	0·7071	1·1785	−0·3333
2	0	2·3570	−0·6667
3	0·7071	3·5355	−1

FIG. 3.4 f.c.c. surfaces. top (210); centre (211); lower (221).

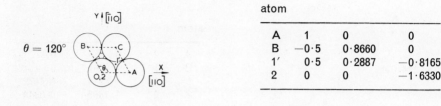

atom			
A	1	0	0
B	−0·5	0·8660	0
1′	0·5	0·2887	−0·8165
2	0	0	−1·6330

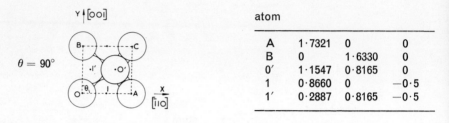

atom			
A	1·7321	0	0
B	0	1·6330	0
0′	1·1547	0·8165	0
1	0·8660	0	−0·5
1′	0·2887	0·8165	−0·5

Fig. 3.5 h.c.p. surfaces. top (001), $0 \leqslant \delta < \frac{1}{2}$; lower (110).

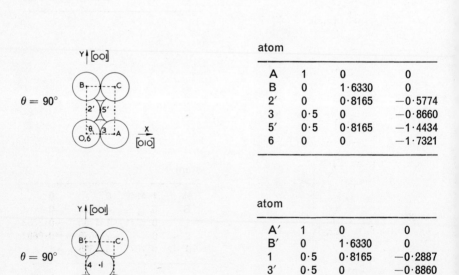

atom			
A	1	0	0
B	0	1·6330	0
2′	0	0·8165	−0·5774
3	0·5	0	−0·8660
5′	0·5	0·8165	−1·4434
6	0	0	−1·7321

atom			
A′	1	0	0
B′	0	1·6330	0
1	0·5	0·8165	−0·2887
3′	0·5	0	−0·8860
4	0	0·8165	−1·1547
6′	0	0	−1·7321

Fig. 3.6 h.c.p. surfaces, (100). top, $0 \leqslant \delta < 1/3$; lower, $1/3 \leqslant \delta < 1$.

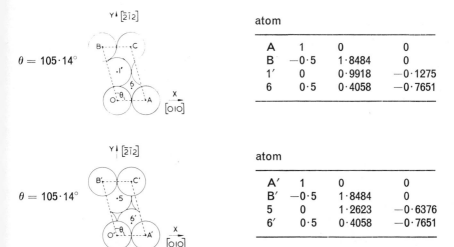

FIG. 3.7 h.c.p. surfaces, (101). top, $0 \leqslant \delta < 5/6$; lower, $5/6 \leqslant \delta < 1$.

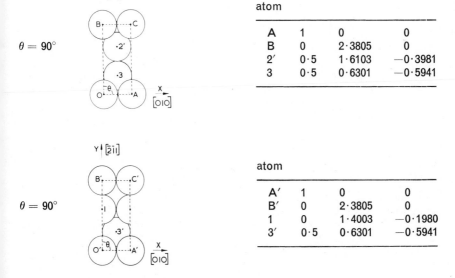

FIG. 3.8 h.c.p. surfaces, (102). top, $0 \leqslant \delta < 1/3$; lower, $1/3 \leqslant \delta < 1$.

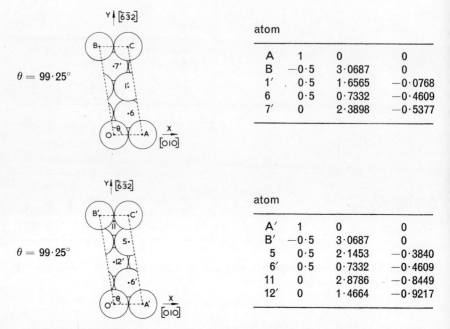

FIG. 3.9 h.c.p. surfaces, (103). top, $0 \leqslant \delta < 5/6$; lower, $5/6 \leqslant \delta < 1$.

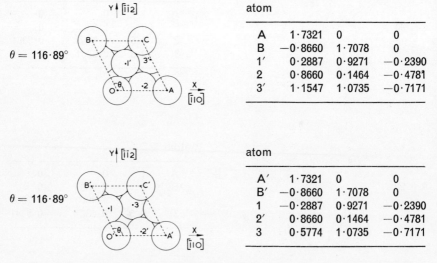

FIG. 3.10 h.c.p. surfaces, (111). top, $0 \leqslant \delta < 1/2$; lower, $1/2 \leqslant \delta < 1$.

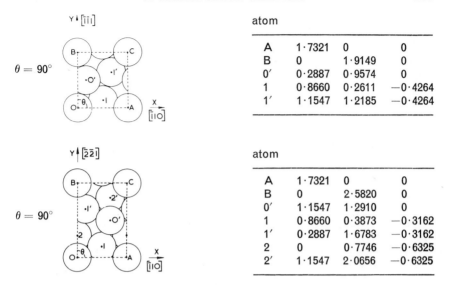

FIG. 3.11 h.c.p. surfaces. top (112); lower (114).

and can be said to possess broken bonds. Each type of surface atom can thus be quite simply specified either by the number of broken bonds or by the number of nearest neighbours it possesses; furthermore, each unit cell of surface will be characterized by a number of broken bonds and, in the approximation that the cohesive energy of the crystal is considered to be made up from nearest neighbour interactions only, the total number of broken bonds per unit area is a measure of the surface energy. In a qualitative sense one would also expect the number of broken bonds or nearest neighbours of a surface atom to be related to its chemical reactivity and to its propensity for surface migration and for evaporation.

We may define N_i to be the number of atoms in a primitive unit cell of surface which have i nearest neighbours and n_i to be the number of atoms per unit area of surface with i nearest neighbours. Then

$$n_i = N_i/A \qquad (3.1)$$

where A is the area per primitive unit cell of surface, and is given by

$$\begin{aligned}
\text{b.c.c.} \quad & A = \tfrac{1}{2} Q a^2 \, (h^2 + k^2 + l^2)^{1/2} \\
\text{f.c.c.} \quad & A = \tfrac{1}{4} Q a^2 \, (h^2 + k^2 + l^2)^{1/2} \\
\text{h.c.p.} \quad & A = ac \, (h^2 + hk + k^2 + 3l^2 a^2/4c^2)^{1/2}
\end{aligned} \qquad (3.2)$$

where $Q = 1$ for f.c.c. with h, k, l all odd; for b.c.c. with $h + k + l$ even, and $Q = 2$ otherwise.

TABLE 3.1 Values for N_i, the number of surface atoms per primitive surface unit cell with i nearest neighbours

Type of surface plane		N_i
b.c.c.	(100)	$i = 4$, $N = 1$
	(110)	$i = 6$, $N = 1$
	(111)	$i = 7$, $N = 2$; $i = 4$, $N = 1$
	(210)	$i = 6$, $N = 2$; $i = 4$, $N = 1$
	(211)	$i = 7$, $N = 1$; $i = 5$, $N = 1$
	(221)	$i = 7$, $N = 2$; $i = 6$, $N = 2$; $i = 4$, $N = 1$
f.c.c.	(100)	$i = 8$, $N = 1$
	(110)	$i = 11$, $N = 1$; $i = 7$, $N = 1$
	(111)	$i = 9$, $N = 1$
	(210)	$i = 11$, $N = 1$; $i = 9$, $N = 1$; $i = 6$, $N = 1$
	(211)	$i = 10$, $N = 1$; $i = 9$, $N = 1$; $i = 7$, $N = 1$
	(221)	$i = 11$, $N = 1$; $i = 9$, $N = 2$; $i = 7$, $N = 1$
h.c.p.*	(001), i.e. (0001)	$i = 9$, $N = 1$
	$(10\bar{1}0)$, $0 \leq \delta < 1/3$	$i = 6$, $N^l = 1$; $i = 10$, $N^m = 1$
	$(10\bar{1}0)$, $1/3 \leq \delta < 1$	$i = 10$, $N^l = 1$; $i = 8$, $N^m = 1$
	$(10\bar{1}1)$, $0 \leq \delta < 5/6$	$i = 8$, $N^l = 1$; $i = 9$, $N^m = 1$
	$(10\bar{1}1)$, $5/6 \leq \delta < 1$	$i = 10$, $N^l = 1$; $i = 11$, $N^m = 1$; $i = 6$, $N^m = 1$
	$(10\bar{1}2)$, $0 \leq \delta < 1/3$	$i = 10$, $N^l = 1$; $i = 7$, $N^l = 1$; $i = 9$, $N^m = 1$
	$(10\bar{1}2)$, $1/3 \leq \delta < 1$	$i = 8$, $N^l = 1$; $i = 11$, $N^m = 1$; $i = 7$, $N^m = 1$
	$(10\bar{1}3)$, $0 \leq \delta < 5/6$	$i = 10$, $N^l = 1$; $i = 7$, $N^l = 1$; $i = 11$, $N^m = 1$; $i = 7$, $N^m = 1$
	$(10\bar{1}3)$, $5/6 \leq \delta < 1$	$i = 8$, $N^l = 1$; $i = 11$, $N^m = 1$; $i = 9$, $N^m = 1$; $i = 7$, $N^m = 1$
	$(11\bar{2}0)$	$i = 11$, $N^l = 1$; $i = 7$, $N^m = 1$
	$(11\bar{2}1)$, $0 \leq \delta < 1/2$	$i = 10$, $N^l = 1$; $i = 6$, $N^l = 1$; $i = 11$, $N^m = 1$; $i = 8$, $N^m = 1$
	$(11\bar{2}1)$, $1/2 \leq \delta < 1$	$i = 11$, $N^l = 1$; $i = 8$, $N^l = 1$; $i = 10$, $N^m = 1$; $i = 6$, $N^m = 1$
	$(11\bar{2}2)$	$i = 10$, $N^l = 1$; $i = 7$, $N^l = 1$; $i = 10$, $N^m = 1$; $i = 7$, $N^m = 1$
	$(11\bar{2}4)$	$i = 11$, $N^l = 1$; $i = 9$, $N^l = 1$; $i = 6$, $N^m = 1$; $i = 11$, $N^m = 1$; $i = 9$, $N^m = 1$; $i = 6$, $N^m = 1$.

* For convenience, the four index notation is also given.

Table 3.1 lists values for N_i for various planes. It will be clear that the number of broken bonds (j) and the nearest neighbour co-ordination number (i) are related by

$$i + j = z \tag{3.3}$$

where z is the number of nearest neighbours of an atom in the interior of the crystal. For f.c.c. and h.c.p. z = 12, for b.c.c., z = 8.

The h.c.p. lattice possesses two crystallographically distinct types of atoms—types l and m (say)—so the exact location of the plane by means of which the crystal is divided so as to expose the surface then requires further definition. For this purpose, a parameter δ is used in Figs 3.1–3.11 and Table 3.1, and this defines the dividing surface (hkl) by

$$hx + ky + lz \leqslant \delta, \text{ with } 0 \leqslant \delta < 1 \tag{3.4}$$

where for the h.c.p lattice the zero for the type l atoms is set at 0, 0, 0, and for type m at 1/3, 2/3, 1/2.

It will be clear from the diagram in Figs 3.1–3.11 that high index planes can be considered as made up from a regular array of terraces and monatomic steps, the latter often containing regularly placed kinks, and the terraces amounting to facets of low index planes.

2. Unsupported Bulk Metal*

We include in this section a discussion of metal specimens used as catalysts in such forms as wire, foil, slabs, buttons, ribbons, etc. Single crystal specimens come in forms such as these, and much of the recent work using unsupported bulk metal catalysts, has, in fact, made use of single crystals prepared to expose known surface planes.

The macroscopic surface roughness of a metal specimen depends on the technique used to produce the surface finish. For mechanical methods of finishing the following list will give an approximate idea of the degree of micro-roughness that is likely. The micro-roughness is expressed in terms of h_{av}, the average departure of the surface profile from the ideal surface plane: lapped steel, about 30 nm; good quality grinding, 50–750 nm; average quality grinding, 200–1500 nm. For comparison, Figs 3.12 and 3.13 show, respectively scanning electron micrographs of tungsten (110) and nickel (111) surfaces prepared by electropolishing, and cleaned to yield optimum LEED patterns by a combination of ion bombardment and thermal desorption, followed by annealing. At this magnification ($\times$ 2000),

* Suppliers of ultrahigh purity metals in the form of wire, foil, single crystals, buttons etc. include: Johnson Matthey; Engelhard Industries; Materials Research Corpn.

the surfaces are relatively featureless, although deviations from exact planarity on a scale beyond this level of resolution, are to be expected (*vide infra*).

Polycrystalline specimens have a long history of use as laboratory catalysts and various examples will be found cited in Bond's compendium.[61] Polycrystalline specimens which have been formed by drawing (wire) or rolling (foil) usually have some degree of preferred crystal orientation as a result of this mechanical treatment. For example, the crystallites in drawn tungsten wire tend to be aligned with [110] axes along the wire axis;

FIG. 3.12   Scanning electron micrograph of tungsten (110) surface, prepared by electropolishing, and cleaned to yield optimum LEED pattern by a combination of ion bombardment and thermal desorption, followed by annealing.

recrystallization occurs on heating to a high temperature and, if suitably controlled, can result in the formation of single crystals extending over relatively long lengths of wire.[104]

Most metal single crystals are obtained from the melt as chunks or rods by solidification or pulling. Occasionally, use has been made of crystal growth from the vapour (e.g. zinc) or by chemical reaction in the vapour phase (e.g. tungsten from tungsten hexachloride[106]). Usually to expose the desired face requires cutting or machining. This may be done with a minimum of structural damage by spark erosion or acid sawing and this is usually followed by lapping and finally by electropolishing. Various formulae for electropolishing are available in the literature.[67] The cut specimen will generally be in the form of a slab, foil or button, but only the highly refractory metals have sufficient strength for a slab or foil of thickness <0·5 mm to have adequate mechanical stability.

It will be obvious that with single crystal metal specimens it may be quite difficult to ensure that the specimen exposes only or substantially one type of surface plane when the whole specimen surface is considered. This can be approximated with very thin slabs or foils if both the major faces are of the same type and if the edge area is relatively small. If this sort of specimen is not feasible, a possible alternative is to cut the whole

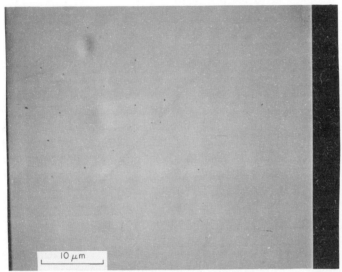

FIG. 3.13 Scanning electron micrograph of nickel (111) surface prepared by electropolishing, and cleaned to yield optimum LEED pattern by a combination of ion bombardment and thermal desorption, followed by annealing.

crystal to the appropriate symmetry (e.g. a cube or rectangular prism with cubic metals exposes six identical faces). This will always be quite difficult and tedious to do, and with h.c.p. metals it will only be possible at all for a relatively limited range of surface types (e.g. (111), $l = 2,4, \ldots$). A possible alternative technique is to mask the unwanted part of the specimen surface with a coating of negligible catalytic reactivity. Inasmuch as the use of clean single crystal catalysts presupposes the use of UHV techniques, a coating needs to have suitable thermal and vacuum properties and should not contribute contaminant to the working face of the crystal either by diffusion or evaporation. Possible coatings include oxide of the metal itself, other metals, oxides of other metals. One would often generate the coating before the working face is finally cleaned, and this might well be carried out as a separate operation and in a separate apparatus from that for final cleaning and catalytic use. A choice of coating material and the procedure for its use can only be made in the light of specific experimental requirements, but the selection of a coating with sufficient stability is fraught with difficulty, and impurity monitoring of the surface essential.

A related technique to this has been described[68] in which the face of the crystal to be used as a catalyst rests on a ground glass annular flange, thus isolating (except for gas leakage across the contact junctions) the face within the flange from the rest of the crystal. The apparatus as originally described would, for most purposes, now require to be modified to allow the crystal to be cleaned and sited on the flange without breaking the vacuum but this would not offer much difficulty using UHV-compatible transfer techniques. A more sophisticated method of studying a catalytic reaction originating at a limited region of crystal surface involves a spatial differentiation of the reaction products. For instance, provided the pressure is in the molecular flow region it is possible to arrange a detector such as a mass spectrometer to accept and identify only those molecules which are desorbed from a specified surface, to which reactant may also be jetted.

In general terms, three methods of surface cleaning are available, namely thermal desorption, ion bombardment and chemical reaction. In practice, the last two are used in combination with the first.

There are relatively few metals which can be reasonably cleaned by thermal desorption of impurity alone, since all metals require to be heated to in excess of 2000 K for this purpose, and for most metals this is in excess of the melting point. Heating to lesser temperatures than this will, of course, serve to desorb weakly held impurities and this is always done as a matter of routine when other techniques such as ion bombardment are to be used. Metals for which thermal desorption can be effectively used to obtain a high level of surface cleanliness are tungsten, molybdenum, tantalum, niobium, rhenium, osmium and iridium. It is important to use metal of the highest possible purity because many impurities concentrate at the surface by diffusion from the bulk, and the complete release of all dissolved impurity may well prove to be an extremely protracted process.

Carbon and sulphur are very troublesome impurities. Even when present at quite low levels, sulphur tends to accumulate in significant amounts at a metal surface by diffusion from the bulk, particularly at annealing temperatures. Carbon is of very low volatility (vapour pressure about 10^{-5} Pa (about 10^{-7} Torr) at 2050 K) and thus also tends to accumulate at the surface. It may be removed from the surface by reaction with oxygen. One method consists of treating the surface with oxygen at about 10^{-4} Pa (about 10^{-6} Torr) with the metal at quite low temperatures (up to a few hundred °C) for periods of a few minutes to a few tens of minutes, followed by evacuation and a relatively quick thermal flash to remove adsorbed oxygen; this cycle may require to be repeated since some more carbon tends to accumulate at the surface during the flash. Alternatively, one may soak the specimen in oxygen at about 10^{-4} Pa (about 10^{-6} Torr) at >2300 K for some hours, followed by thermal removal of adsorbed oxygen; this depletes the metal of carbon to some depth below the surface. Carbon is a particularly serious impurity with tungsten, molybdenum and tantalum, much less so with rhenium, niobium, iridium and osmium, and there is now evidence that even tungsten which has been subjected to very rigorous thermal cleaning at 2500–3000 K is not completely free of surface carbon, but removal of the last vestiges requires oxygen treatment.[37] Attention always needs to be given to the problem of transfer of impurity from the specimen supports to the specimen itself when high temperature treatments are involved. Where possible the supports should be of the same metal as the specimen. By making a support into a loop, it may be outgassed independently of the specimen.

Specimen heating may be achieved by direct resistive heating, by electron bombardment, by inductive heating, or by the use of an intense beam of light. If direct resistive heating is used, the cross-sectional area of the specimen needs to be limited to about 1–2 mm² if the heating current is not to reach unmanageable values. The heating used should be a.c. rather than d.c. since the latter is known (with tungsten wire) to induce surface faceting.[104,105] This process, which only occurs below about 2200 K, appears to be due to surface migration of tungsten ions towards the negative end of the wire, with preferential diffusion resulting in the exposure of $\{110\}$ and some $\{112\}$ and $\{111\}$ planes.

If electron bombardment is used, care is required to ensure that the electron source is not itself a source of impurity: commercial oxide emitters containing alkali or alkaline earth metal oxides should be avoided. It is an advantage if the electron gun is not situated in line-of-sight to the face being cleaned; the beam may be directed onto the rear of the specimen, while an electrostatically deflected electron beam has also been used.

For metals whose melting point is too low to allow thermal cleaning to be efficient, the most satisfactory alternative is ion bombardment with an inert gas such as argon. This method is, of course, applicable to refractory metals too.

Two rather distinct techniques have been used for inert gas ion bombardment. The first sort generates inert gas ions by means of an electron beam, and the ions are then directed towards the specimen by a voltage applied to the specimen itself, or by means of a separate accelerating electrode. Of these alternatives for ion acceleration, the latter alternative is to be preferred since it is possible then to design an ion gun which provides a degree of ion beam collimation and this allows bombardment of components such as specimen supports to be minimized. Commercial ion bombardment guns which are available as part of LEED and electron stimulated Auger analysis systems are usually of this type. A variety of more or less obvious designs is possible, and the interested reader is referred to some of the scientific and commercial literature for further details.[38,40] To avoid contamination it is desirable that the electron emitter should not be in line-of-sight with the surface being cleaned. The second method avoids the use of an electron beam by generating inert gas ions in a glow discharge. This method should be avoided because it is virtually impossible to obtain stable operation with ion energies less than about 1 keV, and at this energy surface damage to the specimen is unacceptably severe. Furthermore the cleaning efficiency of immersing a specimen in a glow discharge is highly questionable because of the amount of impurity generated by adventitious sputtering.

Cleaning a surface by ion bombardment depends of course on the removal of material from the surface. At the same time surface damage usually occurs. The latter involves both embedding inert gas atoms in the surface and altering the surface morphology of the specimen. There are, in fact, separate ion energy thresholds for metal removal and for the onset of surface damage, with the former threshold being at a lower energy than the latter.[42] However, for all the inert gases the rate of metal removal below the damage threshold is too low (about 1% of the rate for 200 eV ions) for this regime to be practically useful. One is left to strike a balance between rate of removal and extent of damage, and in practice ion energies in the range 200–500 eV are used. For comparison, the damage threshold for argon ions on gold is in the vicinity of 40 eV, and for xenon ions on silver or gold is in the region of 15–25 eV[42,43] In practical terms, for a transition metal specimen a cleaning dose of about 10^{17} ions mm^{-2} is typical. With ordinary ion gun sources one can expect to obtain beam currents at the specimen in the region of 0.05–$1\,\mu$A mm^{-2} depending on gun design, and the total bombardment time will be of the order of minutes to tens of minutes. A pressure of $10^{-1} - 10^{-2}$ Pa (about $10^{-3} - 10^{-4}$ Torr) of pure gas (preferably gettered) is used. It is possible to use any of the inert gases, but in practice only argon is worth considering since it is the cheapest and none of the others offers an improvement in the extent of surface damage relative to the metal removal rate.[42]

The effect of ion bombardment on surface morphology has been studied in some detail.[42–45,109–115] One effect is a removal of large features such as scratches, ridges and grooves, so on a relatively gross topographical scale there is a general smoothing of the surface. However, surface roughening

3. MASSIVE METAL CATALYSTS

and damage are apparent on a finer scale. If the specimen is polycrystalline, the surface will expose various surface planes owing to the range of orientations of individual crystallites. Since the rate of metal removal depends on the crystallographic identity of the exposed face, the result is that some crystallites lose metal faster than others, so steps develop between adjacent crystallites. In addition, grain boundary grooving may occur. The surface plane which a crystallite exhibits after bombardment may not be the same as the one originally present, and this effect is enhanced for conditions of non-normal ion beam incidence. The general result of these processes is a roughening of the surface. Normal ion beam incidence is to be preferred if surface roughening is to be minimized. If a single crystal specimen is bombarded these sources of surface roughening are absent, but some surface damage still occurs. At the energies most appropriate for cleaning, the main features of damage are the formation of facets and crystallites at the surface, and the introduction of point defects and dislocations in the surface regions. This sort of surface damage also occurs, of course, within each exposed face of a polycrystalline specimen.

The undesirable effects resulting from surface damage can be eliminated by choice of the optimum bombardment conditions to minimize damage in the first place (minimum workable ion energy) and by vacuum annealing after bombardment; the latter is always essential. Annealing is typically done at temperatures up to about half the melting point (K) of the metal for a few minutes. In practice, a bombardment–annealing cycle may need to be repeated several times. Bombardment of an alloy gives a uniform surface layer but one for which the composition may differ from the bulk.

When a single crystal is cut, the accuracy of orientation is seldom better than $\pm 0 \cdot 5°$ averaged over the whole cut surface, and it is often rather worse than this. In fact, such a cut surface will consist of regions of slightly varying orientation including, no doubt, some with exactly zero deviation from the nominal. One can expect, therefore, that when a crystal is cut to expose a low index plane, the process of cleaning by bombardment and annealing is likely to generate areas of orientation equal to the nominal together with higher index facets or at least varying concentrations of steps. For example a recent photoelectric work-function study[11] has shown that nickel surfaces prepared by cutting single crystals with an accuracy of about $\pm 1°$, followed by ion bombardment and annealing to optimum LEED resolution contained the nominal surface in the following proportions: (111), 90%; (100), 95%; (110), 95%. One concludes that LEED is not a particularly sensitive tool for detecting surface imperfections and this view is in agreement with the conclusions reached by considering the coherence width of the electron beam.[121]

If argon ions are generated by a gun, at a normal working distance from the gun the beam will not bombard an area greater than about 5–50 mm² or so from a gun of conventional size and design. If gun-collimation is not used, somewhat larger specimens can be handled, but even so with conventional laboratory-scale equipment limitations on ion current density and uniformity of bombardment restrict the useful specimen area to about 500–1000 mm². In practice the technique is limited to unidirectional bombardment, so one side of a specimen only can be cleaned at a time if a single ion gun is used.

Even when much trouble has been taken to prepare the surface of a metallic crystal with the greatest degree of planarity, the surface will, in fact, not be flat in a crystallographically ideal sense, except over quite small areas. A real surface will contain terraces linked by steps, and the

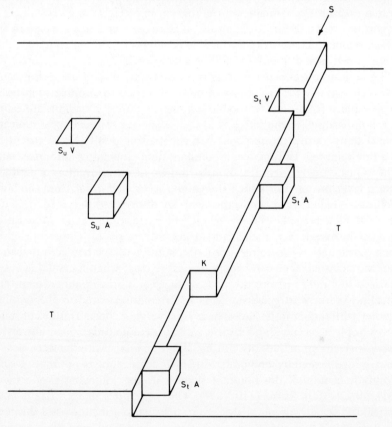

FIG. 3.14 Schematic representation of surface imperfections; showing step, S; terrace, T; kink in step, K; surface vacancy S_uV; step vacancy, S_tV; atom self-adsorbed on surface, S_uA; atom self-adsorbed at step, S_tA.

steps will contain kinks. Furthermore, an atom may leave its position in a flat area to generate a self-adsorbed atom and a surface vacancy. These possible features are illustrated schematically in Fig. 3.14. The importance of monatomic steps and kinks in the process of crystal growth and evaporation has quite a long history (e.g. refs 46–48), and the theory of growth and evaporation is so well established[49–51] that the existence of these surface imperfections is beyond doubt. Steps and kinks of dimensions greater than monatomic also undoubtedly occur, but in lesser number the greater the size. It is important to recognize that real metal surfaces will practically never approach an equilibrium configuration with respect to terraces, steps and kinks, but their topography is largely dictated from their history. A possible exception to this would be a surface maintained close to the melting point for a considerable time, since it is only at a temperature such as this that the rate of approach to equilibrium will be large enough to be important. Nevertheless, it is useful to have some idea of the equilibrium theory, since this is clearly a limiting situation.

The equilibrium theory for the formation of step and kink imperfections in an otherwise ideal surface has been given by Burton, Cabrera and Frank[39,49] and by Dunning.[41] The discussion is restricted to imperfections in an atomically flat plane.

Let us imagine that atoms leave their surface positions to become self-adsorbed with the creation of surface vacancies; there is an attractive potential between the atoms and there will be an equilibrium distribution of varying sizes of aggregates of both self-adsorbed atoms and surface vacancies. An aggregate of either sort clearly has a monatomic step boundary, which will contain kinks. An imperfect surface of this sort exists at three levels (the bottom of a vacancy, the original surface, the top of a self-adsorbed atom). More levels (deeper holes and higher hills) are obviously possible, but five levels is the most for which an analysis is available. The results of Burton, Cabrera and Frank's analysis are summarized in Fig. 3.15 in terms of the surface roughness parameter (s). The latter is defined to be the average number of missing nearest neighbour interactions parallel to the surface per surface site. The model assumes a (100) face of a simple cubic lattice. This treatment uses a generalization of Bethe's method of dealing with order–disorder and assumes nearest neighbour interactions only. The results are rather similar to those obtained by an application of the Bragg-Williams method,[70] so that they represent a relatively crude approximation. The data show that the surface roughness parameter does not differ appreciably from zero until a temperature of about $0 \cdot 25 \phi/k$, where ϕ is the interaction energy between a pair of nearest neighbours. Since for many metals the melting point T_m is roughly equal to about $0 \cdot 2 \phi/k$ (with ϕ evaluated as one-sixth of the heat of vaporization), one would conclude from these data that the equilibrium degree of surface roughening is negligible at all temperatures up to the melting point.

These computations, however, certainly underestimate the degree of roughness. A computation to a better approximation is only available for the two-level problem such as corresponds to the distribution of a partly complete layer of self-adsorbed atoms on the complete layer beneath.

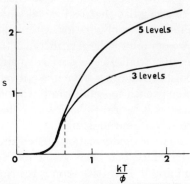

Fig. 3.15 Surface roughness, s, as a function of kT/ϕ for a model (100) surface of a simple cubic crystal. s is defined to be the average number of missing nearest neighbour interactions parallel to the surface per surface site: ϕ is the interaction energy for a pair of nearest neighbours. Curves marked 3 and 5 are for three-level and five-level problems respectively. Reproduced with permission from Burton, W. K., Cabrera, N. and Frank, F. C. *Phil. Trans. Roy. Soc.* **A243**, 299 (1951).

A comparison of the results of an exact solution (still retaining only nearest-neighbour interactions) with those from the Bragg-Williams approximation[41] suggests that one might expect the surface roughness first to become significant at about the same value of kT/ϕ for the two cases, although in the more realistic approximation the computed roughness increases much more rapidly with temperature than in the Bragg-Williams approximation. It seems safe to conclude that for metals, the equilibrium concentration of surface imperfections of this sort is negligible up to temperatures close to the melting point.

However, the energy required for an atom in a monatomic kink position to move away to become, in effect, an atom self-adsorbed at a monatomic step, only requires the breaking of one nearest-neighbour bond. Since this process is, therefore, relatively easy it is likely to be significant at readily accessible temperatures well below the melting point of the metal. By this process, a kink atom in a f.c.c. metal which has six nearest-neighbours ($i = 6$) is converted to an atom with $i = 5$, while in a b.c.c. metal a kink atom with $i = 4$ is converted into one with $i = 3$. The kink is, of course, regenerated except when the atom which moves is the last one which defines the kink.

Steps can also occur in association with emergent screw dislocations (Fig. 3.16). The dislocation density in a metallic crystal can vary within wide limits depending on its history: for carefully prepared and thoroughly annealed single crystals, densities in the range $10–10^3$ mm^{-2} are not uncommon. Attempts to correlate catalytic activity with dislocation density (via the effect of dislocations on surface topography) have either failed or been inconclusive. Thus, no correlation was found by Jaeger[19] for formic acid decomposition over silver film catalysts while the results of Uhara and co-workers,[124,125] and more recently Criado et al.[126] with copper and nickel catalysts are inconclusive because of inadequate control over experimental variables, particularly surface contamination. In most cases one expects the influence of emergent dislocations on catalytic activity to be of negligible, or at best marginal importance, compared to the influence of other forms of surface imperfections.

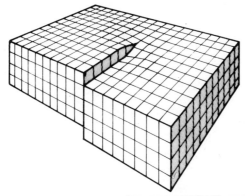

FIG. 3.16 Schematic representation of screw dislocation emerging at a surface.

We have seen that low index surface planes of most metals have a structure close to that for the corresponding ideal surface. The question arises if this is also true for high index surface planes. One's expectations contain an element of uncertainty because of the known tendency for high index surfaces to undergo faceting or thermal etching. The tendency for this to occur is dependent on the conditions in which the specimen is maintained; nevertheless, conditions have been found in a number of cases where the actual structure of a nominal high index surface does not depart far from the ideal. Surface examination has often been made by LEED (e.g. refs 127–129), and we may take as an important example the behaviour of platinum reported by Somorjai et al.[127]. In this work, single crystals were cut at 6·5° from (111) towards (11$\bar{1}$), 9·5° from (111) towards (100), and 9·0° from (100) towards (111), followed by a standard cleaning procedure which involved ion bombardment and thermal treatment in vacuum at temperatures up to about 1120 K. These surfaces can formally be represented by the indices (997), (755) and (911) respectively. The general conclusion emerged that the LEED diffraction patterns from these surfaces corresponded reasonably well with those to be expected if the surface structures were ideal, and two examples of the structures are given in Fig. 3.17. Orientations such as these which are

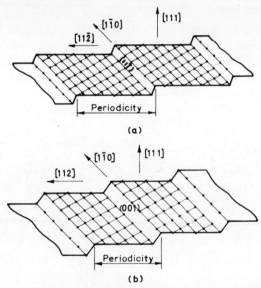

FIG. 3.17 Schematic representations of atomic structures of vicinal surfaces on platinum. (a) surface cut 6·5° from (111) towards (11$\bar{1}$), nominally (997); (b) surface cut 9·5° from (111) towards (100), nominally (755). Reproduced with permission from Lang, B., Joyner, R. W. and Somorjai, G. A. *Surface Sci.* **30**, 440 (1972).

relatively close to an atomically flat low index plane (i.e. vicinal) are characterized by an ordered terrace and edge structure with each terrace exposing a surface structure of the nearby low index face: the terrace width increases and the edge concentration decreases as the orientation approaches that of the low index face. Although the representations shown in Fig. 3.17 show straight terrace edges without kinks, for other orientations kinks will be present. In fact, a close examination of the LEED pattern from the nominally (755) surface shows that regularly spaced kinks are present in the terrace edges, due presumably to a small misorientation away from the nominal. A generally similar conclusion to that outlined above for platinum has been reached for vicinal faces on copper crystals which had been cleaned by ion bombardment and heat treated in vacuum at 770–870 K.[128]

The accuracy of the LEED technique for assessing the degree of order in a surface is fairly limited, since the technique averages over an appreciable area, and one can certainly expect there to be a distribution of step and terrace parameters. Nevertheless, the technique is quite adequate to demonstrate conclusively that these vicinal surfaces were not heavily faceted into hill and valley structures.

Despite the stability of vicinal surfaces under conditions which are not too severe, there is ample evidence that heating a metal to a sufficiently high temperature in vacuum may cause thermal etching of the surface, and this generally results in the formation of pits and facets. This will not be unexpected if the temperature is high enough for evaporation of the metal to occur at an appreciable rate, and examples of this situation are to be found with zinc at 670 K,[52] and magnesium at 770 K,[53] where in both cases the metal vapour pressure is about $6 \cdot 7$ Pa (about 5×10^{-2} Torr) and faceting was observed in times ranging from a few minutes to a few tens of minutes. However, faceting is well known under conditions where evaporation is entirely negligible, and examples are iridium at 1570 K for 3 h,[54] where the iridium vapour pressure is about 10^{-9} Pa (about 10^{-11} Torr), nickel at 1170 K for 50 h,[55] where the nickel vapour pressure is about 3×10^{-7} Pa (about 10^{-9} Torr), and titanium (β) at 1170 K for 3 h,[56] where the titanium vapour pressure is about 10^{-8} Pa (about 10^{-10} Torr). Where evaporation is important, faceting results from the anisotropic dependence of the evaporation rate on the crystallographic orientation of the surface plane. Faceting by evaporation may be particularly prominent if surface impurity is present. If this impurity, which may be an adsorbed molecule or, more likely, at evaporation temperatures a small particle of another phase, inhibits evaporation in its vicinity, the surface remote from the impurity will continue to evaporate at a higher rate than the surface near to it, and this will lead to the formation of hillocks. An example is given by Hirth and Winterbottom[57] for a (111) silver surface heated in vacuum for 7 h at 1070 K and is illustrated in Fig. 3.18.

When evaporation is insignificant, faceting must occur by surface diffusion. The driving force is the requirement to minimize the total surface free energy. If at certain crystallographic orientations of the crystal surface the surface free energy is low, a surface plane of some other orientation of higher energy may reduce its total energy by breaking up into facets containing low energy orientations. In order to retain the average orientation of the faceted surface the same as that before faceting, the facet often contains a complex plane, that is one of higher indices, as well as the low index plane, and the situation is schematically illustrated

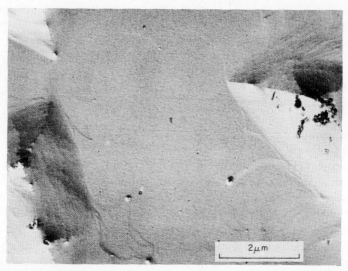

FIG. 3.18 Impurity generated faceted features on a (111) silver surface, produced by heating in vacuum at 1070 K. Reproduced with permission from Hirth, J. P. *In* "Metal Surfaces", American Society for Metals, Metals Park, Ohio (1963), p. 199.

in Fig. 3.19. This faceting appears to be most important when the orientation of the unfaceted surface lies only within a few degrees of a low energy plane, that is for vicinal surfaces. Some examples are: the faceting of nickel surfaces at 1170 K lying within $1 \cdot 7°$ of (100) and within $0 \cdot 5°$ of (111),[58] and of a platinum surface at 1580 K lying within 10° of (100).[59]

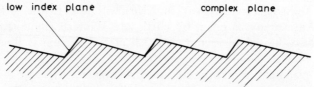

FIG. 3.19 Schematic representation of faceting process.

Inasmuch as the surface energy anisotropy of metals tends to decrease with increasing temperature (cf. platinum[130]), the lower temperature limit to thermal faceting clearly implies a kinetic limitation.

Thermal faceting is often more pronounced if the surface is heated not in a vacuum but in the presence of a reactive gas. This arises from the effect of chemisorption on the orientation anisotropy of evaporation, surface energy and surface diffusion. Oxygen is the gas with which enhanced thermal faceting has been most often studied, but it has also been observed with the halogens and sulphur, and the susceptible metals cover a wide range, including silver, gold, copper, iron, nickel, palladium, platinum, iridium and tungsten. The subject has been reviewed by Moore,[60] while more recent work with tungsten is due to Taylor[107] and Lea and Mee.[108] When chemisorbed gas is present, the propensity of various planes to facet, and the type of planes so produced, may be quite different to those in vacuum. Thus, Lea and Mee observed faceting of tungsten (100) and (110) planes in the presence of chemisorbed oxygen, although these planes are stable in vacuum. Thermal faceting in the presence of a reactive gas is clearly related to the changes which are not infrequently observed in the surface morphology of metal catalysts during the course of catalytic reactions (sometimes referred to as catalytic etching). Thus, Gwathmey[68] observed faceting of both (100) and (110) copper planes during catalysis of the hydrogen/oxygen reaction at 670 K. Baddour et al.[141] found that the infrared spectrum of carbon monoxide adsorbed on palladium changed as a result of catalysis of carbon monoxide oxidation at about 450 K, and this correlated with a change in the kinetic parameters of the oxidation reaction. They interpreted this as being due to a rearrangement of the palladium catalyst surface.

Reactive faceting has been observed with evaporated silver films used as catalysts for ethylene oxidation or heated in oxygen at 500 K:[142] the surface area increased by about 30% and small silver crystals (<50 nm) were eliminated. At 1120–1290 K in hydrogen the surface mobility of platinum is greatly increased,[143] so that sintering of a platinum powder then occurs at a greatly increased rate. Gross morphological changes in platinum wire or gauze catalysts during the oxidation of ammonia (1020–1220 K) are well known,[135] and include extensive faceting. These changes which occur in the platinum catalyst are much more severe under catalytic conditions than with either of the reactants alone at comparable nominal temperatures, presumably because the release of heat of reaction at the catalyst surface generates local transient temperatures in excess of the nominal average. Not all catalytic reactions appear to lead to gross changes in surface morphology of the catalyst and, on the whole, there is no substantial evidence for this with those reactions involving only hydrocarbons and hydrogen, at least so far as massive metal catalysts are concerned. Nevertheless, it should be understood that even when gross morphological changes are absent, the surface of the metal may be subject to a sort of reconstruction process which is limited to the first one or two unit cells of metal beneath the original surface, and which accompanies the adsorption and/or incorporation of reactant molecules. In this sense even hydrocarbon adsorption can sometimes effect a reconstruction, as in the formation of a reconstructed overlayer from the adsorption of ethylene or benzene on nickel (111).[144]

Ultimately, of course, reaction between gas and metal may lead to the formation of an identifiable reaction product phase, and when this happens morphological changes in the gas/solid interface are usually observed. This may well be so even though the product phase exhibits specific crystallographic relationships to the metal when examined by LEED. As an example we may note the data for the initial formation of WO_3 crystals on W(110) in the range 650–1100 K.[136] Above 1000 K, the WO_3 crystals are oriented with WO_3 (11$\bar{1}$) parallel to W(110). These crystals are faceted to expose {100} type planes, and the simplest form of the crystal is thus a triangular pyramid. The WO_3 crystals exist in three pairs of twin related domains with W[100] parallel to (i) WO_3[1$\bar{1}$0] and [$\bar{1}$10], (ii) WO_3[011] and [0$\bar{1}\bar{1}$], and (iii) WO_3[$\bar{1}$0$\bar{1}$] and [101]. Oxidation at 850 K yields WO_3 crystals oriented with WO_3(100) parallel to W(110) and in two rotational domains with (approximately) (i) WO_3[0$\bar{1}\bar{1}$] parallel to W[1$\bar{1}\bar{1}$], and (ii) WO_3[0$\bar{1}$1] parallel to W[1$\bar{1}$1]. Although these LEED data clearly imply that the morphology of the oxide/gas interface is quite different to the planar nature of the metal surface, this becomes much more apparent when the oxide is examined by transmission electron microscopy, since this technique reveals that much of the WO_3 exists with a whisker-like morphology. No doubt the actual nature of the oxide surface morphology will vary from system to system, but it is likely that if the oxide is crystalline, it will very seldom if ever exist as a simple parallel-sided slab.

In many cases, a thermally faceted surface has a striated structure and Fig. 3.20 shows a typical example of this sort. In addition, pits and hillocks are sometimes formed. An example is the formation of tetrahedral facets exposing (211) faces, which are formed by heating a tungsten (111) surface which carries adsorbed oxygen.[107] Faceting in a reactive gas is possible up to much greater maximum angles between the initial surface orientation and the nearest low index plane than is possible in vacuum, and figures in the range 10–30° are observed. If this angle is 20° or so, or above (a not uncommon situation), it means that faceting is possible for most orientations of the initial surface, since most will lie within this allowed range of orientations from one or other of the low index planes. The implications for faceting in a reactive gas of the crystal surfaces exposed in a polycrystalline specimen are obvious. In addition to faceting on individual grain surfaces, a polycrystalline specimen will usually show grain boundary grooving, and this may well be the dominant behaviour.

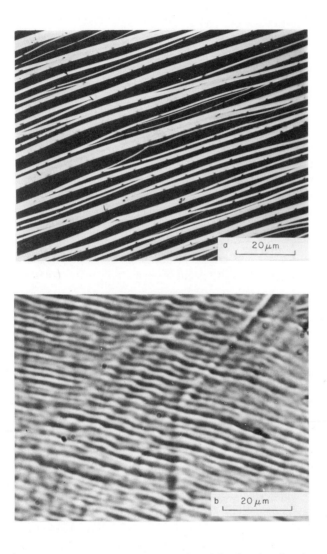

FIG. 3.20 (a) Faceted surface of silver produced by heating in air at 1170 K. Each ridge has a (111) plane on one side and planes of complex orientation on the other. Reproduced with permission from Moore, A. J. W. *Acta Met.* **6**, 293 (1958). (b) Faceting of tungsten surface of orientation close to (110) after heating in vacuum at about 2500 K. Reproduced with permission from Lea, C. and Mee, C. H. B. *Surface Sci.* **25**, 332 (1971).

Despite the success with platinum, the evidence suggests that generally it may well prove quite difficult to retain a high index surface plane on a cut crystal, because the various preparative and cleaning procedures—electropolishing, thermal desorption, ion bombardment—to say nothing of the subsequent catalytic reaction, are likely to lead to extensive faceting. In fact, very little catalytic work has yet been reported with well defined high index surface planes, possibly for this reason.

It will be clear from the previous discussion that, no matter what cleaning procedure has been used, one cannot expect the surface of a polycrystalline metal specimen to be completely flat; in fact, measurements with a number of refractory metals after thorough cleaning involving high temperature treatment have yielded a ratio of actual to apparent surface area in the vicinity of $1 \cdot 3$–$1 \cdot 5$.

There are present in a rough surface, atoms with a lower number of nearest neighbours than are possessed by atoms in a low index surface plane: atoms at corners and terrace kinks are obvious examples. The possibility that such atoms of low co-ordination can function as catalytic reaction sites with special properties has already been referred to in Chapter 1, and the occurrence of atoms of this sort is again discussed in Chapter 5 in relation to the properties of small metal particles.

3. Evaporated Metal Films

We do not intend to include details of film growth mechanisms in the present discussion. For this, reference may be made to recent reviews by Geus[1] and Sanders.[2] Here we confine our attention to films which are thick enough to be "continuous". This means a thickness in excess of 50 nm or so, and in practice the thickness is usually in excess of 100 nm. Evaporated metal films which are thin enough to consist of isolated metal crystals distributed over the support will be discussed in Chapter 4 which is devoted to dispersed metal catalysts. Continuous evaporated metal films are considered here as massive metal in the sense that such a film has both form and structure that show reasonable resemblance to what would be expected for a normal foil of similar thickness. It is certainly true that this thickness is ample for a specimen to have a band structure indistinguishable from that for metal of more usual macroscopic dimensions. However, this is not to imply that in all respects an evaporated film is indistinguishable from a normal specimen of bulk metal. In addition to surface topography which is very dependent on the specimen's history, evaporated films can exhibit a degree of intercrystalline porosity which affects both the apparent bulk density and the electrical resistivity, while continuous films are still often thin enough for the electrical resistivity to be influenced by gas adsorption.

In principle, one has available a range of possible film structures lying between the extreme limits of a perfectly epitaxed single crystal film, and a polycrystalline film of completely random crystal orientation. The actual structure will, in practice, depend on the nature of the metal and of the support surface, and on the conditions of preparation including support temperature, deposition rate, and impurity levels in the metal and the vacuum system. Supports for metal film deposition have already been discussed in Chapter 2, and this includes a description of their surface topography. Gross topographic features in the support are reproduced in the metal surface and this will add to the general roughness of the metal.

We shall deal first with polycrystalline metal films, and most of the discussion assumes that we are dealing with metals of relatively high melting point such as the transition metals.

It is often found that the ratio R (measured for instance by gas adsorption methods) of actual metal surface area accessible to the gas phase, to the geometric film area, exceeds unity. This arises from non-planarity of the outermost film surface both on an atomic and a more macroscopic scale, and from porosity of the film due to gaps between the crystals. These gaps are typically up to about 2 nm wide. However, for film thicknesses >50 nm, this gap structure is never such as completely to isolate metal crystals one from the other, and almost all of the support is, in fact, covered by metal. Except for films deposited at very low temperatures, at very high deposition rates or at exceptionally great thicknesses, films are one crystal thick, and on this basis a section through a random polycrystalline film with intercrystal gaps may be schematically represented as in Fig. 3.21. In practice, catalytic work mostly uses thick films in the thickness range 100–200 nm to which the model in Fig. 3.21 applies, and it is easily shown[8] that intercrystal gaps in these films will not influence catalytic reaction kinetics via diffusional effects provided the half-life of the reaction exceeds about 10–20 s, which will usually be the case.

Provided the film is sufficiently thin for electrons to penetrate (less than about 200 nm for 100 keV electrons), the main features of film structure can be revealed by transmission electron microscopy and diffraction,

FIG. 3.21 Schematic representation of polycrystalline film deposited at 273 K.

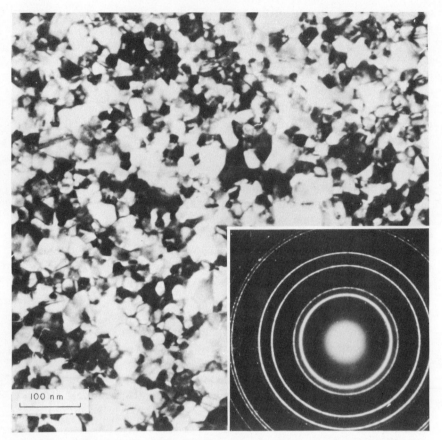

FIG. 3.22 Transmission electron micrograph and electron diffraction pattern for a polycrystalline platinum film deposited at 273 K on glass.

while at any film thickness the surface may be examined by replication and shadowing or by scanning electron microscopy. Thus transmission microscopy can reveal individual crystal widths and also can display at least the larger of the intercrystal gaps, while preferred crystal orientation can be detected in diffraction. Figure 3.22 shows the transmission micrograph and diffraction pattern for a typical polycrystalline platinum film of random crystal orientation.

Films deposited on a glass support are almost always polycrystalline, and Table 3.2 lists some typical parameters for polycrystalline films deposited on glass at 273 K [3,4] with random crystallite orientation. Data of this sort are not unique, but the conditions of film deposition listed in Table 3.2 are typical of many that have been used for catalytic purposes.

TABLE 3.2 Values for average crystal diameter and film porosity for some typical film catalysts deposited on glass at $\sim 10^{-4}$ Pa

Metal	Specific Film Weight (μg mm^{-2})	Substrate Temperature During Deposition (K)	Average Crystal Width (nm)	R
tungsten	0·67	273	7	18·0
	0·69	673	11	6·3
rhodium	0·34	273	8	11·2
palladium	0·22	273	16	3·4
	0·30	573	$\sim$300	$\sim$1·2
platinum	0·34	273	18	4·7
	0·35	673	$\sim$120	$\sim$1·4
iron	0·37	273	24	10·9
nickel	0·65	273	46	7·9
	0·65	673	$\sim$200	$\sim$1·3

There is a trend (with some irregularities) for crystal widths to be larger the lower the melting point of the metal, and there is a tendency for widths to be larger at larger thicknesses. Furthermore, crystal widths are larger the higher the support temperature during deposition; this is also accompanied by lower values of R, due both to the elimination of intercrystal gaps, and a smoothing of the general film surface: this is illustrated by the data in Table 3.2.

It is difficult to assess with high precision the orientations of the crystal planes exposed to the gas phase in low temperature (273 K) polycrystalline films. The assumption has sometimes been made (for instance Brennan et al.[7]), that for f.c.c. metals the surface consists of an equal exposure of (111), (100) and (110) planes, with a similar assumption for b.c.c. metals with regard to (110), (100) and (211) planes. However, for low temperature polycrystalline transition metal films in the thickness range 100–200 nm, high index planes are undoubtedly present to an appreciable extent, and this is the more probable the more refractory the metal. We may take nickel as an example. The photoelectric work-function of a nickel film deposited at 77 K is 4·5–4·6 eV,[1,10] compared with values for the three low index planes[11,9] of 5·35 (111), 5·22 (100) and 5·04 (110) eV, clearly showing the presence of low work-function high index faces in the surface. Or again, nickel films deposited at 273 K have a measured photoelectric work-function of 5·0 eV,[12] from which the same conclusion follows, save that the proportion of high index planes in the surface is here rather less. The relatively high degree of surface crystallographic heterogeneity of 273 K nickel films is also illustrated by the nature of the rare gas adsorption isotherm[13] to which a patch analysis is inapplicable, although this method is more readily applicable to high temperature films of lower heterogeneity.[8, 13, 4] Nevertheless, it must also be said that in low temperature films grown to great thickness and thus consisting of very wide crystals, low index planes are probably dominant. For instance, in a polycrystalline nickel film deposited on glass to a thickness of $3 \cdot 1$ μm and with an average crystal width in the region of $0 \cdot 2$ μm, surface replicas clearly show regular faceted crystal shapes with dominant (111) and (100) surface planes,[15] as expected from both thermodynamic[16] and kinetic arguments.[1] Work-function measurements with films of other metals show a similar trend to that described above for nickel. Thus with platinum films sintered in the range 77 K to 570 K the work-function increased from $5 \cdot 44$ to $5 \cdot 72$ eV,[18] and with ruthenium films sintered in the range 77 K to 700 K the work-function increased from $4 \cdot 52$ to $5 \cdot 10$ eV.[31]

R values for films are reduced by sintering at an elevated temperature after deposition at (say) 273 K or by deposition on a heated substrate. For instance, some support temperatures above which there is little further change in R are as follows (cf. ref. 5): for platinum, palladium and nickel, about 470 K; rhodium, about 620 K; tungsten $\geqslant$ 670 K.

FIG. 3.23 Schematic representation of a high temperature polycrystalline film.

Figure 3.23 shows a schematic cross-section through a film such as would be expected to be produced at temperatures higher than those quoted above. The main features are the complete elimination of intercrystal gaps and a great reduction in the general surface roughness compared to Fig. 3.21. The elimination of intercrystal gaps is clearly evident from transmission micrographs and from the electrical properties of the films.[3, 6] Moreover, an electron micrograph of a shadowed surface replica of a nickel film deposited on glass at 670 K is given in Fig. 3.24, and this may be compared with Fig. 3.25 for a nickel film of comparable thickness deposited on glass at 273 K.

Polycrystalline films deposited on amorphous supports are of lower crystallographic surface heterogeneity the higher the temperature of annealing subsequent to deposition or the higher the support temperature

3. MASSIVE METAL CATALYSTS   135

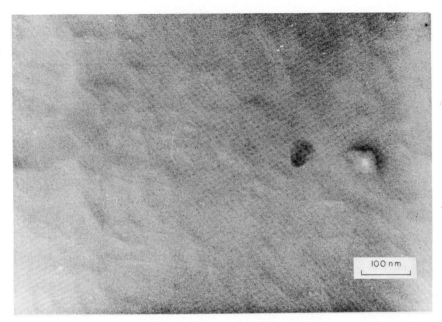

FIG. 3.24 Electron micrograph of shadowed surface replica of polycrystalline nickel film deposited on glass at 670 K.

during deposition. This is clearly implicit in the schematic representations in Fig. 3.21 and Fig. 3.23, and photoelectric work-function data serve to emphasize the point. As mentioned above, nickel films deposited at 77 K give work-functions of about 4·5–4·6 eV, at 273 K give about 5·0 eV, while annealing films such as these at higher temperatures leads to work-function increases into the range of values characteristic of the low index planes. Thus Suhrmann and Wedler[10] observed a value of 4·58 eV for a film deposited at 77 K, 5·09 eV after sintering at 370 K, and 5·25 eV after sintering at 470 K. In a similar way, Baker et al.[9] found 4·54 eV for a film deposited at 77 K, and after sintering at 520 K the photoelectric yield curve could be analysed in terms of three patches which, in terms of their work-functions, may be identified as (111), 28%; (100), 67% and (110), 5%.

In addition to a variation in the proportions of the various types of crystal planes exposed, the thermal history of a film would also be expected to influence the extent to which features such as small asperities, perhaps consisting of clusters of no more than a few atoms may exist on the surface as residues from crystallite nucleation and growth. One would expect this sort of feature to be less important the higher the temperature during film deposition or sintering, while at a given temperature one would expect their importance to increase the more refractory the metal. These expectations are in agreement with reality.[113] This type of surface feature can be of catalytic consequence since it inevitably contains atoms with relatively low numbers of nearest neighbours.

FIG. 3.25 Electron micrograph of shadowed surface replica of polycrystalline nickel film deposited on glass at 273 K.

The question arises of the extent to which, in polycrystalline films, reactant gas has access to the support. It is clear that in high temperature films the total absence of intercrystal gaps means that such access of gas is completely absent. In the case of films deposited at 273 K one may estimate from the measured roughness factor and from transmission electron microscopic evidence that, of the total substrate area, more than 90% is in direct contact with metal; in any case, the support at the base of a gap is almost certainly covered with a thin layer of metal. Thus, even in this case the gas cannot have more than trivial access to the support. This conclusion is of some importance if one is concerned about possible participation of the support as a seat of catalytic activity.

Deposition on glass or other amorphous supports at higher temperatures may result in some degree of preferred crystal orientation. This becomes immediately evident in transmission electron diffraction, although surface replication will still reveal a surface very similar to a film of random crystal orientation deposited at the same temperature. Some typical examples from the literature are: silver on fused quartz above 820 K,[11] gold on fused quartz above 620 K,[22] or above 750 K,[23] and Pd on fused quartz above 800 K.[22] In the author's experience, some preferred orientation is sometimes obtained in the deposition of nickel, platinum and palladium on glass at 620–670 K either in HV or UHV conditions, and on rare occasions may even be observed at deposition temperatures approaching room temperature. In all these cases, the orientational tendency for cubic metals is for $<111>$ normal to the support. With iron deposited on glass, some $<110>$ and $<111>$ preferred orientation has been observed.[24-29] The tendency towards preferred orientation is greater at larger film thicknesses, although it can undoubtedly occur in the initial stages of film growth.[30] In general, however, the occurrence and extent of preferred orientation on glass is of poor reproducibility, and when preferred crystal orientation is deliberately required, glass is not the best choice as a support.

Film deposition on a single crystal support can, in principle, lead to the formation of an epitaxed single crystal film. The art of epitaxial film growth has been extensively described elsewhere.[1,17]

Only two crystalline supports have had much use for the preparation of metal film catalysts. These are mica and rocksalt, but other substances such as single crystals of other alkali halides, magnesia, molybdenum disulphide or graphite are also possible. The cleavage faces of alkali halides or magnesia provide convenient supports for generating epitaxed films of f.c.c. or b.c.c. metals exposing a (100) face, while mica, molybdenum disulphide and graphite supports favour (111) epitaxy with f.c.c. Some examples are to be found in references 117–120.

With some metals it is possible to obtain a high degree of single crystal epitaxy on mica. For instance, silver deposited and annealed at elevated temperatures on mica gives (111) oriented single crystal films: provided the films are deposited with the mica at 570–670 K and then annealed at 720–920 K the films are free of grain boundaries and non-coherent twin boundaries, although they still contain 10^6–10^9 mm^{-2} dislocations, and

40–300 mm^{-1} stacking faults and coherent twin boundaries.[19] If prepared at somewhat lower temperatures these films contain twins parallel to the support whose non-coherent boundaries cut the surface to generate grooves and to create small patches of surface with an orientation different from the ideal.[34] However, single crystal film catalysts have not been prepared on mica from transition metals such as those which are of most importance for catalytic purposes, probably because of temperature limitations imposed by glass apparatus. With f.c.c. transition metals, deposition on mica at 620–670 K in HV or UHV leads to polycrystalline deposits in which each crystal is oriented with a <111> axis normal to the support, but with the crystals oriented with rotational disorder about this axis (cf. ref. 20). A typical transmission micrograph and diffraction pattern

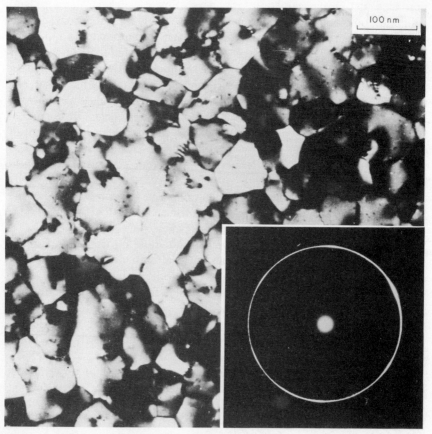

FIG. 3.26 Transmission electron micrograph and electron diffraction pattern for a polycrystalline platinum film deposited on mica at 620 K.

is shown in Fig. 3.26 and a micrograph of a surface replica in Fig. 3.27. In some cases, rotational disorder is not completely random. Complete preferred crystal orientation is not obtained with total reproducibility. The microscopic results show that the exposed surface of each crystal is overall relatively flat, so that the whole exposed film surface must be close to (111). Presumably, some higher order planes are exposed to a relatively small extent in the immediate vicinity of the grain boundaries: nevertheless, the proportion of (111) surface exposed is estimated to be not less than 90% and this is very similar to the estimate for the proportion of (111) surface exposed in completely epitaxed silver films.[19] This estimate is also in agreement with that obtained from a patch-model analysis of photoelectric work-function data for nickel films deposited on mica at 600 K.[9]

FIG. 3.27   Electron micrograph of shadowed surface replica of film from Fig. 3.26.

Although the surface of such a high temperature film may appear relatively flat and featureless to shadowed replication, decoration shows clearly that the surface is not completely smooth on a quasi-atomic scale. Thus, Fig. 3.28 shows a micrograph of a decorated (111) silver surface of a well epitaxed single crystal film on mica, and the presence of surface steps is obvious. Metals more refractory than silver would be expected to present a somewhat rougher and more imperfect surface than that in Fig. 3.28 under similar temperature conditions of preparation. When deposited on mica at 273 K surface replicas show surface roughness comparable to that of films on a glass substrate, and the degree of preferred crystal orientation is also usually negligible.

For comparison with this (111) epitaxy of silver, Jaeger[19] also found that silver deposition onto a (100) cleaved rocksalt surface followed by annealing at 670 K gave a well epitaxed single crystal silver film exposing a (100) surface, with comparable or somewhat greater densities of dislocations, stacking faults and coherent twin boundaries to those found for (111) epitaxed films on mica. The emergence of stacking faults at the surface generates surface grooves.

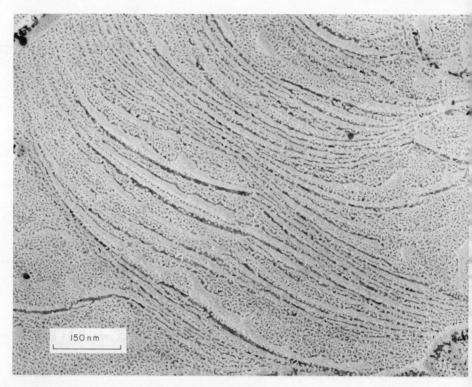

Fig. 3.28 Decoration of (111) surface of silver film deposited on mica. Courtesy J. V. Sanders.

A wide range of metals has been grown epitaxially on a (100) rocksalt face, including (f.c.c.) gold, silver, aluminium, nickel, copper and (b.c.c.) chromium, iron.[21] All can readily give an orientation of (100) metal planes parallel to rocksalt (100), but gold, silver, copper and aluminium can also give (111) metal planes parallel to rocksalt (100) depending on the conditions during metal deposition and during rocksalt cleavage, and this also affects the quality of epitaxy. Various workers[35,36] have also described epitaxy of f.c.c. metals on potassium halides.

3. MASSIVE METAL CATALYSTS

For epitaxy on alkali halide substrates, both adsorbed water vapour and cracking products from hydrocarbon pump oil have been found to affect the result, and adsorbed water vapour is particularly potent.[32,33] At a given substrate temperature, an optimum impurity concentration is needed to obtain the best degree of epitaxy. However, the interplay between these factors makes it extremely difficult to specify optimum conditions for epitaxy, and the matter still remains to be treated in a practical sense as something of an empirical art.

One of the problems of using epitaxed metal film catalysts deposited on cleaved or cut faces of substances such as the alkali halides or magnesia is that the available substrate area is quite limited; a limit of about $2 \times 10^{-3}\,m^2$ would be usual.

Because of the convenience of using for catalytic purposes a film

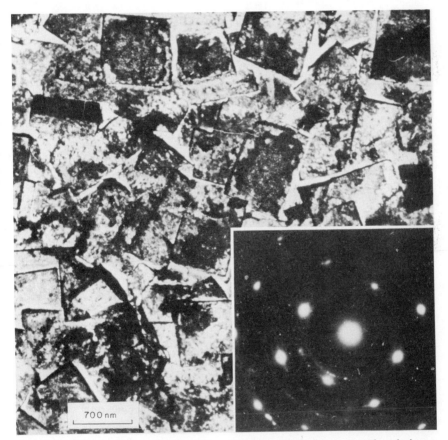

FIG. 3.29 Transmission electron micrograph and diffraction pattern for platinum film deposited on evaporated rocksalt layer at 540 K.

FIG. 3.30 Transmission electron micrograph and diffraction pattern for nickel film deposited on evaporated rocksalt layer at 520 K. Reproduced with permission from Baker, B. G. and Bruce, L. A. *Trans. Faraday Soc.*, **64**, 2533 (1968).

FIG. 3.31 Electron micrograph of shadowed replica of surface of nickel film grown on evaporated rocksalt. Reproduced with permission from Baker, B. G. and Bruce, L. A. *Trans. Faraday Soc.* **64**, 2533 (1968)

deposited on a relatively large area of support, a technique has been developed[20] for producing an evaporated layer of rocksalt as a support for subsequent film deposition. To avoid problems due to sintering, thermal etching and incipient evaporation of the rocksalt layer, and to maintain adequate vacuum conditions for surface cleanliness of the metal film, the support temperature is limited to about 520 K during metal deposition or subsequent annealing. Although a support temperature of 520 K will produce reasonably well epitaxed films of silver, with metals of higher melting point and greater cohesive energy, epitaxy is more difficult. Thus, platinum films produced in this way are polycrystalline on a given rocksalt face;[4, 20] nevertheless, the platinum crystals have strong preferred orientation with the (100) plane parallel to the rocksalt, and with most of the platinum crystals similarly oriented with respect to each other about each <100> axis normal to the support. This is illustrated in Fig. 3.29 which shows a typical transmission micrograph and a diffraction pattern from a platinum film deposited in this way; the diffraction pattern refers to an area of metal within that on a single rocksalt crystal face. On the other hand, with nickel under similar conditions, single crystal epitaxy occurs on a rocksalt face as shown by Fig. 3.30. Nevertheless, this nickel surface is far from flat, as shown by the shadowed surface replica of Fig. 3.31 where the dimpled structure of the surface remaining from the coalescence stage of film growth is clearly seen. With this type of film, the exposed surface is far from perfect, due both to the microcrystallinity of the rocksalt support and to the imperfections in the surface of the epitaxed film. For a nickel film so prepared, the proportion (100) surface exposed, as judged from rare gas adsorption data[14] is no more than 70% and is probably rather less than this with platinum.

4. Alloy Catalysts

Many of the important features of surface topography of alloys are similar to those of metals, and the previous discussion need not be repeated. We shall be mostly concerned with features by which alloy surfaces differ qualitatively from those of pure metals.

It has long been clear that alloys offer exciting possibilities as catalysts because variation in alloy composition offers a ready method for the more or less continuous alteration of metallic properties.

Alloy catalysts in massive form have, like pure metals themselves, been used mainly as laboratory catalysts in model systems, and for this reason they have usually been designed with an eye to interpretative insight and simplicity. No single crystal alloy catalysts have yet been used, the work having been confined to polycrystalline specimens.

Much of the philosophic framework for the study of alloy catalysts arose from the empirical correlations which were found between the activity of metal catalysts for a number of hydrogenation, dehydrogenation and exchange reactions, and certain metallic properties such as magnetic susceptibility which could be interpreted in terms of the extent of d-band occupancy or proportion of d-character of the metallic bond. Alloying a paramagnetic or ferromagnetic metal such as a group VIII metal, with a diamagnetic metal such as a group 1B metal changes the magnetic susceptibility, and this has been interpreted in terms of changes in d-band occupancy or proportion of d-character of the metallic bond. This has been discussed earlier in Chapter 1.

A number of alloy systems have been used as catalysts in the form of thick continuous evaporated films. Compared to catalysts in the form of foils and wires, alloy films have the advantage of a low degree of adventitious surface contamination, which should be reduced to negligible proportions if UHV techniques are used. They can also be prepared with reasonably large surface areas. However, these benefits are obtained at the expense of rather greater uncertainty in alloy equilibration because of the temperature limitations imposed by the apparatus. With glass supports, film equilibration temperatures are limited to <670 K.

The magnitude of the equilibration problem depends to some extent on the method and conditions of film preparation. Thus, if equilibrium requires a single phase, this will be favoured by simultaneous evaporation: if equilibrium requires two phases, this will be favoured by sequential evaporation. However, in this situation, the order of evaporation is also important since, as we shall subsequently see, there is a tendency for two phase alloy films to consist of particles in each of which there is an outer layer consisting of the phase with the lower surface energy, and this usually corresponds to the phase which is richest in the component of lower melting point.

Consider the case of the simultaneous deposition of the components of an alloy which at equilibrium would consist of two phases. Provided the support temperature is kept low (e.g. about room temperature with most transition metals) one would expect that the kinetic energy of the impinging metal atoms in combination with the lattice energy liberated by each condensing atom, and the heat of radiation emitted from the evaporation sources, will enable the growing film to acquire an atomic arrangement which is somewhere between the metastable state of a random distribution and the thermodynamically stable state of two separate phases so arranged to minimize the total free energy (including surface energy). Presumably, there are clusters present which represent regions having more or less short range order tending towards the structures of the two phases.

During annealing of the film a coalescence of like clusters takes place by relatively rapid surface diffusion, grain boundary migration, and relatively slow volume diffusion. Because the components may well have markedly different mobilities, a pronounced Kirkendall effect is to be expected. Although volume diffusion is certainly much slower than surface or grain boundary diffusion, it is nevertheless not of negligible importance since, in the later stages of annealing when each film particle has its duplex structure, the final stages of composition equilibration has to involve volume diffusion. The consequence is that the later stages of equilibration will usually be relatively protracted. On the other hand, if deposition occurs onto a support at an elevated temperature, there will be much greater atom mobility and the two separate phases can grow right from the initial nucleation state, but the generation of the correct film morphology (e.g. duplex particle structure) will still require extensive diffusion.

One might expect to be able to assess the experimental requirements for equilibrating alloy films in terms of known diffusion data. However, the best that can be said is that these data offer only a rough guide, and an independent examination of the specimen structure is mandatory. The problem is the complexity of the processes which occur during film growth and, furthermore, the relevant diffusion data (e.g. ref. 71) are only available for relatively few systems of catalytic interest. The diffusion coefficients of individual components in alloys are strongly dependent on the alloy composition. Because of this, the most satisfactory diffusion data to use are those for interdiffusion, where the latter term refers to the diffusive growth of alloy at the junction region of the two pure components. Such interdiffusion data for alloy systems of interest as film catalysts are recorded in Table 3.3. This table also includes the RMS interdiffusion distance, L, expected from heating for 1 h at the indicated temperatures, computed from the diffusion coefficient, D, from the Einstein relation

$$L = \sqrt{2Dt} \qquad (3.5)$$

where t is the time: this assumes diffusion in one direction, that is, normal to the interface.

One needs to bear in mind the following facts: film thicknesses are generally in the range 100–200 nm or so, and crystallite widths for polycrystalline films are of a similar order to this thickness, or smaller. One would expect much metal transport to occur by surface and grain boundary diffusion which are both considerably faster than volume diffusion. For instance the activation energies for the diffusion of silver on palladium and palladium on silver are 55 and 105 kJ mol^{-1} respectively.[72] The data of Table 3.3 indicate that of the systems listed, only with palladium–copper is serious equilibration difficulty to be expected if film annealing is carried out under the indicated conditions. This is confirmed by the experimental behaviour of individual systems which will be discussed presently.

TABLE 3.3 Volume interdiffusion data for some alloy systems
$D = D_0 \exp(-Q/RT)$

System	Q (kJ mol^{-1})	D_0 (mm^{2}s^{-1})	Temperature range* (K)	Calculated diffusion distance in 1 h* (nm)	Reference
Ni, Cu	124	$4 \cdot 2 \times 10$	870–1270	$8 \cdot 6 \times 10^3$(673 K) $7 \cdot 5 \times 10$(473 K)	73
Pd, Cu	224	$4 \cdot 8 \times 10$	870–1170	$1 \cdot 2$(673 K)	74
Pd, Ag	103	$1 \cdot 5 \times 10^{-4}$	870–1170	$1 \cdot 1 \times 10^2$(673 K) $2 \cdot 1 \times 10$(573 K)	74
Pd, Au	153	$3 \cdot 2 \times 10^{-2}$	870–1320	$1 \cdot 9 \times 10^3$(673 K)	74
	46	$1 \cdot 2 \times 10^{-3}$	670–870	$4 \cdot 6 \times 10^4$(673 K)	75
Pt, Au	163	$1 \cdot 5 \times 10^{-1}$	970–1270	$4 \cdot 3 \times 10^2$(673 K) $3 \cdot 1 \times 10$(573 K)	76

* The temperature range indicates the range over which the diffusion measurements were made: the temperatures used for the calculated diffusion distances were chosen to correspond to values actually used in the preparation of alloy films and often involve a short extrapolation beyond the temperature range for the diffusion data.

Much attention has been focused on the nickel–copper system (e.g. refs 77–84). The standard reference works[85] list nickel–copper as an alloy which is single phase over the entire composition range, in which there are no solid state transformations or intermediate phases, and little or no ordering. The liquidus–solidus curves are shown in Fig. 3.32. However, this system exemplifies the problems which exist in trying to assess the nature of the equilibrium phase diagram at temperatures a long way below the solidus line. The problem is, of course, that transformations in the solid are slow and it is always difficult to be sure that true equilibrium has been reached; moreover, the determination of surface composition is a difficult task. In the case of the nickel–copper system, the metal activities in the alloys have been measured electrochemically.[86, 87] These and other data have been analysed by Sachtler[80] and the result is summarized in Fig. 3.33 for the free energy change for alloy formation at 473 K. This result inevitably means that at 473 K there must exist a two phase region in the phase diagram extending between about 2 and 80 atom % copper.

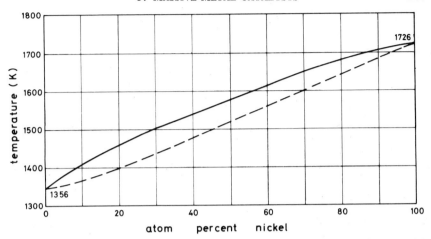

FIG. 3.32 Solidus–liquidus curves for the nickel–copper system. After Hansen, M. and Elliot, R. P. (cf. ref. 85).

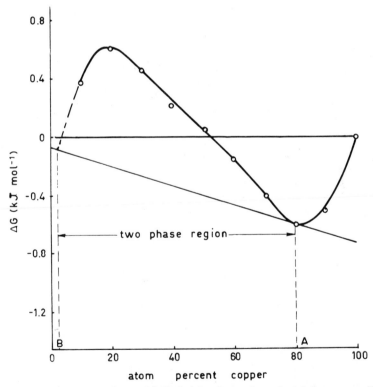

FIG. 3.33 Free energy change (ΔG) for the formation of nickel–copper alloy at 473 K Reproduced with permission from Sachtler, W. M. H. and Jongepier, R. *Catal.* **4**, 665 (1965).

The question then arises: if one actually had a two phase specimen, how would these phases be distributed in relation to the surface and, in any case how does the surface composition of a phase compare with the bulk composition? At the moment, there are two sets of data which require mutual reconciliation. On the one hand there are some measurements of the structure of nickel–copper alloy films by Sachtler and co-workers[79, 80, 83] who used photoelectric work-function measurements to assess the nature of the surface. These films were prepared by sequential evaporation under UHV conditions, followed by homogenizing by annealing at about 473 K. As indicated by Fig. 3.33, at 473 K there is a large two-phase field consist-

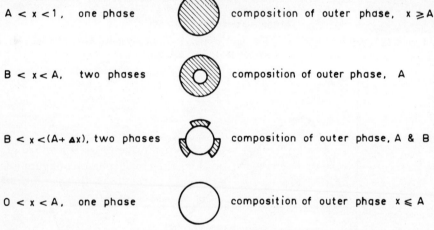

FIG. 3.34 Schematic representation of particle types in nickel–copper evaporated films. Hatched regions are nickel-rich phase, open regions are copper-rich phase. The composition, x, is atom % copper. The compositions A and B refer to the values indicated in Fig. 3.33. After Sachtler, W. M. H. and Jongepier, R. *J. Catla.* **4**, 665 (1965).

ing of alloy A (80 atom % copper, 20 atom % nickel) and alloy B (2 atom % copper, 98 atom % nickel). From the fact that the film work-function was constant over a wide range of overall compositions, it appears that each crystallite consists of a core of alloy B, surrounded by a layer of alloy A. Thus, over a wide range of compositions, the film exposes a surface of constant composition (alloy A), and it is only when the proportion of A becomes small, so that A is present as patches on the surface of B, that some surface of alloy B becomes exposed. The situation is illustrated schematically in Fig. 3.34. One should add that in terms of surface energy (cf. Table A3, Appendix 1) one would expect this to be minimized by an arrangement in which the copper-rich alloy A formed the outside layer. On the other hand, there are measurements of the surface composition of nickel–copper alloy slab specimens, both polycrystals and single crystals with bulk compositions in the range 16–17 atom % nickel,[88,137-139] using electron-stimulated Auger spectroscopy. In all cases, the measured surface composition was identical with the bulk composition. The quantitative application of this technique to alloys is not as straightforward as might at first be thought owing to the difficulty of establishing meaningful calibrations, nevertheless it is hard to avoid the conclusion that with these specimens the differences between bulk and surface compositions cannot be very substantial. This difference between the results from alloy films and alloy slab specimens is probably a result of lack of equilibration with the latter. Lack of equilibration in this sort of specimen has already been well recognized[138] and is, in any case, evidenced by the absence of phase separation under temperature and composition conditions for which it is mandatory if true equilibrium is reached. Clearly equilibration occurs much more easily with evaporated alloy films than with thick slab specimens, and this may well be due to a higher defect concentration in the former which increases the rate of interdiffusion, as well as the much shorter diffusion path necessary in films. Furthermore the slab specimens were cleaned by ion bombardment followed by annealing at temperatures $\geqslant 670$ K, while the critical temperature for phase separation is in the range below 570 K; since the specimens were cooled quickly, the single phase configuration above the critical temperature is frozen in. It thus appears that this lack of phase equilibration is paralleled by a lack of equilibrium with respect to surface composition. Clearly the surface composition of nickel–copper alloy specimens is much dependent on their form and on their thermal history.

Although Sachtler's work with nickel–copper alloy film provided no actual evidence at all to indicate whether the surface of alloy A exposed had the same composition as that of the bulk of that phase, it would be a reasonable inference on surface energy grounds to expect some surface enrichment with copper which has the lower surface energy.

Platinum–gold alloy films behave quite analogously to nickel–copper. In the platinum–gold system a wide miscibility gap is well known for temperatures below 1531 K and the phase diagram is shown in Fig. 3.35. With platinum–gold alloy films[89] equilibration required simultaneous evaporation (UHV) followed by sintering at 570 K. Work-function data have been interpreted to indicate a constant surface composition in the range 10–90 atom % platinum, and it is likely that the outermost phase of each particle is the gold-rich phase which, after equilibration at 570 K probably contains in the region of 10–20 atom % platinum.

Platinum–ruthenium films have also been studied,[89,131] having been prepared by sequential evaporation followed by sintering at 770 K (UHV). X-ray data show mutual solubility of the components except in the range 30–50 atom % platinum. At <30 atom % platinum a hexagonal lattice structure is retained (cf. ruthenium), while at >50 atom % platinum the alloy has an f.c.c. lattice structure (cf. platinum). In the single phase regions the surface composition appears to change continuously as the bulk

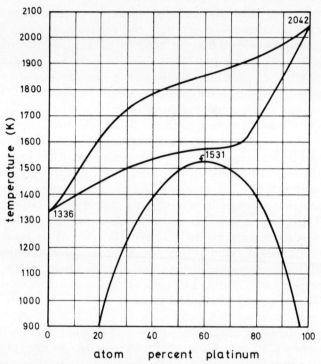

Fig. 3.35 Phase diagram for the platinum–gold system. The lines have been drawn to be the best representation of the data from various sources which are presented by Hansen, M. and Elliot, R. P. (cf. ref. 85).

composition changes, and on surface energy grounds one would expect some surface enrichment with platinum.

Moss et al.[90,91] have also concluded that two phases are present in rhodium–palladium alloy films, in agreement with the phase diagram (Fig. 3.36) after annealing at 670 K (about 10^{-4} Pa, about 10^{-6} Torr) using simultaneous deposition. Nevertheless, it is clear that in many cases these films did not reach equilibrium. Moss and Gibbins[92] suggest making a virtue out of necessity by using this difficulty of equilibration to generate surface compositions that are inaccessible under equilibrium conditions. This is a highly hazardous proposition for controlled experimental work, unless one can use some independent method for assessing surface composition, such as Auger spectroscopy.

The one system which appears to be unambiguously single phase and which has been used for the preparation of evaporated alloy films, is

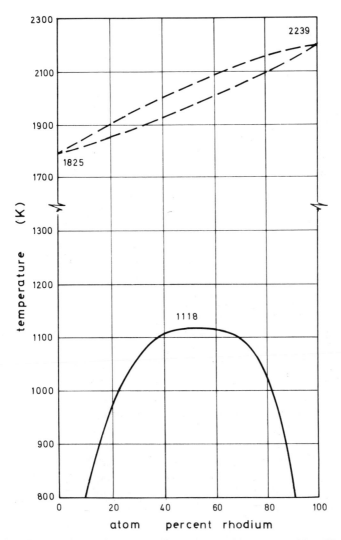

FIG. 3.36 Phase diagram for the rhodium–palladium system. After Hansen, M. and Elliot, R. P. (cf. ref. 85).

palladium–silver. The phase diagram is shown in Fig. 3.37. In agreement with this, alloy films in this system, prepared by simultaneous evaporation and sintering at 670 K (about 10^{-4} Pa, about 10^{-6} Torr) appear to have a surface composition which changes continuously with changing bulk concentration.[89] On surface energy grounds one would expect some surface enrichment by silver.

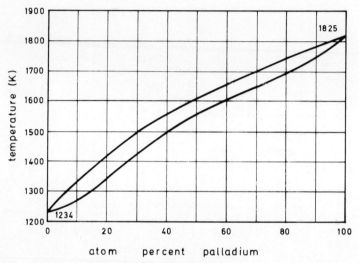

FIG. 3.37 Phase diagram for the palladium–silver system. After Hansen, M. and Elliot, R. P. (cf. ref. 85).

In broad terms, the gross microstructure of alloy films lies between the extremes for films of the pure components. An example of the microstructure is shown in Fig. 3.38 for the palladium–silver system.

Moss and co-workers[93] have shown that the ease of achieving alloy film equilibration can be strongly dependent on the vacuum conditions used during film deposition. Thus, while with palladium–silver films, deposition in vacua of the order of 10^{-4} Pa (about 10^{-6} Torr) requires a deposition temperature and a subsequent annealing temperature of 670 K to achieve equilibrium, if ultrahigh vacuum is used, a temperature of 273 K is adequate. The difference here is a matter of some importance since deposition and annealing at a high substrate temperature results in film sintering with crystal growth and loss of surface area, while the use of ultrahigh vacuum and lower substrate temperatures leads to the maintenance of an uncontaminated film surface. The resulting difference in film microstructure is illustrated in Fig. 3.39.

3. MASSIVE METAL CATALYSTS   153

FIG. 3.38 Transmission electron micrograph of palladium–silver alloy film. 62 atom % palladium, deposited under high vacuum at 273 K. Reproduced with permission from Whalley, L., Thomas, D. H. and Moss, R. L. *J. Catal.* **22**, 302 (1971), and Crown Copyright.

Quite a wide range of alloys have been used as catalysts in foil, wire or slab form, but explicit quantitative information about the surface composition is largely lacking. In some cases such as palladium–silver[97] and palladium–gold,[94,95] the phase diagrams indicate only the existence of a continuous range of solid solutions, and it seems likely that the only departure of the surface composition from that of the bulk will be some degree of enrichment with silver or gold. The surface composition of nickel–gold alloy foil has been analysed with Auger electron spectroscopy,[140] with the result that the gold concentration was substantially higher on a clean surface than in the bulk. Thus, for an alloy with a bulk gold atom fraction of 0·5 atom %, the concentration of gold in the surface

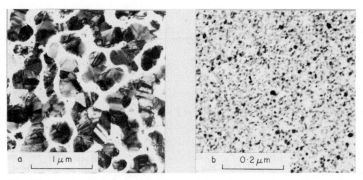

FIG. 3.39 Transmission electron micrographs of palladium–silver alloy films. (a) 45 atom % palladium, deposited/annealed in vacuum at 670 K; (b) 42 atom % palladium deposited in ultra-high vacuum at 273 K. In both cases the alloy was homogeneous. Reproduced with permission from Moss, R. L., Thomas, D. H. and Whalley, L. *Thin Solid Films*, **5**, R19 (1970) and Crown Copyright.

was 50 atom %. In some cases where the phase diagram indicates the formation of intermediate compounds, the alloy compositions have been chosen to correspond to these compounds and so at least the uncertainty of having to deal with a two-phase system was avoided (assuming the specimens really reached equilibrium during preparation); examples are Cu_3Au, $CuAu$, Cu_3Pd, $CuPd$, Cu_3Pt and $CuPt$ studied by Rienacker.[96] Alternatively, Schwab and co-workers[146-151] have used alloys for which the phase diagrams indicate the existence of more than one phase, but avoided this complication by restricting the alloy composition to lie within a single phase field (silver alloyed with one of cadmium, indium, tin, antimony, mercury, thallium, lead or bismuth), while these workers also studied binary phase alloy catalysts in the following systems; copper–silver, copper–tin, copper–magnesium, gold–cadmium, gold iron and silver–aluminium. The more the phase diagram for an alloy departs from an ideal solid solution, the more likely it is that the composition of the surface will differ from that of the bulk.

The driving force for the enrichment of the surface by one of the alloy components is of course that of minimizing the total free energy. The theory of surface enrichment has recently been discussed from the viewpoint of statistical thermodynamics both for cases where disordered alloys consisting of a continuous series of solid solutions are formed,[132] and also for ordered alloys.[133] In both cases the surface enrichment was assumed to result only from an exchange of atoms between the outer layer and the one immediately beneath it, and the energy changes were evaluated from a parametized nearest neighbour interaction energy model. For disordered solid solutions, it was shown that the composition of the surface relative to that of the bulk is given by

$$\ln\left[\frac{x_1^B(1-x_1^S)}{x_1^S(1-x_1^B)}\right] = \frac{\bar{a}}{RT}(\gamma_1 - \gamma_2) + \frac{\Omega}{RT}\left\{p(1-2x_1^S) + q(1-2x_1^B) + r\right\} \quad (3.6)$$

The model assumes some specific surface plane to be exposed: x_1^B and x_1^S are the mole fraction compositions of the bulk and surface respectively, both referred to component 1; γ_1 and γ_2 are the surface free energies of the pure components in the plane in question; Ω is the heat of formation of the alloy; $\bar{a}$ is the mean surface area of the constituent atoms in the plane, p, q and r are constants which are determined by the number of nearest neighbour atoms possessed by an atom in the surface and the bulk. For an f.c.c. bulk structure, the constants are: (111) surface: $p = 1$, $q = -3/4$, $r = -1/2$; (100) surface: $p = 2/3$; $q = -2/3$, $r = -1/3$. An example of the results from the application of equation 3.6 is shown in Fig. 3.40 for gold–silver alloys. As expected, the concentration of the component of lower surface energy (silver) is augmented in the surface, and the extent to which this occurs decreases with increasing temperature and with decreasing ($\gamma_1-\gamma_2$). Since this theory is confined to the top two atomic layers, it is probable that the results will be at best only a semi-quantitative description of reality. A more complete and hence more accurate description would include more atomic layers.

Providing equilibrium can be achieved, the measurement of the surface energy as a function of bulk composition remains the classical method for studying the surface composition of mixtures. With metals there are two

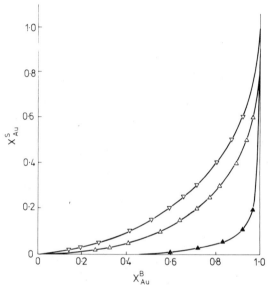

FIG. 3.40 Variation of mole fraction surface concentration of gold, x^S_{Au}, with the mole fraction bulk concentration of gold, x^B_{Au}, for (111) surface of gold–silver alloys. Calculation from equation 3·40, using $\Omega = -6\cdot 65$ kJ mol^{-1}. Curve ▲ is for ($\gamma_{Au} - \gamma_{Ag}$) = 0·6 J m^{-2} at 523 K and corresponds to the best estimates for $\gamma_{Au(111)}$ and $\gamma_{Ag(111)}$ of 2·2 and 1·6 J m^{-2} respectively. For comparison, curves ▽ and △ correspond to ($\gamma_{Au} - \gamma_{Ag}$) = 0·3 J m^{-2} at 773 K and 523 K respectively.
After van Santen, R. A. and Boersma, M. A. M. *J. Catal.* **34**, 13 (1974).

obvious problems. At temperatures much below the melting point it will be difficult to establish equilibrium, and in any case, the measurement of surface energy is difficult. Secondly, great care is needed to avoid contamination of the surface: UHV techniques or their equivalent are mandatory. We know of no alloy surface energy data which meet these criteria with certainty, particularly in respect to surface contaminant. Such data as are available have been summarized by Hondros and McLean[116] and indicate solute accumulation at the free surfaces of the following alloy systems: gold in copper (1123 K), nickel in iron, chromium in iron, manganese in iron (all 1473 K). However, there are some reliable data for the closely related phenomenon of component accumulation at grain boundaries. At an experimental level, a study of grain boundaries avoids the contamination problem which plagues work with free surfaces. Some grain boundary accumulation data for alloy polycrystals are summarized in Table 3.4. Since the analysis is dependent on the use of the Gibbs adsorption isotherm, these data refer to very low solute concentrations, typically <1 atom %. Since a solute monolayer contains about $1 \cdot 2 - 1 \cdot 5 \times 10^{19}$ atoms m^{-2}, it will be seen that the grain boundary concentrations given in Table 3.4 range from about $2 \cdot 5$ % (Au in Cu) to 45 % (Cr in Fe) of a monolayer.

TABLE 3.4 Grain boundary accumulation of alloy solute

System		Temperature (K)	Excess grain boundary solute concentration (10^{18} atom m^{-2})	Reference
solvent	solute			
copper	gold	1123	0·3	98
iron	nickel	1473	4·4	99
iron	chromium	1473	6·1	99

One might also note the considerable body of work which has demonstrated the segregation of various group VIII metals at the free surface of ferrous alloys (particularly stainless steel), titanium and chromium (cf. ref. 100). This is significant at quite low solute concentrations (<1 atom %) and it has important consequences for the electrochemical and corrosive properties of the surfaces. In the case of palladium in titanium, surface segregation has been directly demonstrated radiochemically.[101]

Attempts have also been made to characterize alloy surfaces by electrochemical methods.[122,123] The voltametric technique, using a dilute sulphuric acid electrolyte, itself often alters the composition of the alloy surface by the preferential removal of one component, so the result then depends on the number of times the electrode is subjected to a voltametric sweep. Nevertheless, it was concluded that with platinum–gold alloys which had been quenched from the melt to give a homogeneous phase, the initial surface was highly enriched in gold, although the result of repeated voltametric sweeps was to generate, on the surface, islands or crystallites of the two phases which would be expected for equilibrium at about room temperature. With palladium–gold it was found that the electrochemical treatment resulted in surface enrichment with gold, that is preferential removal of palladium, while with platinum–rhodium there was enrichment with platinum, and there was no preferential enrichment with palladium–rhodium.

This dependence of alloy surface composition on the electrochemical history of the specimen is, of course, not a unique situation. Cases are known where the surface composition of an alloy is dependent on the history of its interaction with chemisorbable gas, and this possibility has important consequences for understanding the behaviour of alloy catalysts, because one cannot automatically assume that the composition of the surface will remain unaffected by reaction with the gas. This phenomenon has particularly been demonstrated in the interaction of carbon monoxide with some alloy surfaces: in this environment there is platinum enrichment of the surface of platinum–gold alloy, and palladium enrichment of the surface of palladium–silver;[31] and it has also been reported for oxygen adsorption on nickel–gold alloys which leads to nickel enrichment.[140] The effect arises when one component in the alloy can chemisorb the gas and the other component cannot, or when there is a large difference in the heat of adsorption. One can expect with some confidence that further work will show this sort of behaviour to be of widespread occurrence in alloy/gas interactions. Since catalysis always involves chemisorption of at least one of the reactant species, one must conclude that when an alloy is functioning as a catalyst, the relative concentrations of the metallic components at the surface may bear little relation to the corresponding concentrations measured in vacuum, and there is a clear need for the surface composition to be determined under reaction conditions if the catalytic properties of the alloy are to be properly understood.

Cleaning the surface of an alloy catalyst which is in the form of a wire, foil or slab presents a fairly difficult problem. In almost all cases, the technique adopted has consisted of hydrogen reduction at temperatures up to about 620 K. In the case of nickel–copper it was found that higher temperatures tend to result in copper evaporation, presumably via the relatively unstable hydride. Alternatively, a high frequency hydrogen discharge has been used. It is most unlikely that surfaces so prepared will be atomically clean. On the other hand, cleaning by argon ion bombardment can lead to preferential loss of one component from the surface region, thus changing its composition. In the case of nickel–copper for instance, preferential loss of copper occurred to a depth of about 0·8 nm. The degree of differentiation is also dependent on the argon ion energy, and is more marked below 150 eV.[102] Although one should be able to reduce or eliminate this by post-bombardment annealing, it represents an added uncertainty with the use of ion-bombarded alloys.

References

1. Geus, J. W. *In* "Chemisorption and Reactions on Metallic Films" (J. R. Anderson, ed.), Academic Press, London (1971), Vol. 1, p. 129.
2. Sanders, J. V. *In* "Chemisorption and Reactions on Metallic Films" (J. R. Anderson, ed.), Academic Press, London (1971), Vol. 1, p. 1.
3. Anderson, J. R., Baker, B. G. and Sanders, J. V. *J. Catal.* **1**, 443 (1962).
4. Macdonald, R. J. Ph.D. Thesis, Flinders University, Adelaide, Australia (1970).
5. McConkey, B. H. Ph.D. Thesis, University of Melbourne, Melbourne, Australia (1965).
6. Anderson, J. R. and McConkey, B. H. *Proc. Inst. Rad. Elec. Eng. (Australia)* **1967**, 132.
7. Brennan, D., Hayward, D. O. and Trapnell, B. M. W., *Proc. Roy. Soc.* **A256**, 81 (1960).
8. Anderson, J. R. and Baker, B. G. *In* "Chemisorption and Reactions on Metallic Films" (J. R. Anderson, ed.), Academic Press, London (1971), Vol. 2, p.1.
9. Baker, B. G., Johnson, B. B. and Marie, G. *Surface Sci.* **24**, 572 (1971).
10. Suhrmann, R. and Wedler, G. *Z. Angew. Phys.* **14**, 70 (1962).
11. Maire, G., Anderson, J. R. and Johnson, B. B. *Proc. Roy. Soc.* **A320**, 227 (1970).
12. Anderson, J. S. and Klemperer, D. F. *Proc. Roy. Soc.* **A256**, 350 (1960).
13. Baker, B. G. and Fox, P. G. *Trans. Faraday Soc.* **61**, 2001 (1965).
14. Baker, B. G. and Bruce, L. A. *Trans. Faraday Soc.* **64**, 2533 (1968).
15. Suhrmann, R., Gerdes, R. and Wedler, G. *Z. Naturforsch.* **18a**, 1208 (1963)
16. Drechsler, M. and Nicholas, J. F. *J. Phys. Chem. Solids* **28**, 2609 (1967). Nicholas, J. F. *Austral. J. Phys.* **21**, 21 (1968).
17. "Single Crystal Films" (M. H. Francombe and H. Sato, eds), Pergamon Press, London (1964).
18. Bouwman, R., van Keulen, H. P. and Sachtler, W. M. H. *Ber. Bunsen Ges.* **74**, 32 (1970).
19. Jaeger, H. *J. Catal.* **9**, 237 (1967).
20. Anderson, J. R. and Avery, N. R. *J. Catal.* **5**, 446 (1966).
21. Sella, C. and Trillat, J. J. *In* "Single Crystal Films" (M. H. Francombe and H. Sato, eds), Pergamon Press, London (1964), p. 201.
22. Rudiger, O. *Ann. Phys.* **30**, 505 (1937).
23. Bruck, L. *Ann. Phys.* **26**, 233 (1936).
24. Evans, D. M. and Wilman, H. *Acta Cryst.* **5**, 731 (1952).
25. Ono, K. and Mizushima, Y. *J. Appl. Phys. Japan* **25**, 349 (1956).
26. Umansky, M. M. and Krylov, W. A. *Zhur. Eksp. i. Teoret. Fiz.* **6**. 684 (1936).
27. Knorr, T. G. and Hoffman, R. W. *Phys. Rev.* **113**, 1039 (1959).
28. Yelon, A., Asik, J. R. and Hoffman, R. W. *J. Appl. Phys.* **33**, 949 (1962).
29. Adamsky, R. F. *J. Appl. Phys.* **31**, 2895 (1960).

30. Bauer, E. *In* "Single Crystal Films" (M. H. Francombe and H. Sato, eds), Pergamon Press, London (1964), p. 43.
31. Bouwman, R. Ph.D. Thesis, University of Leiden, The Netherlands (1970).
32. Harsdorff, M. and Raether, H. *Z. Naturforsch.* **19a**, 1497 (1964).
33. Krohn, M. and Barna, A. *In* "Proceedings of the Second Colloquium on Thin Films, Budapest, 1967" (E. Hahn, ed.), p. 45 Van den Hoeck and Rupprecht, Gottingen.
34. Bagg, J., Jaeger, H. and Sanders, J. V. *J. Catal.* **2**, 449 (1963).
35. Kunz, K. M., Green, A. K. and Bauer, E. *Phys. Status Solidi* **18**, 441 (1966).
36. Ogawa, S., Ino, S., Kato, T. and Ota, H. *J. Phys. Soc. Japan* **21**, 1965 (1966).
37. Hopkins, B. J. and Rivière, J. C. *Proc. Phys. Soc.* **81**, 590 (1963).
38. Farnsworth, H. E. *In* "The Solid–Gas Interface" (E. A. Flood, ed.), Dekker, New York (1967), Vol. 1, p. 431.
39. Burton, W. K. and Cabrera, N. *Discussions Faraday Soc.* **5**, 33, 40 (1949).
40. Technical details of ion gun specifications are available from, for instance, Varian or Vacuum Generators.
41. Dunning, W. J. *In* "The Solid–Gas Interface" (E. A. Flood, ed.), Dekker, New York (1967), Vol. 1., p. 271.
42. Ogilvie, G. J., Sanders, J. V. and Thomson, A. A. *J. Phys. Chem. Solids* **24**, 247 (1963).
43. Ogilvie, G. J. *Austral. J. Phys.* **22**, 169 (1969).
44. Ogilvie, G. J. *J. Phys. Chem. Solids* **10**, 222 (1959).
45. Ogilvie, G. J. *Austral. J. Phys.* **13**, 402 (1960).
46. Gibbs J. W. "Collected Works, Vol. 1, Thermodynamics", Yale University Press, New Haven (1948).
47. Kossel, W. *Nach. Ges. Wiss. Gottingen* **1927**, 135.
48. Stranski, I. N. *Z. Phys. Chem.* **136**, 259 (1928); **11**, 421 (1931).
49. Burton, W. K., Cabrera, N. and Frank, F. C. *Phil. Trans. Roy. Soc.* **A243**, 299 (1950).
50. Hirth, J. P. and Pound, G. M. *J. Chem. Phys.* **26**, 1216 (1957).
51. Knacke, O., Stranski, I. N. and Wolff, G. *Z. Electrochem.* **56**, 476 (1952); *Z. Phys. Chem.* (N. F.) **198**, 157 (1951).
52. Rias, C. B. and Bromberg, M. I. *Kristallografiya*, **4**, 594 (1959).
53. Grall, L. *Rev. Met.* **52**, 603 (1955).
54. Lozinskii, M. G. and Fedotov, S. G. *Izv. Akad, Nauk. SSSR.* No. 5, 109 (1955).
55. Blakely, J. M. and Mykura, H. *Acta Met.* **9**, 595 (1961).
56. Bennett, W. D. *J. Metals* **7**, 322 (1955).
57. Work by Hirth, J. P. and Winterbottom, W. L., quoted by Hirth, J. P. *In* "Metal Surfaces", American Society for Metals, Metals Park, Ohio (1963), p. 199.
58. Mykura, H. *Acta Met.* **9**, 570 (1961).
59. Blakely, J. M. and Mykura, H. *Acta Met.* **10**, 565 (1962).

60. Moore, A. J. W. *In* "Metal Surfaces", American Society for Metals, Metals Park, Ohio (1963), p. 155.
61. Bond, G. C. "Catalysis by Metals", Academic Press, London (1962).
62. Nicholas, J. F. "An Atlas of Models of Crystal Surfaces", Gordon and Breach, New York (1965).
63. MacRae, A. U. *Science* **139**, 379 (1963). MacRae, A. U. and Germer, L. *Phys. Rev. Letters* **8**, 489 (1962).
64. Nicholas, J. F. CSIRO Division of Tribophysics, unpublished.
65. MacRae, A. U. *Surface Sci.* **2**, 522 (1964). Lyon, H. B. and Somorjai, G. A *J. Chem. Phys.* **44**, 3701 (1966). Goodman, R. M., Farrell, H. H. and Somorjai, G. A. *J. Chem. Phys.* **48**, 1046 (1968). Jones, E. R., McKinney, J. T. and Webb, M. B. *Phys. Rev.* **151**, 476 (1966).
66. Palmberg, P. W. and Rhodin, T. N. *Phys. Rev.* **161**, 586 (1967). Fedak, D. G. and Gjostein, N. A. *Acta Met.* **15**, 825 (1967). *Surface Sci.* **8**, 77 (1967).
67. Tegart, W. J. McG. "The Electrolytic and Chemical Polishing of Metals", Pergamon Press, London (1959).
68. Gwathmey, A. T. and Cunningham, R. E. *Advances in Catalysis* **10**, 59 (1958).
69. Burton, J. J. and Jura, G. *In* "The Structure and Chemistry of Solid Surfaces" (G. A. Somorjai, ed.), Wiley, New York (1969), p. 21-1.
70. Mullins, W. W. *Acta Met.* **7**, 746 (1959).
71. "Diffusion Data", Diffusion Information Center, Cleveland, U.S.A.
72. Pines, B. Ya., Grebennik, I. P. and Zyman, Z. Z. *Fiz. Met. Metalloved.* **27**, 307 (1969)
73. Burminskaya, L. N. and Pachkov, P. O. *Zavod. Lab.* **34**, 206 (1968).
74. Neukam, O. *Galvanotechnik* **61**, 626 (1970).
75. Boiko, B. T., Palatnik, L. S., Lebedeva, M. V. *Fiz. Met. Metalloved.* **25**, 845 (1968).
76. Kincera, J., Fiedler, R. and Ciha, K. *Cesk. Casopis. Fys.* **A17**, 262 (1967).
77. Dowden, D. A. and Reynolds, P. W. *Discussions Faraday Soc.* **8**, 184 (1950).
78. Takeuchi, T., Sakaguchi, M., Miyoshi, I. and Takabatake, T. *Bull. Chem. Soc. Japan* **35**, 1390 (1962).
79. Sachtler, W. M. H. and Dorgelo, G. J. H. *J. Catal.* **4**, 654 (1965).
80. Sachtler, W. M. H. and Jongepier, R. *J. Catal.* **4**, 665 (1965).
81. Campbell, J. S. and Emmett, P. H. *J. Catal.* **7**, 252 (1967).
82. van der Plank, P. and Sachtler, W. M. H. *J. Catal.* **12**, 35 (1968).
83. Sachtler, W. M. H. and van der Plank, P. *J. Catal.* **18**, 62 (1969).
84. Ponec, V. and Sachtler, W. M. H. *J. Catal.* **24**, 250 (1972).
85. Hansen, M. "Constitution of Binary Alloys", McGraw-Hill, New York, 1958. Elliot, R. P. "Constitution of Binary Alloys, First Supplement", McGraw-Hill, New York, 1965. Smithells, C. J. "Metals Reference Book", Butterworths, London, 1967.
86. Vecher, A. A. and Gerasimov, Ya. I. *Russ. J. Phys. Chem.* **37**, 254, 258 (1963).

87. Rapp, R. A. and Maak, E. *Acta Met.* **10**, 62, 69 (1962).
88. Quinto, D. T., Sundaram, V. S. and Robertson, W. D. *Surface Sci.* **28**, 504 (1971).
89. Bouwman, R. Ph.D. Thesis, University of Leiden, 1970.
90. Moss, R. L., Gibbens, H. R. and Thomas, D. H. *J. Catal.* **16**, 117 (1970).
91. Moss, R. L., Gibbens, H. R. and Thomas, D. H. *J. Catal.* **16**, 181 (1970).
92. Moss, R. L. and Gibbens, H. R. *J. Catal.* **24**, 48 (1972).
93. Moss, R. L., Thomas, D. H. and Whalley, L. *Thin Solid Films*, **5**, R19 (1970).
94. Couper, A. and Eley, D. D. *Discussions Faraday Soc.* **8**, 172 (1950).
95. Eley, D. D. *J. Res. Inst. Catal. Hokkaido Univ.* **16**, 101 (1968).
96. Rienacker, G. *Z. Electrochem.* **47**, 805 (1941).
97. Rienacker, G. and Engels, S. *Z. Anorg. Allg. Chem.* **336**, 259 (1965).
98. Hilliard, J. E., Cohen, M. and Averback, B. L. *Acta Met.* **8**, 26 (1960).
99. Hondros, E. D. In "Interfaces" (R. C. Gifkins, ed.), Butterworths, Sydney (1969), p. 77.
100. Llopis, J. *Catal. Rev.* **2**, 161 (1969).
101. Tomashov, N. D., Shchelepnikov, M. N. and Ivanov, Ya. M. *Zashch. Metal.* **1**, 122 (1965).
102. Tarng, M. L. and Wehner, G. K. *J. Vac. Sci. Tech.* **8**, 23 (1971).
103. Tabor, D., Wilson, J. M. and Bastow, T. J. *Surface Sci.* **26**, 471 (1971).
104. Johnson, R. P. *Phys. Rev.* **54**, 459 (1938).
105. Adam, P. and Wever, H. *Surface Sci.* **21**, 307 (1970).
106. Weise, G. and Owsian, G. *J. Less-Common Metals* **22**, 99 (1970).
107. Taylor, N. *Surface Sci.* **2**, 544 (1964).
108. Lea, C. and Mee, C. H. B. *Surface Sci.* **25**, 332 (1971).
109. Magnusen, G. D., Meckel, B. B. and Harkins, P. A. *J. Appl. Phys.* **32**, 369 (1961).
110. Mollenstedt, G. and Duker, H. *Optik* **10**, 192 (1953).
111. Carter, G. and Colligon, J. S. "Ion Bombardment of Solids", Elsevier, New York (1968).
112. Wehner, G. K. *Adv. in Electronics* and *Electron Physics* **7**, 239 (1955).
113. Klemperer, D. F. and Snaith, J. C. *Surface Sci.* **28**, 209 (1971).
114. Hauffe, W. *Phys. Status Solidi.* **36**, K83 (1969).
115. Hauffe, W. *Phys. Status Solidi.* (a) **4**, 111 (1971).
116. Hondros, E. D. and McLean, D. Society of Chemical Industry Monograph No. 28, London (1968).
117. Darby, T. P. and Wayman, C. M. *Phys. Status Solidi.* (a), **1**, 729 (1970).
118. Stowell, M. J. and Law, T. J. *Phys. Status Solidi* **25**, 139 (1968).
119. Sato, H., Toth, R. S. and Astrue, R. W. *J. Appl. Phys. Suppl.* **33**, 1113 (1962).
120. Thirsk, H. R. *Proc. Phys. Soc.* **63B**, 833 (1950).
121. Heckingbottom, R. *Surface Sci.* **17**, 394 (1969).
122. Woods, R. *Electrochem. Acta* **16**, 655 (1971).
123. Rand, D. A. J. and Woods, R. *J. Electroanalytical Chem.* **36**, 57 (1972).

124. Uhara, I., Yanagimoto, S., Tani, K., Adachi, G. and Teratani, S. *J. Phys. Chem.* **66**, 2691 (1962).
125. Uhara, I., Hikino, T., Numata, Y., Hamada, H. and Kageyama, Y. *J. Phys. Chem.* **66**, 1374 (1962).
126. Criado, J. M., Herrera, E. J. and Trillo, J. M. *In* "Proceedings 5th International Congress on Catalysis" (J. W. Hightower, ed.), North-Holland, Amsterdam (1973), p. 541.
127. Lang, B., Joyner, R. W. and Somorjai, G. A. *Surface Sci.* **30**, 440 (1972).
128. Perdereau, J. and Rhead, G. E. *Surface Sci.* **24**, 555 (1971).
129. Houston, J. E. and Park, R. L. *Surface Sci.* **21**, 209 (1970); **26**, 269 (1971).
130. McLean, M. and Mykura, H. *Surface Sci.* **5**, 466 (1966).
131. Bouwman, R. and Sachtler, W. M. H. *J. Catal.* **26**, 63 (1972).
132. van Santen, R. A. and Boersma, M. A. M. *J. Catal.* **34**, 13 (1974).
133. van Santen, R. A. and Sachtler, N. M. H. *J. Catal.* **33**, 202 (1974).
134. Grant, J. T. *Surface Sci.* **18**, 228 (1969).
135. Dixon, J. K. and Longfield, J. E. *In* "Catalysis" (P. H. Emmett, ed.), Reinhold, New York (1960), Vol. 7, p. 281.
136. Avery, N. R. *Surface Sci.* **33**, 107 (1973).
137. Ertl, G. and Kuppers, J. *Surface Sci.* **24**, 104 (1971).
138. Ertl, G. and Kuppers, J. *J. Vac. Sci. Tech.* **9**, 829 (1971).
139. Ono, M., Takasu, Y., Nakayama, K. and Yamashima, T. *Surface Sci.* **26**, 313 (1971).
140. Willams, F. L. and Boudart, M. *33rd Annual Conference on Physical Electronics*, Berkeley (1973).
141. Baddour, R. F., Modell, M. and Goldsmith, R. L. *J. Phys. Chem.* **74**, 1787 (1970).
142. Presland, A. E. B., Price, G. L. and Trimm, D. L. *J. Catal.* **26**, 313 (1972)
143. Norris, L. F. and Parravano, G. *In* "Reactivity of Solids: Proceedings 6th. International Symposium on the Reactivity of Solids" (J. W. Mitchell, R. C. DeVries, R. W. Roberts and P. Cannon, eds), Wiley-Interscience, New York, 1969, p. 149.
144. McCaroll, J. J., Edmonds, T. and Pitkethly, R. C. *Nature* **223**, 1260 (1969).
145. Martin, M. R. and Somorjai, G. A. *Phys. Rev.* **B, 7**, 3607 (1973).
146. Schwab, G. M. *Discussions Faraday Soc.* **8**, 166 (1950).
147. Schwab, G. M. and Holz, G. *Z. Anorg. Chem.* **252**, 205 (1944).
148. Schwab, G. M. *Trans. Faraday Soc.* **42**, 689 (1946).
149. Schwab, G. M. and Schwab-Agallidis, E. *Ber.* **76**, 1228 (1943).
150. Schwab, G. M. and Karatzas, A. *Z. Elektrochem.* **50**, 242 (1944).
151. Schwab, G. M. and Pesmatjoglou, S. *J. Phys. Chem.* **52**, 1046 (1948).

CHAPTER 4

Dispersed Metal Catalysts

	page
1. SUPPORTED METAL CATALYSTS	164
Platinum	177
Platinum/Silica Catalysts	184
Platinum/Alumina Catalysts	190
Platinum/Silica–Alumina Catalysts	194
Platinum/Zeolite Catalysts	194
Platinum/Carbon Catalysts	196
Noble metals other than platinum	197
Non-noble metals	203
Ultrathin evaporated metal films	216
2. UNSUPPORTED METAL CATALYSTS	218
Metal powders	218
Stabilized porous metals	222
Skeletal metals	228
3. DISPERSED MULTIMETALLIC CATALYSTS	231

Many metallic catalysts which are used for preparative conversions at either the industrial or laboratory scale have a relatively high specific surface area. Such catalysts vary widely in their actual morphology, but they all differ from massive metal in that they consist of metal particles which are to some degree separated from one another. The particles may be widely separated as in conventional supported catalysts, or they may be close together but more or less separated by a small amount of refractory oxide stabilizer as in the classical iron synthetic ammonia catalysts. The skeletal metal catalysts (Raney catalysts) are another example of porous metal catalysts which are usually oxide stabilized to some extent and which are more properly related to dispersed than to massive metal catalysts. Finally, the dispersed metal may be merely in the form of metal powder.

1. Supported Metal Catalysts*

Metal is introduced to the support, usually from aqueous solution or suspension, by processes such as impregnation, adsorption or ion exchange, co-precipitation, or deposition, followed by drying and hydrogen reduction. There is a vast range of empirical variations to preparative recipes, and an extensive catalogue is given by Innes,[1] while Gil'debrand[2] provides a more recent empirical summary. Metal deposition by decomposition of volatile metal compounds has also been described.[227, 228]

After drying, catalysts are often (but not always) subjected to high temperature treatment (calcining) before reduction. During calcining, and to a lesser extent during drying, the metal compounds on the carrier decompose, and to maximize the degree of metal dispersion ultimately obtained, the compounds should be chosen so that decomposition and reduction occur at temperatures that are as low as possible.

With all supported catalysts there is a trend towards metal particles of larger average size the higher the metal concentration on the support. Furthermore, if a catalyst is maintained for periods of time at successively increasing temperatures, it is found that at any given temperature there is an increase in the average metal particle size—a decrease in the metal dispersion—with time, until a time-independent average particle size is reached: this time-independent average size tends to increase with increasing temperature. However, the influence of these factors—temperature, time and metal concentration—is much dependent on other variables such as the nature of the support, particularly its pore structure, and on the atmosphere in contact with the catalyst during heating.

This process of sintering is obviously important during catalyst preparation, but it may also occur during catalyst use if the reaction conditions are relatively severe, and it then contributes to the overall phenomenon of catalyst deactivation. In an engineering and an economic sense, it is often more important to achieve a long catalyst lifetime even at a somewhat reduced level of activity, than to produce a catalyst with very high initial activity but whose activity declines rapidly and extensively during use. This is particularly so if catalyst regeneration or replacement requires plant down-time.

Catalyst deactivation is a complicated phenomenon to which several distinct processes can contribute. A high area support material may undergo a reduction in porosity and surface area, and not only may this render some of the dispersed metal inaccessible, but there will be a more direct

* Many supported catalysts use platinum group metals: among the well-known suppliers of these and their derivatives are Johnson Matthey; Engelhard Industries; Research Organic/Inorganic Chemical Corpn.; Alpha Inorganics; Halewood Chemicals; Strem Chemicals.

loss if the support itself has a catalytic function. Some reference to the sintering of support materials has been made in Chapter 2, but it is here worth recalling that the presence of water vapour is a particularly potent factor with oxide-type supports. Porosity may also be lost by direct pore blockage with adventitious material such as fine particulate matter entrained in the reactant stream, or by the formation of solid reaction products within the pore structure.

With regard to the metal function in a dispersed catalyst, the extent of deactivation as indicated by reaction rate is not necessarily paralleled by a decrease in the total metal surface area. Obviously, if a reaction is site selective—for instance if the preferred site is a metal atom in a corner position rather than in a low index face—it would be possible for the catalyst activity to decrease as a result of a change in the surface topography of the dispersed metal, without a corresponding change in the metal dispersion. Or again, if deactivation were due to the adsorption of a poison at the metal surface, there would be no necessary correspondence between the decline in activity and any change in the metal dispersion.

While one can generally anticipate that the susceptibility of a dispersed metal catalyst to deactivation by agglomeration of the metal particles will be more severe the more highly dispersed the metal, the susceptibility to deactivation by poisoning will generally be less severe because, for given quantities of metal and poison, the fraction of the metal surface covered will be smaller the greater the degree of metal dispersion. Furthermore, it has been suggested in the important case of sulphur poisoning of platinum/zeolite reforming catalysts, that the poison is less strongly adsorbed for very high degrees of metal disperson:[42, 89] however this latter sort of behaviour is unlikely to be general for all systems.

In contrast to this general process of sintering of dispersed catalysts, under certain circumstances heating can result in platinum redispersion: for instance, it has been reported[122] that redispersion occurred by heating in air to 700–1100 K hydrogen-reduced platinum/γ-alumina catalysts (0·21–1·54 wt. % platinum) which had been prepared by chloroplatinic acid impregnation. This redispersion behaviour is a complex phenomenon. It can be associated at least in part with the formation of volatile platinum compounds to which residual chlorine may contribute (e.g. $PtCl_4$ formation) and with the formation of platinum oxide which interacts strongly with an oxide support and so may tend to spread on it.

Some more detailed comments about the thermal stability of dispersed metal systems, particularly some of the process kinetics, are contained in Chapter 5.

Since we are restricting our attention to catalysts in which metal is present in elemental form, it is implicit that we are only concerned with those metals which can be reduced to elemental form, generally by the use of hydrogen at temperatures in the region of 570–770 K, from the metal compounds which are formed on the support. As a practical guide, it is useful to examine the feasibility of hydrogen reduction of the oxides and the chlorides. If we write the reduction reactions for the metal oxide $MO_{x(s)}$ and the chloride $MCl_{x(s)}$ as

$$MO_{x(s)} + xH_{2(g)} \rightleftarrows M_{(s)} + xH_2O_{(g)} \quad (4.1)$$

$$MCl_{x(s)} + \frac{x}{2} H_{2(g)} \rightleftarrows M_{(s)} + xHCl_{(g)} \quad (4.2)$$

we may examine the values of $(p_{H_2O}p_{H_2}^{-1})_{equil}$ and $(p_{HCl}^2 p_{H_2}^{-1})_{equil}$ for comparison with experimentally accessible values to see if reduction can be achieved. The equilibrium data are recorded in Table 4.1. Under practical reduction conditions, we may take p_{H_2} equal to 101 kPa (760 Torr), and p_{H_2O} or p_{HCl} equal to 0·133 Pa (10^{-3} Torr), so $p_{H_2O}p_{H_2}^{-1}$ and $p_{HCl}^2 p_{H_2}^{-1}$ take values of $1·3 \times 10^{-6}$ (dimensionless) and $1·8 \times 10^{-7}$ Pa respectively; when the equilibrium ratios are greater than the experimental ratios, reduction can proceed under the conditions assumed. One may conclude on these criteria that it should be possible to generate reduced metallic catalysts with the group VIII metals, copper, silver, gold, rhenium, molybdenum and tungsten. Although the chlorides are more easily reduced than the oxides, the latter can never be entirely avoided during catalyst processing. Hydrogen reduction of other transition metals, other than those named above, is improbable under conditions which can be tolerated for catalyst integrity.

If the metal oxide readily reacts chemically with the support, a compound may be generated which is much more resistant to reduction than the metal oxide itself. This is the situation when one attempts to disperse MoO_3 or WO_3 on supports such as high area alumina or silica. Provided the molybdenum or tungsten content is limited to 15–20 wt.%, in neither case is any trioxide present on high area alumina[185,186] after calcining in air at about 770 K. Aluminium molybdate is apparently not formed either,[185] and the analogous result is also presumably true for the tungsten system. It seems likely that at these concentrations the molybdenum or tungsten is present as a surface oxide, perhaps only about a monolayer thick on the alumina surface. Certainly, in neither case can reduction to the metal be effected by hydrogen at temperatures up to 820 K, although in the case of molybdenum[185,187] but not tungsten,[186] some reduction to valence states below the hexavalent does occur. When silica gel is used as a support, although some WO_3 can be detected by X-ray diffraction after calcining, there is still no evidence for reduction to metallic tungsten, although some reduction to valence states below the hexavalent occurs.[186] Some other examples of interaction between dispersed metal oxides and alumina or silica supports will be encountered subsequently.

TABLE 4.1 Thermodynamics of metal oxide and metal chloride reduction at 673 K

	$(p_{H_2O} p_{H_2}^{-1})_{equil}$	$(p^2_{HC\ell} p_{H_2}^{-1})_{equil}$ (Pa)
nickel	5×10^2(NiO)	6×10^3(NiCl$_2$)
iron	1×10^{-1}(FeO); 7×10^{-1}(Fe$_2$O$_3$)	3(FeCl$_2$); 6×10^7(FeCl$_3$)
cobalt	5×10^1(CoO)	3×10^2(CoCl$_2$)
rhenium	$\sim 10^4$(ReO$_2$); $\sim 10^5$(ReO$_3$)	
ruthenium	$\sim 10^{12}$(RuO$_2$)	$\sim 10^{17}$(RuCl$_3$)
rhodium	$\sim 10^{13}$(RhO)	$\sim 10^{14}$(RhCl$_2$); $\sim 10^{15}$(RhCl$_3$)
iridium	$\sim 10^{13}$(IrO$_2$)	$\sim 10^{16}$(IrCl$_2$)
palladium	$\sim 10^{14}$(PdO)	
platinum		$\sim 10^{18}$(PtCl$_2$)
copper	2×10^6(Cu$_2$O); 2×10^8(CuO)	4×10^{11}(CuCl$_2$)
silver	3×10^{17}(Ag$_2$O)	4×10^6(AgCl)
tungsten	1×10^{-1}(WO$_3$)	
molybdenum	2×10^{-2}(MoO$_2$); 4×10^1(MoO$_3$)	
chromium	3×10^{-9}(Cr$_2$O$_3$)	2×10^{-4}(CrCl$_2$); 6×10^{-1}(CrCl$_3$)
vanadium	2×10^{-11}(VO); 6×10^{-4}(V$_2$O$_5$)	2×10^{-7}(VCl$_2$); 4×10^{-2}(VCl$_3$)
tantalum	4×10^{-12}(Ta$_2$O$_5$)	
titanium	2×10^{-19}(TiO); 4×10^{-16}(TiO$_2$)	$\sim 10^{-11}$(TiCl$_2$); $\sim 10^{-10}$(TiCl$_3$)
manganese	2×10^{-10}(MnO); 1×10^1(MnO$_2$)	1×10^{-9}(MnCl$_2$)

The decomposition of a metal compound to yield the metal oxide which can be then reduced to the metal in hydrogen is an important procedure in catalyst preparation. It can occur for instance with material which is occluded in the pores of a support, and it can occur in the preparation of unsupported metal powder. Heavy metal nitrates are hydrated and their decomposition is a complex process about whose details little is known. It suffices to say that the most strongly held water of hydration is present as a ligand to the heavy metal ion, and this is lost only at the same time as the nitrate decomposes, so that the gaseous decomposition products generally consist of a complex mixture containing oxides of nitrogen, nitric acid, oxygen and water vapour. Although these ultimately escape, appreciable corrosion of the support may occur. Although anhydrous copper(II) nitrate is volatile without decomposition (sublimes in vacuum at 420–470 K) and some volatility is known with other anhydrous heavy metal nitrates, this can be expected to be of negligible importance during catalyst processing, since decomposition to a metal oxide takes precedence when the hydrated nitrate is decomposed in the presence of oxygen.

With unsupported samples the oxide is often generated by dehydration of the precipitated hydroxide. With supported samples a technique of hydroxide precipitation onto the support is occasionally used, but is not readily applicable to high area supports with a large internal porosity.

In most cases where the metal oxide is generated by a decomposition process in which a gas is evolved, the oxide is produced with some degree of porosity. This is a result of the incomplete collapse of the parent structure in regions from which volatile products were generated. The situation is quite analogous to that described in Chapter 2 for the formation of porosity in some support materials. This porosity is of some importance because it facilitates the subsequent process of reduction to the metal. The degree of collapse is dependent on the impurity content and the particle size of the precursor. However, the details are dependent on the nature of the decomposition reaction. When water vapour is a decomposition product, its speedy removal is vital for maintaining a high internal surface area in the product because of the possibility of hydrothermal reactions. On the other hand, when the gaseous products are relatively inert chemically, this sort of factor does not operate and it is often found that collapse of the internal structure is more important with smaller particles. The same general conclusions also hold for metallic derivatives which decompose to the metal rather than to the metal oxide: nickel formate is a case in point.[173] In the latter sort of situation, the ultimate metal particle size is strongly dependent on sintering of the metal particles initially produced, and it was concluded that the most effective method of generating a high area product lies in the use of oxide stabilizers which inhibit metal particle sintering rather than in attempts to generate particularly high areas by suitably tailoring the defect structure of the starting material. The conclusion is entirely consistent with the established technique for the preparation of stabilized porous metal catalysts (*vide infra*).

Although differing to some extent in matters of detail, the main features of the hydrogen reduction of metal oxide follow a general pattern.[165,166] The $O^{2-}_{(s)}$ ions at the oxide surface react with hydrogen and are converted to $OH^-_{(s)}$, and a corresponding number of electrons are trapped at sites in or near the surface. The reaction can be propagated if the temperature is such that water is eliminated from $OH^-_{(s)}$ groups and is desorbed. The electrons are immediately or eventually collected by cations, and the reduced atoms can nucleate a metallic phase. This initial stage of the reduction process depends in detail on the defect structure of the oxide, since this will control the hydrogen chemisorption process. However, once metal nuclei have been formed a second process for the propagation of the reduction may intervene if the metal is itself capable of chemisorbing hydrogen. Hydrogen will then be chemisorbed dissociatively on the metal, and will be available for reaction with the oxide in the near vicinity of the metal/oxide interface following migration across the metal surface towards the interface: this hydrogen will react more readily with the oxide than will molecular hydrogen.

This model for two consecutive stages in the reduction reaction is well confirmed experimentally, for instance with the reduction of copper (II) oxide[167] and nickel(II) oxide.[168,169] The first stage leading to the formation of metal nuclei is slow and kinetically takes the form of an induction period. The subsequent reaction is relatively rapid and its rate increases while the extent of the metal/oxide interface increases. The model implies that if metal nuclei can be created by another method, the induction period will be eliminated: this has been experimentally demonstrated with nickel(II) oxide[168] and copper(II) oxide.[167] For instance, in the case of nickel(II) oxide, vacuum decomposition of the superficial nickel formate formed from formic acid impregnation results in the generation of nickel particles and a consequent strong acceleration in the rate of oxide reduction.[168,177] Furthermore, impregnation of the nickel oxide with relatively easily reducible compounds of other metals such as platinum or copper has a similar effect.[178] However, the effect of additives on the reducibility of nickel oxide is more complex than these simple comments would indicate. In particular, reduction is more difficult when the nickel oxide is highly dispersed on a support, and this has been attributed to nucleation inhibition as a consequence of the dispersion, with the function of (say) a copper additive being to restore the nucleation ability.[178] At typical hydrogen pressures used for oxide reduction (10–100 kPa; $0 \cdot 1$–$1 \cdot 0$ atm), the initial reaction of hydrogen with nickel oxide is dependent on a $p^x_{H_2}$ with $x \approx 0 \cdot 4$–1 since hydrogen is not very strongly adsorbed on the oxide surface; the succeeding reaction has $x \approx 0$ since hydrogen is strongly adsorbed on the nickel metal.

The presence of water vapour retards the rate of the oxide reduction process. At least in the case of copper(II) oxide it appears that the main effect of the water vapour is on the rate of the initial reaction leading to metal nuclei, and this would indeed be consistent with the model in which the adsorption of molecular hydrogen is more difficult the more heavily hydroxylated the oxide surface.

Charcosset et al.[170] followed the morphology of the oxide and metal phases during the reduction of nickel(II) oxide by surface area measurements and by electron microscopy. As would be expected from the model involving the growth of metal nuclei, the nickel particles tend to be individually smaller than the original oxide particles, and the metal/oxide interface is, as a consequence, heavily dissected by the metal grain boundaries.

Nevertheless, it is usually observed that as the reduction proceeds, the size of an oxide particle shrinks and it becomes surrounded by a polycrystalline layer of metal through the interstices of which gas transport occurs. The tendency for sintering of the reduced metal increases, the higher the processing temperature. On the other hand, the density of metal nuclei produced in the initial reaction increases with increasing temperature, and this tends to result in smaller metal particles.

When the metal to be reduced is present as cations dispersed by exchange onto a support surface, the hydrogen reduction process must be somewhat different. The support surface must function as a proton acceptor to allow for electron transfer to the reducible metal cation: one method of accommodating the protons is by the conversion of $O^{2-}_{(s)}$ to $OH^-_{(s)}$, but surface defects may possibly also function as trapping sites. Aggregation to form a discrete metal particle again requires atom mobility, but on the average over greater distances than in the reduction of the metal oxide itself. The nature of the mobile species depends on the system. Thus, there is some evidence obtained during the hydrogen reduction of a platinum ammine complex adsorbed on zeolite that a mobile neutral complex may be produced before the metal atom is stripped of its ligands (vide infra), and it is at least conceivable that mobile neutral complexes are also important in other systems. On the other hand, the way in which metal particles are generated by, for instance, the reduction of Ni(II) ions which are relatively sparsely distributed in tetrahedral sites on the surface of (say) γ-alumina, is quite uncertain. Certainly nickel atoms in an oxidation state of zero should be highly mobile, but the thermodynamics for the generation of Ni(0) are so unfavourable that this seems a highly improbable pathway. One is left to speculate on the migration of nickel atoms in uncharged species in an intermediate state of reduction, that is Ni(I), perhaps as NiH. We estimate that only with gold would M(0) atom formation be thermodynamically favourable by hydrogen reduction, assuming that the precursors have about the same free energy of formation as the corresponding solid metal oxides.

Since oxide reduction is often highly exothermic, and a substantial part of the reaction is acceleratory, careful control is needed if high temperatures are to be avoided. This is, of course the more serious with unsupported than with support catalysts because of heat transfer limitations with the former.

This model for metal oxide reduction has immediate consequences for designing reduction conditions to maximize the degree of dispersion in reduced metal catalysts. In particular, the conditions in the early stages of reduction are of crucial importance since it is then that the number of metal nuclei is determined. In general, a high density of metal nuclei is favoured by effecting the initial reduction as rapidly as possible: this means keeping the partial pressure of water vapour as low as possible and using as high a reduction temperature as is consistent with the avoidance of catalyst sintering. With regard to temperature, a compromise will be necessary between these opposing factors. The water vapour partial pressure will be minimized by using a high hydrogen flow rate, and low temperature dehydration of the catalyst before reduction may be advantageous: this may be done by using a flow of thoroughly dried inert gas. These conclusions are in good agreement with experimental observations on the influence of reduction conditions on the dispersion of supported platinum catalysts prepared via impregnation methods.[172,6]

The process of impregnation for the introduction of metal to a support is seldom as simple as the name implies because it is frequently accompanied by adsorption from the solution onto the support surface, and the overall process is then a combination of adsorption together with a deposition of solute in the support pores when the solvent is evaporated. We propose to retain the term impregnation for the overall process, but material which is not adsorbed and is mechanically held in the support pores will be referred to as being occluded.

The transport processes by which material is introduced into a porous support have been discussed in some detail in the literature.[123-125] When a porous support first comes in contact with a liquid, capillary forces draw the liquid into the pores. The capillary pressure, p_c, is given by

$$p_c = \frac{4\gamma \cos \theta}{d} \tag{4.3}$$

where γ is the surface tension, θ is the contact angle and d the diameter of the capillary (pore). For an oxide-type material in contact with an aqueous solution which does not contain strongly surface active solute, it is a good approximation to set $\cos \theta = 1$. However, some support materials such as heavily graphitized carbon have surfaces which are, at least in part hydrophobic, and for these $\cos \theta < 1$ and the capillary pressure will be correspondingly smaller. Indeed, the difficulty of wetting the surface of a partly graphitized carbon support with an aqueous impregnating solution has led to the use of non-aqueous solvent.[131]

Many high area support materials have average pore diameters in the range 2–50 nm, and assuming a value for γ of 7×10^{-2} N m^{-1} (70 dyn cm^{-1}), this would result in capillary pressures in the range 140–5·6 MPa (1380–55 atm).

The rate of movement of a liquid meniscus along a capillary is determined by the magnitude of the forces opposing the capillary forces. If the opposing forces are constant and due only to viscous flow of the liquid in the capillary, the time t required for movement through a distance x, is given by

$$t = \frac{4\eta x^2}{\gamma d} \tag{4.4}$$

where η is the viscosity of the liquid, and it has been assumed that $\cos \theta = 1$. Again taking average pore diameters in the range 2–50 nm, with γ the same as before, the time required for movement through a distance of (say) 2 mm lies in the range 114–4·6 seconds. Thus, under practical conditions the penetration of solution into a porous catalyst support which is wet by the solution is quite rapid, and complete penetration throughout a support pellet will occur within a time of a few minutes at most. Closed pores cannot be completely filled unless the contained gas is removed by prior evacuation or by solution in the entering liquid. Closed pores are likely to be small pores: assume $d = 2$ nm with a p_c value of 140 MPa (1380 atm). If one ignores gas solubility for the sake of argument, only some 0·07% of the pore volume would remain unfilled at equilibrium if the gas had not previously been pumped away. However, at these high pressures the gas solubility would be so increased that this would be an important process for gas removal.

Although penetration of a solution throughout a porous support is rapid, a uniform distribution of metal-containing solute throughout the support will not necessarily be easily achieved if adsorption occurs to an important extent, because the adsorbed material will be first deposited near the external surface of the support pellet, and if insufficient time is allowed to get beyond this stage, the processed catalysts will necessarily contain a correspondingly uneven distribution of metal particles. If the concentration of the impregnating solution exterior to the support pellet is always constant (often not a practically realistic situation for economic reasons, since it implies a very large external reservoir), the adsorbed concentration is never in excess of its adsorption equilibrium value, and with the passage of time the adsorbed zone moves inwards through the porous medium until a uniform distribution of adsorbed species is reached. However, if the quantity of adsorbable material in the exterior solution is limited, its concentration may become depleted to a low value during the initial stages of adsorption, and a uniform distribution of the adsorbed species throughout the porous medium can then only be obtained by a redistribution of the initially adsorbed material.

A higher concentration of adsorbed material towards the external surface of the support pellet is a matter of some importance. It will result in a poorer metal dispersion in the reduced catalyst, and such a catalyst will be more susceptible to metal loss by mechanical attrition. On the other hand, having metal non-uniformly distributed in this way may lead to enhanced catalytic activity compared to a uniform distribution if the rate of the catalytic reaction would be mass transport limited by diffusion to the centre of the support pellet.

The problem of transport by a sorption–diffusion mechanism has been analysed by a number of authors: the account given by Weisz[126-128] is perhaps the most satisfactory in relation to practical situations, but even so it is restricted to the case of constant external solution concentration. It was shown that if the fractional completion, f, of the total process is expressed in terms of a generalized time parameter T, the dependence of f on T is not too strongly dependent on the adsorption strength: a plot of f against $T^{1/2}$ is shown in Fig. 4.1 for uptake into spherical and cylindrical porous bodies. The shaded areas define the limited behaviour for the entire class of sorption–diffusion systems bounded by the extreme weak and strong cases. The term weak adsorption is used in the sense that the adsorbed concentration is linearly dependent on the solution concentration, while strong adsorption means that the adsorbed concentration is always at its saturation value and is independent of the solution concentration for all concentrations above zero. In this treatment the generalized time parameter, T, is defined by

$$T = t \left(\frac{DP}{R^2 b}\right) \left(\frac{c_0}{c_f}\right) \quad (4.5)$$

where t is the actual time, D is the true diffusivity of the solute in the solution, P is the porosity fraction of the porous medium, b is the so-called tortuosity factor (about $\sqrt{3}$); R is the length dimension of the porous medium: the ratio c_0/c_f contains information about the strength of adsorption, since c_f is the total amount of adsorbable species in unit volume of the porous medium when it is in equilibrium with a solution of concentration c_0. Equation 4.5 clearly indicates the way t varies with the system para-

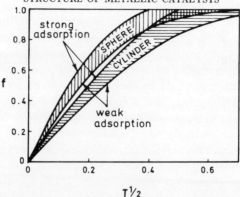

FIG. 4.1 Uptake by sorption-diffusion in a porous adsorbent: f is the fractional completion of the process and T a generalized time parameter. The figure represents the behaviour at a constant adsorbate concentration exterior to the adsorbent, for weak and strong adsorption and with two alternative adsorbent geometries. Reproduced with permission from Weisz, P. B. *Trans. Faraday Soc.* **63**, 1801 (1967).

meters. To achieve a specified f requires a specified T (via Fig. 4.1): to achieve this value of T will require an elapsed time that varies inversely with D, and directly with R^2 and c_f/c_0.

Some results for the variation of f with $T^{1/2}$ have been given by Harriott[125] for cases when the exterior solution concentration is depleted

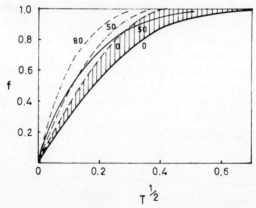

FIG. 4.2 Uptake by sorption-diffusion in a porous adsorbent with depletion of the adsorbate concentration in the external solution: f is the fractional completion of the process and T a generalized time parameter. The figure refers to spherical adsorbent geometry. The full lines are for weak adsorption and the broken lines for strong adsorption. The numbers on the lines indicate the maximum percentage depletion of the solution. The hatched area is the same as that similarly indicated in Fig. 4.1. Reproduced with permission from Harriott, P. *J. Catal.* **14**, 43 (1969).

by the uptake process. The data are given in Fig. 4.2 and refer to a porous body of spherical geometry. The figure also includes for comparison the curves for zero depletion and the shaded area has the same significance as in Fig. 4.1 in relation to the limits set by weak and strong adsorption.

For a given adsorbate–adsorbent system, there are various ways in which some modification of behaviour may be achieved. The first is by competitive adsorption of another adsorbate; the net effect being a relative reduction in the adsorption strength of the desired adsorbate. This general process can be considered to include the effect of decreasing pH which moves the position of equilibrium of reaction 2.6 (Chapter 2) towards the left. The added competitor must, of course, not have undesirable catalytic properties. Maatman[124] has described how the structure of dispersed metal catalysts may be controlled in this way. The relative strength of adsorption also falls with increasing temperature, so this will also lead to a more uniform distribution of adsorbate. A change in the chemical nature of the metal-containing solute species is an obvious way in which solute adsorption may be modified, and illustrations of this, and other variables will be found in later sections.

The fate of the occluded solute depends on how the solvent is removed because of the interaction between capillary forces and the distribution of pore sizes. If all the solvent could be evaporated instantaneously, the solute would simply deposit locally and uniformly throughout the mass. However, evaporation does not occur instantaneously, but starts at the outer edges of the porous particles and proceeds preferentially from those regions containing pores of larger diameter because of the higher vapour pressure of a liquid in larger pores. Furthermore, the liquid which evaporates from small pores is replaced by liquid drawn by capillary forces from large pores. The net result for slow evaporation is an enrichment of solute in the small pores and towards the centre of the porous particle.[123] The general effect will usually be to make worse the dispersion of the supported metal ultimately obtained.

The metal dispersion will usually be improved if solid is deposited from the occluded solution by reactive precipitation rather than by mere solvent evaporation. This may occur to some extent by hydrolysis during solvent removal, or a reagent may be added. Thus, impregnated chloroplatinic acid solution precipitates platinum sulphide with hydrogen treatment, or heating to >373 K with co-impregnated hydrogen peroxide deposits a platinum oxide.[229]

Assuming other factors to be the same, ion adsorption or ion exchange methods give a more even distribution of metal throughout the internal surface of the carrier than impregnation, and as a consequence, the metal crystallites are of smaller average diameter when the former methods are used. If halogen-containing metal compounds are used (e.g. H_2PtCl_6), some halogen may be retained on the carrier surface. This is important with an alumina carrier because it is known that halogen can affect the surface properties. There is also a possibility of halogen being retained on the metal surface.

The fact that metal can be introduced onto a support in two ways, that is by ion adsorption and as solute in occluded solution, is of importance in the preparation of supported bimetallic catalysts. Suppose we have in a solution ions of two metals M_A and M_B. If both ions are adsorbed onto the support with equal strength (or weakness) no extra complications arise in the sense that aggregation occurs from a common sort of precursor in each case—a mixed adsorbed layer in the case of strong adsorption, or mixed occluded solute in the case of very weak adsorption. However, suppose the ions of M_A are strongly adsorbed but those of M_B are only very weakly adsorbed. In this case the support will separate the ions, M_A ions being distributed over the support surface, M_B ions remaining as occluded solute. On processing the catalyst these two precursor states will lead to aggregation by competing processes and the tendency will be to enhance the chance of forming separate particles of M_A and M_B at the expense of $(M_A + M_B)$ particles.

In practical terms, in the presentation of a solution to a catalyst support, one has always to be conscious of the possibility that the support may suffer serious attack by the solution if the pH is too high or too low. The seriousness of this will obviously depend, *inter alia*, on the surface area of the support, and it is common experience that materials in an "active" form (e.g. γ-alumina) are much more reactive than if subjected to a high temperature treatment which converts them to low surface area and to a crystalline modification of relatively low intrinsic reactivity (e.g. α-alumina). Carbon is relatively inert, particularly in a highly graphitized form, but high area aluminas and chromias are susceptible to attack by solutions of high or low pH, silica–alumina and the zeolites are readily attacked by solutions of low pH, while high area silica is subject to attack in solutions of high pH. When this problem exists, it is mostly in connection with the adjustment of the pH of the solution used for exchange or impregnation so as to stabilize some desired metallic ion in solution: a compromise then has to be struck between ion stability and support attack. It hardly needs to be pointed out that when this is done, the added acid or base (or buffering agent) must be so chosen as ultimately to be volatile if contamination of the final catalyst is to be avoided. Nevertheless, when acid or base is used in an unbuffered solution, the problem of attack on the support may still be difficult to avoid completely, even though the pH of the initial solution has been adjusted to a value consistent with support stability, because the acid or base concentration may rise during the drying process. Even when the solution used for impregnation or exchange does not contain added acid or base, the ability of a support to react with acid or base may still be important. For instance, if the support functions as a base, the extent of solute hydrolysis will be enhanced if this is accompanied by the liberation of acid.

In the following discussion, all metal contents are given on a dry weight percent basis.

Platinum

First it is convenient to indicate some of the factors which affect the formation of dispersed platinum catalysts. The two common approaches which have been used are impregnation of the support with an aqueous solution of chloroplatinic acid (H_2PtCl_6) and controlled ion-adsorption or ion-exchange of $[Pt(NH_3)_4]^{2+}$ onto the support. In coprecipitation as a preparative method, control is generally more difficult, but the net result is presumably a combination of adsorption and the generation of an intimate physical mixture of support gel and platinum-containing precipitate.

Presentation of the platinum(II) tetrammine ion, $[Pt(NH_3)_4]^{2+}$ to a silica gel support results in adsorption by ion exchange in which the platinum ion replaces two hydrogen ions from the surface. The process is pH dependent[9] as shown in Fig. 4.3, which also shows that under similar conditions adsorption onto an alumina surface is not possible to any useful degree, apparently because the surface hydroxyl groups do not

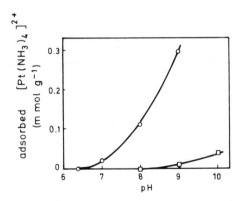

FIG. 4.3 Adsorption from solution containing $[Pt(NH_3)_4]^{2+}$ (as chloride) onto silica gel (Davison 70, 370 m^{2}g^{-1}), and γ-alumina (Aloca F-20, 204 m^{2}g^{-1}), as a function of pH. Room temperature. ○, silica gel; □, γ-alumina. Reproduced with permission from Benesi, H. A., Curtis, R. M. and Studer, H. P. *J. Catal.* **10**, 328 (1968).

have sufficient Brønsted acidity. This is in line with the relatively high value for the pH at zero net surface charge on alumina (about 9). Assuming that the ion $[Pt(NH_3)_4]^{2+}$ remains stable over the pH range of interest (pH > 6), the ion adsorption process may be represented by

$$\{H^+\}_{(s)} + x[Pt(NH_3)_4]^{2+}_{(aq)} \rightleftharpoons \{(1-2x)H^+; x[Pt(NH_3)_4]^{2+}\}_{(s)} + 2xH^+_{(aq)} \tag{4.6}$$

Exchange of $[Pt(NH_3)_4]^{2+}$ into cation exchanged faujasite-type zeolites is straightforward. Zeolites with doubly charged cations as the majority cation (about 80%) have been used, e.g. $Ca^{2+}, Na^+ - 13Y$,[42, 89] Mg^{2+}, $Na^+ - 13Y$,[89] as well as those with only singly charged cations, such as $NH_4^+ - 13Y$,[90] and $Na^+ - 13Y$[89, 90]. Rare earth exchanged zeolites have also been used.[23, 89] The exchange processes may be represented, for instance by,

$$\{Na^+\}_{(s)} + x[Pt(NH_3)_4]^{2+}_{(aq)} \rightleftharpoons \{(1-2x)Na^+; x[Pt(NH_3)_4]^{2+}\}_{(s)} + 2xNa^+_{(aq)} \tag{4.7}$$

$$\{Ca^{2+}\}_{(s)} + x[Pt(NH_3)_4]^{2+}_{(aq)} \rightleftharpoons \{(1-x)Ca^{2+}, x[Pt(NH_3)_4]^{2+}\}_{(s)} + xCa^{2+}_{(aq)} \tag{4.8}$$

In the case of $Ca^{2+}, Na^+ - 13Y$ (80% Ca^{2+}), about 40% the calcium ions are located in S_I and 60% in S_{II} sites,[42] (cf. Chapter 2). Since the minimum van der Waals dimension of the $[Pt(NH_3)_4]^{2+}$ ion is some 0·8 nm across the square, it is certain that S_I sites are inaccessible and exchange must only involve S_{II} sites, so the $[Pt(NH_3)_4]^{2+}$ ion is located at the surface of the main cage. After cation exchange with $[Pt(NH_3)_4]^{2+}$ to give a platinum content of about 0·5 wt. %, only about 1% of the calcium ions on S_{II} sites will have been exchanged, and the $[Pt(NH_3)_4]^{2+}$ cations are on the average some 4 nm apart. Although there will be a spectrum of platinum–platinum distances, very close approach will be improbable.

The platinum cations exchanged onto silica gel or onto zeolite are held relatively strongly as judged by their resistance to removal by washing with decationized water, since this requires hydrolysis with the formation of silanol groups, and $[Pt(NH_3)_4]^{2+}$ and 2 OH^-. The equilibrium constant for this is unfavourable.

The processes of dehydration, calcining and hydrogen reduction have been studied by the use of infrared spectrometry with $[Pt(NH_3)_4]^{2+}$ exchanged into $Ca^{2+}, Na^+ - 13Y$ (80% Ca^{2+}).[89] Initially, the platinum ion contains two water molecules relatively weakly bound to the platinum, so the ion geometry is square bipyramidal. This geometry is retained for the initially exchanged ion in an aqueous environment. However, these water molecules are lost during dehydration of the exchanged zeolite. If, after preliminary drying, the sample is reduced in hydrogen at 4×10^4 Pa (300 Torr) at successively increasing temperatures in the range 320–420 K, it is found that hydrogen consumption occurs simultaneously with decomposition of the tetrammine ion as evidenced from the infrared spectra. The reduction is typically finished by raising the temperature to >570 K, but the resultant platinum dispersion* is then relatively poor ($D_{Pt} \approx 0 \cdot 08$ from hydrogen adsorption). On the other hand, if the $[Pt(NH_3)_4]^{2+}$ exchanged zeolite sample is heated in air, the tetrammine ion does not begin to decompose until 520 K: presumably Pt^{2+} ions largely remain at the cation sites of the zeolite (cf. reaction 4.11). Thus, calcination in air at 620 K, followed by hydrogen reduction at 670 K, gives a very high degree of platinum dispersion ($D_{Pt} \approx 1$). Irrespective of the expectation that isolated platinum atoms with an oxidation number zero would in any case be thermodynamically unlikely, and would be unstable with respect to aggregation by surface diffusion at reduction temperatures, these results clearly show that when $[Pt(NH_3)_4]^{2+}$ is reduced in hydrogen, there must exist highly mobile species before reduction is complete. Insofar as one would expect any ionic species to have a very low mobility across the zeolite surface, one may conclude that the mobile species must be some form of neutral platinum complex, and the following reactions have been suggested:[89]

$$[Pt(NH_3)_4]^{2+} + 2H_2 \rightarrow [Pt(NH_3)_2(H)_2]^{\circ} + 2NH_3 + 2H^+ \qquad (4.9)$$

$$[Pt(NH_3)_2(H)_2]^{\circ} \rightarrow Pt + NH_3 + H_2 \qquad (4.10)$$

Competitive with these reactions would be the process of direct ammonia loss

$$[Pt(NH_3)_4]^{2+} \rightarrow Pt^{2+} + 4NH_3 \qquad (4.11)$$

followed by

$$Pt^{2+} + H_2 \rightarrow Pt + H^+ \qquad (4.12)$$

In reactions 4.9 and 4.12, it is assumed that H^+ is taken up by the zeolite. The complex $[Pt(NH_3)_2(H)_2]^{\circ}$ is suggested as the species of high surface mobility. No complex of this structure has yet been isolated, but in the present circumstances it would only need to be of transitory existence to allow for platinum aggregation. Nevertheless, hydrido complexes of platinum are well known,[91] that with a known structure nearest to

* The dispersion of a metal D_M is defined to be the ratio of the number of surface atoms to the total number of metal atoms, that is $D_M = N_{(S)M}/N_{(T)M}$: cf. Chapter 6, pp. 360–363.

$[Pt(NH_3)_2(H)_2]°$ being the neutral complex $[Pt(L)_2(H)_2]°$ where L ≡ monodentate phosphine. The mechanism suggested above for $[Pt(NH_3)_4]^{2+}$ exchanged zeolite also can plausibly be applied to silica gel. It is known[92] that if silica gel which has been exchanged with platinum ammine solution (the precise identity of the platinum cationic species being unspecified) is subjected only to low temperature drying (about 330 K), then to hydrogen reduction at 570 K, the best platinum dispersion ($D_{Pt} \approx 1$) is achieved if the hydrogen pressure is in the range $0 \cdot 13 – 1 \cdot 3 \times 10^2$ Pa ($0 \cdot 1 – 1 \cdot 0$ Torr); if the hydrogen pressure is 101 kPa (1 atm) the platinum dispersion is appreciably poorer with D_{Pt} about $0 \cdot 8$. It therefore seems likely that reaction 4.9 is reversible and in equilibrium, so that at low hydrogen pressures the concentration of $[Pt(NH_3)_2(H)_2]°$ on the support is reduced, and reactions 4.11 and 4.12 then become competitive for the formation of reduced platinum.

It has been shown[89] that when $[Pt(NH_3)_4]^{2+}$ exchanged zeolite is dehydrated and calcined in air before reduction, the platinum dispersion finally achieved is dependent on the water vapour partial pressure during these processes. Higher water vapour exposures tend to yield poorer dispersions. The effect may originate from the presence of water vapour in the air surrounding the specimen as a whole, or even more potently from attempts to dehydrate and calcine large bulk samples of exchanged zeolite where the thickness of the sample layer impedes the escape of water vapour, thus leading to quite high partial pressures at fairly high temperatures within the sample mass. It is conceivable that the water vapour might in some way enhance the mobility of Pt^{2+}, but it seems much more likely that the effect occurs before decomposition of the tetrammine ion, and the formation of a neutral platinum complex is again possible, for instance,

$$[Pt(NH_3)_4]^{2+} + 2H_2O \rightarrow [Pt(NH_3)_2(OH)_2]° + 2NH_3 + 2H^+ \quad (4.13)$$

Although heating Pt(II) compounds in air or oxygen at temperatures in the vicinity of 570 K will usually generate Pt(IV), calcination of $[Pt(NH_3)_4]^{2+}$ exchanged zeolite under these conditions will not necessarily convert all Pt(II) to Pt(IV). In particular, Pt^{2+} ions strongly held at cation sites on the zeolite are likely to be more resistant to oxidation, but such platinum aggregates as may be formed are certainly likely to be converted to PtO_2 particles.

The use of $[Pt(NH_3)_4]^{2+}$ ion exchange should be possible with other supports provided they have sufficient Brønsted acidity.

We turn now to consider some of the factors relating to the preparation of catalysts by chloroplatinic acid impregnation. The contribution made by adsorption is very much dependent on the nature of the support, and some data are given in Table 4.2.[123]

TABLE 4.2 Adsorption of chloroplatinic acid from aqueous solution at room temperature

Adsorbent	Surface area ($m^2 g^{-1}$)	Saturation adsorption (wt. % Pt, dry basis)	Adsorption equilibrium constant* ($dm^3\ mol^{-1}$)
α-alumina[(a)]	110	2·1	10^4
silica-alumina[(b)]	364	0·26	50
silica gel[(c)]	491	0	0
activated carbon[(d)]	1010	22	7×10^3

(a) Alcoa, F–10.
(b) 10 mol % Al_2O_3, 90 mol % SiO_2.
(c) Davison.
(d) Columbia.
* Defined as $C^{-1}(\theta/(1-\theta))$, where C is the solution concentration in mol dm^{-3}, and θ is the fractional surface coverage.

Clearly, chloroplatinic acid adsorption is negligible with silica, of low but appreciable strength with the silica–alumina, and relatively strong with the high area alumina and carbon. With the latter supports, the adsorption equilibrium constants are such that saturation adsorption is approximated at solution concentrations of $\geqslant 2 \times 10^{-3}$ and $\geqslant 5 \times 10^{-3}$ mol dm^{-3} respectively. Chloroplatinic acid adsorption is strong on all aluminas, and is even stronger on α-alumina than on χ-alumina.[123] The relative strength of chloroplatinic acid adsorption on alumina can be reduced by competitive adsorption of hydrochloric acid or nitric acid, the former being the more effective:[124] this allows a considerable improvement in the ease of achieving a uniform platinum distribution throughout the alumina support pellet.

There is very little definite information as to how chloroplatinic acid is strongly adsorbed on alumina. There is the possibility that the platinate ion may be held to the surface by an anion exchange process involving the loss of surface hydroxyl (cf. Chapter 2, reaction 2.6). Alternatively, or in addition, there is the possibility that a platinate residue could be bound via ligand exchange, in which one (or two) chloro ligands are lost and are replaced by one (or two) surface hydroxyls, for instance

$$S^+OH^-_{(s)} + [PtCl_6]^{2-}_{(aq)} \to S^+[(HO)PtCl_5]^-_{(s)} + Cl^-_{(aq)} \qquad (4.14)$$

Although chloroplatinic acid is relatively stable in aqueous solution, some hydrolysis does occur and involves reactions such as

$$[PtCl_6]^{2-}_{(aq)} + H_2O \rightleftharpoons [PtCl_5(H_2O)]^{-}_{(aq)} + Cl^{-}_{(aq)} \quad (4.15)$$

$$[PtCl_5(H_2O)]^{-}_{(aq)} + H_2O \rightleftharpoons [PtCl_4(H_2O)_2]^{\circ}_{(aq)} + Cl^{-}_{(aq)} \quad (4.16)$$

$$[PtCl_5(H_2O)]^{-}_{(aq)} \rightleftharpoons [PtCl_5(OH)]^{2-}_{(aq)} + H^{+}_{(aq)} \quad (4.17)$$

$$[PtCl_4(H_2O)_2]^{\circ}_{(aq)} \rightleftharpoons [PtCl_4(OH)(H_2O)]^{-}_{(aq)} + H^{+}_{(aq)} \quad (4.18)$$

$$[PtCl_4(OH)(H_2O)]^{-}_{(aq)} \rightleftharpoons [PtCl_4(OH)_2]^{2-}_{(aq)} + H^{+}_{(aq)} \quad (4.19)$$

The equilibrium constants for reactions 4.15–4.19 at about 310 K are $5 \cdot 6 \times 10^{-3}, 2 \times 10^{-4}, \sim 10^{-5}, 6 \cdot 3 \times 10^{-5}$ and $6 \cdot 3 \times 10^{-7}$ respectively. At 328 K, hydrolysis equilibrium is reached in about 30 min, but at room temperature many hours are required. In neutral or weakly acidic solutions and at temperatures below about 328 K, the hydrolysis appears to be limited to the formation of the species $[PtCl_4(OH)_n(H_2O)_{2-n}]^{n-}$ ($0 \leqslant n \leqslant 2$), but at higher pH values and at higher temperatures more extensive hydrolysis occurs, so that on boiling in a dilute alkali solution, it is possible to obtain $[Pt(OH)_6]^{2-}$.

If a support is impregnated with an aqueous chloroplatinic acid solution, drying at room temperature would produce as the first solid to be deposited from the occluded solution, mainly $(H_3O)_2PtCl_6 \cdot 4H_2O$, together with perhaps small amounts of the aquochloro- and hydroxychloroplatinate hydrolysis products, provided the support is inert to acid. However, if the support can react with acid, one would expect the extent of hydrolysis to hydroxychloroplatinate to be increased. Further dehydration requires heating, and this is often done at about 370 K. At this temperature, further hydrolysis occurs, hydrogen chloride being liberated and scavenged by the support if it is basic enough to do so; water is also liberated and the platinum-containing solid on the support will be a complex mixture of hydroxychloroplatinate species, with the chlorine content of this complex ion being generally lower on basic supports. Of course, if the impregnated support is dried at about 370 K from the start, the product will be much the same. If the specimen is now further heated in vacuum or in an inert atmosphere to about 570–670 K, further hydrogen chloride evolution occurs together with chlorine, and the product is platinum(II) chloride plus platinum(IV) oxide. Above about 770 K, the platinum(II) chloride decomposes to yield platinum metal and chlorine. On the other hand, if after dehydration at about 370 K, the material is heated in air at (say) 770 K, most of the platinum will be present as platinum(IV) oxide, together with some residual bound chlorine. All of the platinum-containing species are readily reduced to metallic platinum by gaseous hydrogen at (say)

570 K, and this reaction commences as low as room temperature. Reduction may thus be initiated either after drying or after calcining. There is some evidence that rapid heating of the impregnated support for purposes of dehydration leads to a more highly dispersed platinum deposit than does slow drying and dehydration. There are at least two factors which need to be considered. One has been alluded to previously and is the tendency for liquid to accumulate in the smaller pores by virtue of the interplay between capillary forces and the pore size distribution. However, we have also seen that the transport of liquid in a small porous body such as a catalyst support pellet under the action of capillary forces is quite fast, and is probably too fast to be influenced by accessible variations in the drying time. The second factor is merely the decrepitation of the individual platinate deposits which exist in the support pores, which leads finally to smaller individual platinum particles.

Platinum(IV) oxide (PtO_2) is not completely inert towards other di-, tri- and tetravalent metal oxides, and the position has recently been summarized by Hoekstra et al.[93] We restrict our attention here to those oxides which have been considered in Chapter 2 as support materials. There is no evidence for extensive compound formation or solid solution with silica or alumina (1470 K; 4 GPa). There is extensive solid solution formation with titania (1020–1470 K; 4 GPa), and some with chromia, but no evidence of compound formation. Magnesia reacts with platinum(IV) oxide to form the spinel Mg_2PtO_4, and this decomposes at about 1120 K to give metallic platinum. These results need to be borne in mind if titania, magnesia or chromia is used as a support for a platinum catalyst.

To achieve a uniform distribution of platinum throughout the support grain may take a considerable contact time between the support and the solution when adsorption or exchange of the metal ions occurs. The actual time required depends on the experimental conditions, including the grain size and pore structure of the support, but figures in the range 18 to 72 hours have been quoted.[94-96] It has been reported[123] that for chloroplatinic acid adsorbed in the peripheral region of an α-alumina support pellet, redistribution was still incomplete after being allowed to stand for 23 h at room temperature with the original liquid remaining in the pores. On the other hand, with χ-alumina, redistribution was approximately complete in 3 h. Some degree of control may be achieved via the factors alluded to previously, but in practice economic factors also intrude to limit the permissible equilibration time.

There is evidence[92, 94] that treatment of a silica gel with a solution containing $[Pt(NH_3)_4]^{2+}$ causes attack on the gel surface with the formation of some silicic acid and colloidal silica. Failure to remove this corrosion product by thorough washing leads to a substantial reduction in platinum dispersion. Why this should be so is not clear. Some attack on high area alumina is also known to occur during contact with chloroplatinic acid solution.

In addition to methods based on adsorption and impregnation, supported dispersed platinum–metal catalysts can be prepared by reductive precipitation of the metal. The difficulty is that if reduction takes place easily from solution, it becomes very difficult to confine the precipitated metal particles to a reasonably uniform distribution on the support surface. However, Zeliger[222] has reported that this method can be used successfully with silica or asbestos supports if gaseous hydrogen is used as the reducing agent with an aqueous solution of the platinum–metal chloride: reduction appears to occur at the support surface only.

Platinum/Silica Catalysts

A range of platinum/silica catalysts has been studied by Moss and co-workers,[3-7] and by van Hardevelt and van Montfoort.[8] The support used by Moss was mostly a Davison 70 silica gel (specific surface area about 300 m^2g^{-1}, average pore diameter about $10 \cdot 4$ nm, average spherical particle diameter about $10 \cdot 0$ nm), while van Hardeveld and van Montfoort used an "Aerosil" silica powder (specific surface area about 170 m^2g^{-1}). The smallest average platinum particle size was obtained by ion exchange with $[Pt(NH_3)_4]^{2+}$: after reduction at 570 K, the average diameter was $1 \cdot 4$ nm for a catalyst containing $2 \cdot 45$ wt. % platinum. This is in very good agreement with the result obtained by Benesi et al.[9] For comparison, a catalyst containing $2 \cdot 5$–3 wt. % platinum, but prepared by impregnation with a solution of chloroplatinic acid gave an average particle diameter of about $4 \cdot 5$ nm after reduction at 480 K: the corresponding thin section electron micrograph is shown in Fig. 4.4. Figure 4.5 shows the particle size distribution obtained for the sample corresponding to Fig. 4.4 and the dominance of small particles with diameter $<5 \cdot 0$ nm is clearly seen. The nature of the platinum particle distribution is probably sensitive to quite small variations in the preparative procedure since Adams et al.[10] prepared a $2 \cdot 5$ wt. % platinum catalyst, again using Davison 70 silica gel and using preparative conditions quite similar to those described by Moss and co-workers, and found a rather different particle size distribution. The distribution obtained by Adams et al. is contained in Fig. 4.5 for comparison.

On the "Aerosil" support, van Hardeveld and van Montfoort found that the highest degree of platinum dispersion was obtained by homogeneous hydrolysis (i.e. slow hydrolysis) from a solution of chloroplatinic acid: after hydrogen reduction at 770 K the average platinum particle diameter was about $1 \cdot 2$ nm (for $2 \cdot 9$ wt. % platinum) and this compares well with the degree of dispersion obtained by Moss with an ion exchange method. However, impregnation with chloroplatinic acid and hydrogen reduction at 770 K gave a relatively poor dispersion with the average diameter >20 nm (for 5·2 wt. % platinium).

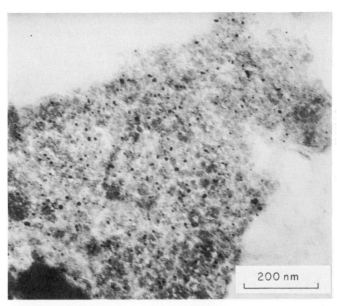

FIG. 4.4 Electron micrograph of thin section of 2·5 wt. % platinum/silica catalyst, prepared by chloroplatinic acid impregnation of Davison 70 silica gel, hydrogen reduced at about 480 K. Reproduced with permission of R. L. Moss, Warren Spring Laboratory, England and British Crown Copyright. The black spots are platinum particles.

Moss and co-workers[6,7] combined X-ray diffraction line broadening, electron microscopy and carbon monoxide adsorption data to study a range of platinum/silica catalysts, and showed in some detail the influence of parameters such as platinum content, method of preparation, reduction temperature, air calcination and surface area of the silica support. Catalysts prepared by impregnation of Davison 70 silica with chloroplatinic acid solution to give platinum contents in the range 0·15–11·5 wt. %, showed two distinct regimes for the dependence of platinum particle size on platinum content. In the range from 0·15 to about 3 wt. % platinum, the average platinum particle diameter was approximately constant at about 3·6 nm, while the number of platinum particles per gram of catalyst increased in the range $0·11 \times 10^{16} - \sim 3 \times 10^{16}$. In the range of about 3–11·5 wt. % platinum, the number of platinum crystallites per gram of

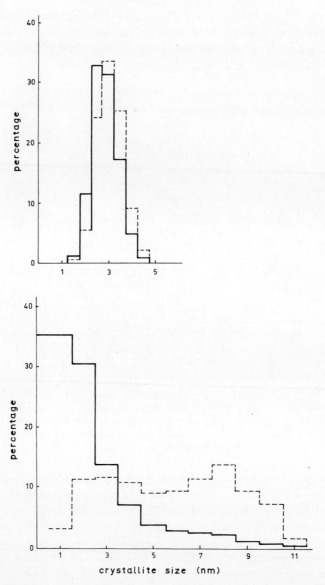

FIG. 4.5 Particle size distributions for platinum/silica gel catalysts. Lower; 3 wt.% platinum, data from Moss, R. L. *Platinum Metals Rev.* **11**, 1 (1967). Upper; 2·5 wt.% platinum, data from Adams, C. R., Benesi, H. A., Curtis, R. M. and Meisenheimer, R. G. *J. Catal.* **1**, 336 (1962). Both catalysts prepared by chloroplatinic acid impregnation of Davison 70 silica gel, hydrogen reduced at 480 K. In both cases the full line refers to a number distribution, the broken line to a surface area distribution.

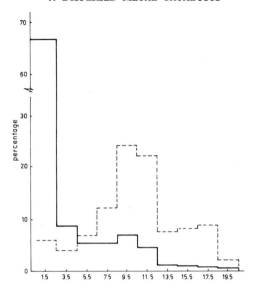
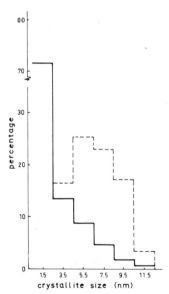

FIG. 4.6 Particle size distributions for 10 wt.% platinum/silica gel catalysts prepared by impregnating Davison 70 silica gel with chloroplatinic acid. Upper: catalyst hydrogen reduced at 770 K; lower: catalyst hydrogen reduced at 410 K. In both cases, the full line refers to a number distribution, the broken line to a surface area distribution. Data from Dorling, T. A., Eastlake, M. J. and Moss, R. L. *J. Catal.* **14**, 23 (1969).

catalyst remained approximately constant at about $3 \cdot 3 \times 10^{16}$, while the average platinum particle diameter increased to about 6–7 nm (the catalysts were dried for 16 h at 390 K and reduced in hydrogen for 2 h at 480 K). This behaviour is undoubtedly related to the way in which the pore structure of the silica gel controls the availability of platinum for crystal growth, so that the behaviour varies in a way that depends on the structure of the silica gel. In general, the average crystallite diameter tends to decrease as the specific surface area increases, that is, as the average pore size decreases.

On the other hand, catalysts prepared by ion exchange with $[Pt(NH_3)_4]^{2+}$ from aqueous solution onto Davison 70 silica showed the average platinum particle diameter to be approximately constant at $1 \cdot 4$–$1 \cdot 6$ nm for platinum contents in the range $0 \cdot 1$–$4 \cdot 45$ wt. %. This agrees with the conclusion of Poltorak and Boronin.[11] With this method of preparation, the average platinum particle size is independent of the porosity of the support.

Moss and co-workers also found that the platinum particle size increased somewhat at increased reduction temperatures. Thus, impregnated catalysts in the composition range 1–10 wt. % platinum suffered an area-reduction in the vicinity of 10–30 % on increasing the reduction temperature from 410 to 770 K. This is also reflected in the distribution of particle

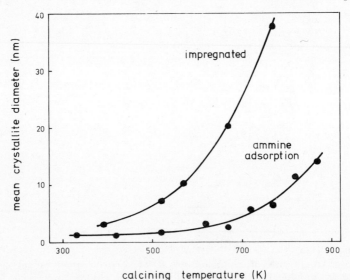

FIG. 4.7 Dependence of crystallite size on temperature of air-calcining prior to hydrogen reduction for platinum ($2 \cdot 5$ wt%) on Davison 70 silica gel. Reproduced with permission from Dorling, T. A., Lynch, B. W. J. and Moss, R. L. *J. Catal.* **20**, 190 (1971).

sizes: Fig. 4.6 shows, for 10 wt.% platinum catalysts reduced at 410 K and 770 K, the fraction of total metal surface area attributable to particles of various sizes and obtained from electron microscopic particle size distributions. Calcining the catalysts in air prior to reduction had a relatively strong effect on the particle size finally achieved after reduction. This is illustrated in Fig. 4.7. The volatility of $PtCl_4$ has been suggested[7] as the reason for the relatively high rate of crystallite growth with impregnated specimens. This treatment can conveniently be used to generate catalysts of controlled average particle size: this is particularly so

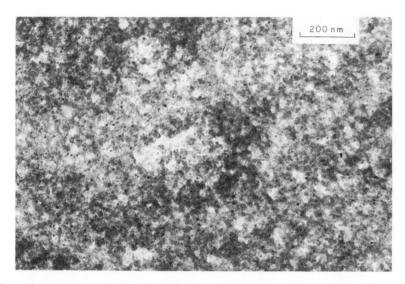

FIG. 4.8 Electron micrograph of platinum/silica catalyst, prepared by adsorption of $[Pt(NH_3)_4]^{2+}$ on Davison 70 silica gel. Calcined at 620 K in air prior to hydrogen reduction at 570 K. Reproduced with permission from R. L. Moss, Warren Spring Laboratory, England and British Crown Copyright. The black spots are platinum particles.

for catalysts prepared by adsorption where it has been demonstrated[4] that the even distribution of crystallites over the support remains unaffected by the calcining process. A typical case is illustrated by the electron micrograph in Fig. 4.8 which shows a catalyst prepared by ammine adsorption which had been calcined in air at 620 K prior to hydrogen reduction at 570 K. In general, ion exchange with $[Pt(NH_3)_4]^{2+}$ produces a more even distribution of platinum particles than does impregnation with chloroplatinic acid (cf. Figs 4.5 and 4.8). The result of calcining depends on the structure of the silica gel support. The relative area reduction is greater the lower the gel porosity, and again this is probably a consequence of the influence of porosity on the transport of platinum required for crystal growth, particularly in relation to the effect of residual water vapour.

The reduction temperature has one other influence on the catalyst which is of considerable importance: it may influence the amount of halogen retained by the catalyst. Thus, an impregnated catalyst containing 3·3 wt.% platinum had retained 4·2% of its initial chlorine after reduction for 2 h at 480 K and 0·2% after 2 h at 770 K.[7] This residual chlorine may have catalytic consequences. For instance, Dorling et al.[4] found that a 10 wt.% platinum catalyst prepared by impregnation with chloroplatinic acid and reduced at 350, 480 and 770 K retained 0·16, 0·13 and 0·034% chlorine respectively (the blank for the untreated silica was 0·029%), and that the activity of this catalyst for the test reaction of ethylene hydrogenation was dependent on residual chlorine. Inasmuch as the silica substrate surface is catalytically inert irrespective of the level of residual chlorine, this effect apparently results from the presence of chemisorbed chlorine on the platinum surface.

Platinum/Alumina Catalysts

A detailed study of a range of platinum/alumina catalysts has been made by Wilson and Hall.[23] These were prepared by chloroplatinic acid impregnation. A 2·83 wt.% platinum content on alumina (140 m^{2}g^{-1}, probably γ-alumina), reduced in hydrogen at 750 K had an average platinum

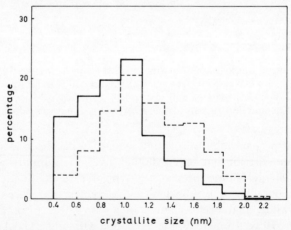

FIG. 4.9 Particle size distributions for 2·83 wt.% platinum/γ-alumina catalyst, prepared by chloroplatinic acid impregnation, hydrogen reduced at about 750 K. Full line, number distribution; broken line, surface area distribution. Data from Wilson, G. R. and Hall, W. K. J. Catal. 17, 190 (1970).

particle diameter of about 1·1 nm and the corresponding particle size distribution is shown in Fig. 4.9. This average diameter is quite close to the values of < 1·0–2·0 nm obtained by a number of previous workers[24-27] for catalysts containing between 0·1 and 1 wt.% platinum on γ- or η-alumina, prepared by impregnation and reduced at 770 K. The platinum dispersion is independent of whether the support is γ- or η-alumina, other factors being the same.[28] It was also shown that increasing the temperature of hydrogen reduction in the range 750–950 K resulted in a modest increase in the average particle diameter from about 1·1 to 2·1 nm. However, the gaseous environment coupled with temperature is important in controlling platinum particle growth. Thus, although heating in an inert nitrogen atmosphere at 870 K for 4 h had only a very small effect on particle size, heating at 820–890 K in oxygen

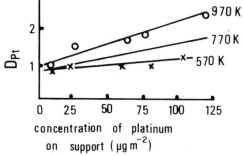

FIG. 4.10 Dependence of platinum dispersion $D_{Pt}(=N_{(S)Pt}/N_{(T)Pt})$ on platinum concentration in support, and on temperature of hydrogen treatment. Supports γ-alumina, 100–240 m²g⁻¹. Catalysts prepared by impregnation with aqueous solution of chloroplatinic acid. After Zaidman, N. M. et al. *Kinetics and Catal.* **10**, 313 (1969).

or air caused a large increase. It was also shown that the average platinum particle size increased by about 25 % on increasing the platinum content from 0·75 to 2·83 wt.%. Zaidman et al.[107] showed that heating platinum/alumina catalysts (prepared by chloroplatinic acid impregnation of γ-alumina) in hydrogen also resulted in an appreciable growth in the average metal particle size, the effect being the more pronounced the greater the platinum concentration on the support. Figure 4.10 summarizes their data for catalysts with 0·1–2·0 wt.% platinum, in which is shown the platinum dispersion D_{Pt}, assuming a hydrogen chemisorption stoichiometry of two (cf. p. 295), obtained in each case as a time-independent value after 6 h hydrogen treatment at the indicated temperature.

A thin section electron micrograph of a 2·5 wt.% platinum catalyst on a low area γ-alumina, obtained by Moss[3] is shown in Fig. 4.11.

On the whole, the degree of platinum dispersion obtained by chloro-

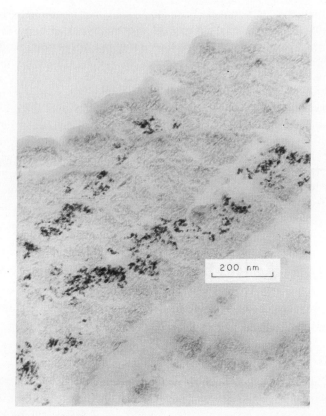

FIG. 4.11 Electron micrograph of thin section of 2·5 wt.% platinum/γ-alumina catalyst, prepared by chloroplatinic acid impregnation, hydrogen reduced at about 480 K. Reproduced by permission R. L. Moss, Warren Spring Laboratory, England, and British Crown Copyright. The black spots are platinum particles.

platinic acid impregnation on alumina is better than on silica of comparable porosity, and this is no doubt due to the fact that on alumina a substantial proportion of the chloroplatinic acid is adsorbed.

An example of how the uniformity of the platinum distribution may be influenced by the nature of the platinum species in the solution in relation to the strength of adsorption is afforded by the data due to Roth and Reichard.[111] A medium area alumina (probably mainly γ-alumina) with a specific surface area of 74 m²g⁻¹ was used in the form of $\frac{1}{8}$ in × $\frac{1}{8}$ in. tablets. Two impregnation techniques were used to give 0·3 wt.% platinum. One employed an ammoniacal solution of platinum dinitrito diammine $[Pt(NO_2)_2(NH_3)_2]°$, the second an aqueous solution of chloroplatinic acid. The impregnated samples were dried at 390 K and calcined at 770 K, and the platinum concentration profile across the support tablet was determined by electron microprobe analysis. The results are shown in Fig. 4.12 and it is obvious that whereas impregnation with $[Pt(NO_2)_2(NH_3)_2]°$ solution leads to a nearly uniform platinum distribution, the chloroplatinic acid treatment leads, as expected, to a distribution in which the platinum is highly concentrated towards the outer regions of the alumina tablet due to strong chloroplatinic acid adsorption. By contrast, the neutral complex, $[Pt(NO_2)_2(NH_3)_2]°$ is presumably not adsorbed.

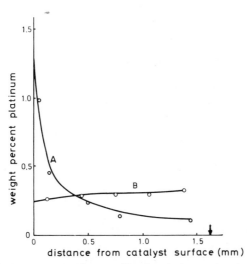

FIG. 4.12 Platinum concentration profiles in platinum/alumina catalyst tablets. Support mainly γ-alumina (74 m²g⁻¹), platinum content 0·3 wt.% overall. Catalysts prepared by impregnation, dried 390 K, calcined 770 K. Curve A, impregnation with aqueous solution of chloroplatinic acid; curve B, impregnation with aqueous solution of platinum dinitrito diammine. The arrow indicates the centre of the tablet. After Roth, J. F. and Reichard, T. E. *J. Res. Inst. Catal. Hokkaido Univ.* **20**, 85 (1972).

The work of Kral[114] has shown that for a series of 5 wt.% platinum catalysts processed under the same conditions (calcined at 770 K, hydrogen reduced at 470 K), the platinum particle size is dependent on the pore size of the alumina support (rather than on the specific surface area as such). Thus an increase in the average pore diameter by a factor of about 2·9 resulted in an increase in the average particle diameter by a factor of about 2·1. The situation is therefore generally similar to that found with silica gel supports.

McHenry *et al.*[44] showed that some of the platinum in reduced (770 K) platinum/alumina catalysts prepared by chloroplatinic acid impregnation, was soluble in reagents such as hydrofluoric acid or acetylacetone. Solubility of the platinum in hydrofluoric acid can be simply a consequence of exposure of the catalyst to oxygen, the surface platinum atoms carrying chemisorbed oxygen being those which dissolve.[98,106] However, incomplete reduction, leaving some platinum present on the alumina surface as platinum(IV) complexes, presumably can also be responsible for the presence of soluble platinum: following the original suggestion of McHenry et al.[44] this may consist of platinum atoms co-ordinated in part with the alumina and in part with hydroxyl or chlorine ligands. In general, we believe these models to be more reasonable than one which would identify the soluble fraction with monatomic platinum atoms with an oxidation number of zero, the existence of which we regard as most improbable under these conditions on energetic grounds (cf. p. 170 and Chapter 5).

Platinum/Silica–Alumina Catalysts

Figueras et al.[34] observed by electron microscopy platinum particles in the range <2–10 nm for catalysts containing $0 \cdot 6$–$1 \cdot 0$ wt. % platinum and prepared by chloroplatinic acid impregnation, with hydrogen reduction at 670 K. However, there is a tendency for the average metal particle size for silica–alumina supports to be larger than with silica or alumina supports. Thus, Cusumano et al.[35] found for 2 wt. % platinum (prepared by chloroplatinic acid impregnation and hydrogen reduction at 720 K particle sizes of $1 \cdot 0$ nm and $8 \cdot 5$ nm on alumina (925 m^{2}g^{-1}) and silica–alumina (Davison DA-1, 13 % Al$_2$O$_3$, 450 m^{2}g^{-1}) respectively. This may be because silica–alumina has an enhanced surface heterogeneity by virtue of its surface aluminium sites, with the latter providing regions in the vicinity of which metal tends to congregate, thus facilitating crystal growth. As with other supports, there is a trend for larger average metal crystallite sizes to be formed at higher temperatures of calcination and hydrogen reduction.

Platinum/Zeolite Catalysts

In the introduction of platinum to zeolite supports, use is made of their cation exchange properties, and [Pt(NH$_3$)$_4$]$^{2+}$ is the most commonly used cation. The important question is the nature of the platinum in the zeolite after the platinum-containing material has been subjected to dehydration, calcining and hydrogen reduction.

Lewis[41] examined a catalyst containing $0 \cdot 5$ wt. % platinum which had been exchanged onto Ca^{2+}–13Y zeolite, followed by hydrogen reduction at 570 K. A variety of techniques were used, including X-ray absorption edge spectroscopy, X-ray diffraction line broadening, together with hydrogen adsorption. It was concluded that about 60% of the platinum was present as particles of average diameter about 1·0 nm, probably situated in the main cage of the zeolite: the rest of the platinum was of average particle size about 6 nm, presumably residing exterior to the zeolite pore structure. The absorption edge data showed that, within the limits of detection, all the platinum was reduced to a zero-valent state. On the other hand, Rabo et al.[42] exchanged $[Pt(NH_3)_4]^{2+}$ into a Ca^{2+}–13Y zeolite to the extent of 0·5 wt.% platinum, followed by drying, calcining in air at 770 K and hydrogen reduction at 570 K, and showed that the platinum was in a very high state of dispersion. It is clear that the platinum dispersion achieved by Rabo et al.[42] was appreciably better than that obtained by Lewis.[41]

We have already alluded in an earlier section in which were discussed some of the chemical processes occurring during the processing of a platinum catalyst, to the way in which the dispersion of platinum on (inter alia) zeolite supports may be influenced by processing variables: in particular, if the maximum dispersion is to be obtained, the need for decomposition of the exchanged $[Pt(NH_3)_4]^{2+}$ ion in a non-hydrogen atmosphere (e.g. air), and the need for a low water vapour partial pressure. The results obtained by Rabo et al.[42] and by Lewis[41] are entirely consistent with this, and with the later results of Dalla Betta and Boudart.[89]

The question arises if, as asserted by Rabo et al.,[42] the platinum on their highly dispersed catalyst was in monatomic form with zero oxidation number, Pt(0). Whether it is reasonable to believe that Pt(0) atoms can be generated and exist under these conditions on a catalyst support has been previously commented on (cf. p. 170) and is further discussed on energetic grounds in Chapter 5: the conclusion emerges that they are improbable. Nevertheless, the platinum in this catalyst must certainly have existed in very small aggregates, the average size of which was no more than a few atoms. This is in qualitative agreement with the conclusion reached by Dalla Betta and Boudart[89] from an examination of the $OH_{(s)}/Pt$ stoichiometry in relation to deuterium exchange with $OH_{(s)}$ in a highly dispersed platinum on zeolite catalyst of a type very similar to that described by Rabo et al:[42] it was concluded that the upper limit to the platinum aggregate size was six platinum atoms.

Rabo et al. also described a catalyst prepared by impregnation of a Ca^{2+}–13Y zeolite with chloroplatinic acid solution to give 0·5 wt.% platinum, followed by drying, calcining in air at 770 K and hydrogen reduction at 570 K. The platinum dispersion, D_{Pt}, appeared to be about 60% of that for the catalyst prepared by cation exchange.

Platinum/Carbon Catalysts

Impregnation methods using chloroplatinic acid have generally been used. There appears to have yet been little attempt to make use of the ion exchange properties associated with the oxygen-containing residues on carbon surfaces, although ion-exchange adsorption of $[Pt(NH_3)_4]^{2+}$ should be possible with the surface carboxylic acid groups of an oxidized carbon support, in the manner described for the palladium ammine ion.[14]

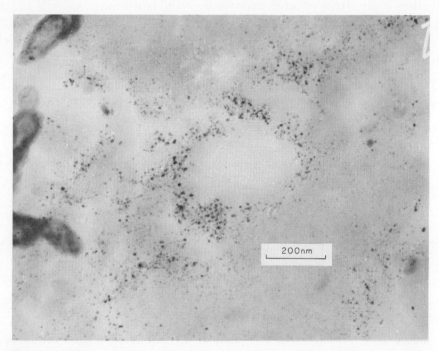

FIG. 4.13 Electron micrograph of thin section of 5 wt.% platinum/carbon catalyst, prepared by chloroplatinic acid impregnation; calcined in air at 570 K, hydrogen reduced at about 480 K. The carbon was a commercial charcoal. Reproduced by permission R. L. Moss, Warren Spring Laboratory, England and British Crown Copyright. The black spots are platinum particles.

Moss[3] used a thin section technique with electron microscopy to examine a 5 wt.% platinum/charcoal catalyst (Fig. 4.13) which was prepared by impregnation with chloroplatinic acid solution, followed by reduction. It was found that even after heating in air at 570 K, the platinum crystallites were extremely small, and judging from Moss's electron micrographs the platinum crystallites appear to be at least as finely dispersed on charcoal as on silica or alumina, or possibly more so. This retention of a very high degree of dispersion is probably the result of the use of a charcoal with an extremely large internal surface so that particle growth becomes improbable, as well as to relatively strong adsorption of chloroplatinic acid by charcoal.

Noble Metals other than Platinum

We refer in this section to the metals palladium, rhodium, iridium, ruthenium, osmium and gold. With none of these metals has the chemistry or technology of catalyst preparation been studied in as much detail as with platinum. Nevertheless, what is known about platinum serves as a useful general guide to what may be expected.

Except for gold these metals may be introduced to the support by cation exchange or by impregnation. Gold* does not exist in stable cations, and impregnation methods are used. Table 4.3 summarizes some reagents which have been used, or which seem likely to be of potential utility.

The same general factors govern the susceptibility of a support for ion exchange as have already been discussed previously. Impregnation methods have most often used aqueous solutions of chloro-complex ions: however, the aqueous chemistry of these ions is often quite complicated, so that the exact nature of the ionic species in solution may not be well defined.

The $[IrCl_6]^{3-}$ ion is relatively stable and survives in aqueous solution without the addition of hydrochloric acid to suppress hydrolysis (cf. ref. 17). On the other hand, $[RhCl_6]^{3-}$ is relatively unstable in aqueous solution and a range of chloroaquoions exist. Solution of $RuCl_3$ or its hydrate in water results in rapid hydrolysis with the precipitation of a hydrated ruthenium hydroxide: this requires acidification with, for instance, hydrochloric acid for its suppression. However, osmium trichloride solutions are extremely stable to hydrolysis. Palladium(II) chloride and the ion $[PdCl_4]^{2-}$ are both subject to hydrolysis in water which is more extensive than for the $[PtCl_6]^{2-}$ ion.

With ruthenium and osmium, although cationic complexes such as $[Ru(NH_3)_6]^{3+}$, $[Ru(NH_3)_4(NO)(OH)]^{2+}$ and $[Os(NH_3)_6]^{3+}$ are well known, they are labile in alkaline solutions with the loss of ammonia and their conversion into hydroxyanionic complexes; thus, they are only available for cation exchange provided this can be achieved at a pH

---

* It has been reported[220] that supported metal catalysts which contain gold should never be prepared by impregnation of a support with solutions which contain both gold salts and NH_4OH: the dried catalyst contains extremely shock-sensitive gold–nitrogen compounds which may explode with the lightest touch.

TABLE 4.3 Examples of reagents for preparation of supported catalysts from noble metals other than platinum

	Reagent		References to examples of use
Palladium	$PdCl_2$ (or hydrate)	a, c	17, 101, 97
	$(NH_4)_2[PdCl_4]$	a, c	
	$Pd(NO_3)_2 \cdot 2H_2O$	a, c	100
	$[Pd(NH_3)_4]^{2+}$	b, e, f	21
Rhodium	$RhCl_3 \cdot xH_2O$	a, c	17, 19, 101
	$(NH_4)_3[RhCl_6] \cdot 1\tfrac{1}{2}H_2O$	a, c	115
	$Rh(NO_3)_2 \cdot 2H_2O$	a, c	
	$[Rh(NH_3)_6]^{3+}$	b, d	
Iridium	$IrCl_3 \cdot xH_2O$	a, c	
	$IrCl_4 \cdot xH_2O$	a, c	101
	$H_2[IrCl_6]$ (or hydrate)	a, c	17, 114
	$(NH_4)_2[IrCl_6]$	a, c	115
	$(NH_4)_3[IrCl_6] \cdot H_2O$	a, c	
	$[Ir(NH_3)_5Cl]^{2+}$	b, e	
	$[Ir(NH_3)_6]^{3+}$	b, d	
Ruthenium	$RuCl_3$ (or hydrate)	a, c	17, 101
	$(NH_4)_2[RuCl_6]$	a, c	115
	$(NH_4)_2[RuCl_5H_2O]$	a, c	
	$[Ru(NH_3)_4(NO)(OH)]^{2+}$	b, e, g	
	$[Ru(NH_3)_6]^{3+}$	b, d, g	
Osmium	$OsCl_3$ (or hydrate)	a, c	18
	$(NH_4)_2[OsCl_6]$	a, c	115
	$[Os(NH_3)_6]^{3+}$	b, d, g	
Gold	$AuCl_3$	a, c	115
	$H[AuCl_4]$ (or hydrate)	a, c	99
	$(NH_4)[AuCl_4]$	a, c	
	$H[Au(NO_3)_4] \cdot H_2O$	a, c	

a, for impregnation methods
b, for cation exchange methods
c, commercially available in solid form
d, for instance, as the nitrate or chloride
e, commercially available in solid form as dichloride
f, commercially available in solid form as dinitrate
g, labile at pH > 8.

below about 8. The cationic hexammine complexes of iridium(III) and rhodium(III) and the tetrammine complex of palladium(II) are all stable in alkaline solutions and are potentially useful for cation exchange procedures over a wide pH range.

The methods of catalyst processing are generally similar to those used with platinum.

Some of the main features of the reactions of the hydrated chlorides of rhodium, iridium and ruthenium upon heating in hydrogen or in air have been described by Newkirk and McKee.[179] When $RhCl_3 \cdot xH_2O$ (x ≈ 2·74) was heated in hydrogen, weight loss was first observed at about 310 K and, as judged by gravimetric measurements, reduction to metallic rhodium was complete above about 380 K: with $IrCl_3 \cdot xH_2O$ (x ≈ 3) and $RuCl_3$ (hydrate) the corresponding temperatures were 320 K, 470 K and 330 K, 620 K respectively. When $RhCl_3 \cdot xH_2O$ or $IrCl_3 \cdot xH_2O$ was heated in air, the thermograms showed steps which, although there was some overlap between the processes, could be mainly identified first with the loss of water to give the anhydrous chloride, followed by conversion to metal oxide. In the case of $RhCl_3 \cdot xH_2O$ the first plateau corresponding to the presence of anhydrous $RhCl_3$ occurred at 620–780 K: the second plateau at 1030–1310 K corresponded to the existence of Rh_2O_3, and at temperatures 1310–1330 K this oxide decomposed to give rhodium metal. These conclusions are in agreement with earlier work.[180,181] Heating $IrCl_3 \cdot xH_2O$ in air gave first a sloping plateau with a mid-point at about 620 K corresponding to anhydrous $IrCl_3$, while a second sloping plateau commenced at about 950 K and corresponded to IrO_2: the oxide decomposed to the metal at 1300–1340 K, although some oxide volatilization occurred above about 950 K. In the case of $RuCl_3$ (hydrate) heating in air gave the first detectable weight loss at about 350 K following which there was a gradual conversion to RuO_2 which was complete at 710 K.

Palladium/alumina catalysts have been reported with average metal particle sizes in the region 1·0–4·5 nm[8, 21, 29] for palladium contents in the range 0·5–15 wt. %; these were prepared by impregnation methods. Heating in vacuum in the range 370–670 K did not have much influence on palladium particle size, but at 770 K, it resulted in an increase in the average palladium particle size by a factor of 3–4. On the other hand, the palladium dispersion was more sensitive to the temperature of hydrogen reduction: increasing the reduction temperature from 370 to 670 K resulted in an increase in the average palladium particle size by a factor of nearly 2^{29}.

On a silica gel support (Davison 70) the best palladium dispersion was obtained by cation exchange using $[Pd(NH_3)_4]^{2+}$—for instance, an average particle diameter of 1.4 nm for 2.2 wt. % palladium, after drying at 390 K and calcining at 770 K. By comparison, the dispersion resulting from impregnation was appreciably poorer—an average particle diameter of 2·2 nm for 2 wt. % palladium, after drying at 390 K and calcining at 770 K. There is a trend to larger average particle sizes with increased palladium loadings—for instance, an average diameter of 3·7 nm for 5 wt.% palladium[29] prepared by impregnation compared with 2·2 nm for 2 wt. %: an average diameter of 10·6 nm has been reported for 10 wt. % palladium.[17]

A variety of palladium/charcoal catalysts has been studied by Pope et al.,[71] and Fig. 4.14 shows an electron micrograph particle size distribution for a 1 wt. % palladium/charcoal catalyst which had been treated in hydrogen at 300 K, followed by evacuation at 370 K. This treatment does not result in serious sintering, but the reduction temperature is so low that a high degree of surface cleanliness would seem improbable. The average palladium particle diameter was estimated at

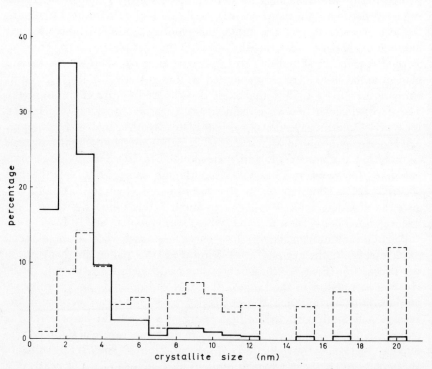

FIG. 4.14 Particle size distribution for 1 wt.% palladium/carbon catalyst, hydrogen reduced at 298 K followed by evacuation at 370 K. Full line, number distribution; broken line, surface area distribution. The carbon was a commercial charcoal. Reproduced with permission from Pope, D., Smith, W. L., Eastlake, M. J. and Moss, R. L. *J. Catal.* **22**, 72 (1971).

about 4·0 nm, taking into account both the electron micrograph data and results also obtained by gas adsorption. Increasing the palladium content in the range 1–20 wt. % was shown not to have much effect on the average particle size. High temperature treatment caused considerable palladium particle growth, so that after heating to 970–1170 K, the average diameter was in the vicinity of 40–60 nm. Morikawa et al.[14] have also reported on the size of palladium particles on a charcoal support (5 wt. %), and quote a range 15–44 nm after hydrogen reduction. Comparison with Pope's data lends emphasis to the strong temperature sensitivity of the average particle size in these materials. Morikawa et al. also noted that some reduction of palladous chloride to metallic palladium occurred by reaction with oxidizable groups already present on the charcoal surface. A 1 wt. % palladium/graphite catalyst has been studied electron microscopically by Brownlie et al.[48] This was prepared by impregnation of powdered synthetic graphite (nuclear grade) with aqueous palladous chloride solution followed by reduction at 470 K, and alternatively by direct deposition of palladium vapour. Although not explicitly stated by Brownlie et al. these support surfaces are almost certainly basal graphite planes. The average crystallite diameter was rather smaller from the impregnation method than from evaporation, electron microscopic values being 6·5 and 14 nm respectively. As might be expected for the basal plane of graphite, the metal mobility is relatively high. Thus, heating above 570 K caused considerable metal diffusion resulting in decoration of topographical features of the graphite surface, and there was extensive particle growth during a catalytic reaction at 310 K. Morikawa et al.[14] have also reported ion-exchange adsorption of palladium(II) ammine ions (probably $[Pd(NH_3)_4]^{2+}$) on a high area carbon support by making use of surface carboxyl groups which were generated by nitric acid treatment.

Carbon molecular sieves have also recently been used as supports for metal catalysts; for instance, the palladium/carbon sieve catalyst described by Trimm and Cooper.[49]

The particle sizes for some typical catalysts containing rhodium, iridium, osmium, ruthenium and gold are given in Table 4.4. The general trends with respect to the effects of metal concentration and calcining temperature are similar to those for platinum. Iridium catalysts prepared by coprecipitation (cogelation) of aluminium hydroxide and iridium hydroxide (5–36 wt. % iridium) gave a somewhat larger average iridium particle size after dehydration and hydrogen reduction than was obtained under comparable conditions by impregnation.[112] It has been reported[223] that in the preparation of ruthenium/γ-alumina catalysts by impregnation with ruthenium chloride solution, the metallic ruthenium obtained after

TABLE 4.4 Average metal particle size for various supported catalysts containing Rh, Ir, Os, Ru, Au

Catalyst*		Average metal particle diameter (nm)	Reference
0·1 wt. % Rh/silica	(a)	1·1	19
0·3 wt. % Rh/silica	(a)	1·1	19
1 wt. % Rh/silica	(a)	1·2	19
5 wt. % Rh/silica	(a)	2·0	19
10 wt. % Rh/silica	(a)	2·3	19
5 wt. % Rh/silica	(b)	4·1	19
5 wt. % Rh/silica	(c)	12·7	19
5 wt. % Rh/silica	(d)	1·5	101
5 wt. % Rh/carbon	(f)	1·3	101
10 wt. % Ir/silica	(a)	1·4	17
5 wt. % Ir/silica	(d)	1·8	101
30 wt. % Ir/γ-alumina	(g)	2·8	102
20 wt. % Ir/γ-alumina	(j)	1·7	112
36 wt. % Ir/γ-alumina	(k)	3·7	112
36 wt. % Ir/γ-alumina	(l)	7·5	112
36 wt. % Ir/γ-alumina	(m)	7·0	112
5 wt. % Ru/silica	(e)	4·2	17
5 wt. % Ru/silica	(d)	3·0	101
5 wt. % Ru/carbon	(f)	2·3	101
10 wt. % Os/silica	(a)	1·9	18
0·7 wt. % Au/magnesia	(h)	5·0	99
0·7 wt. % Au/magnesia	(i)	130	99
5 wt. % Au/magnesia	(h)	15·0	99
5 wt. % Au/magnesia	(i)	110	99

* With the exception of the commercial catalysts whose preparative origins are unknown, all catalysts were prepared by impregnation methods using metal halide in aqueous solution.

(a) support Cabosil HS5, 300 m²g^{-1}: catalyst dried 380 K, H$_2$ reduced 720 K.
(b) support Cabosil HS5, 300 m²g^{-1}: catalyst dried 380 K, calcined 810 K, H$_2$ reduced 720 K.
(c) support Cabosil HS5, 300 m²g^{-1}: catalyst dried 380 K, calcined 1070 K, H$_2$ reduced 720 K.
(d) support silica gel: catalyst dried and calcined 770 K, H$_2$ reduced 720 K.
(e) support silica gel: catalyst vacuum dried, H$_2$ reduced 720 K.
(f) commercial catalyst: H$_2$ reduced 720 K.
(g) commercial catalyst: H$_2$ reduced 620 K.
(h) support reagent grade magnesium oxide powder, 18 m²g^{-1}: catalyst reduced chemically at <370 K.

footnotes continued bottom p. 203.

hydrogen reduction is much better dispersed (average particle size about 2 nm) if the ruthenium chloride is decomposed in hydrogen, than if the decomposition is first carried out in air followed by hydrogen reduction (average particle size 7–14 nm). Particle growth may occur via a relatively volatile ruthenium oxide.

Non-Noble Metals

We are mainly concerned in this section with nickel, cobalt, iron, copper, silver and rhenium. In the cases of nickel, cobalt, iron, copper and silver, the simple aquated cations are stable in aqueous solutions of appropriate pH, so that impregnation of a support may be readily done using an aqueous solution of a metal salt such as the nitrate which will decompose to give metal oxide upon dehydration and calcining. Other salts such as oxalates or formates have sometimes been used, the latter often decomposing to yield the metal itself.

However, with many of these elements, the support is not inert. Cations from the impregnating solution may exchange with cations from the support surface: the latter may be hydrogen ions if the support surface has sufficient Brønsted acidity, or exchange may occur with some other metal cations present in the support surface. However, in addition to this, under the conditions used for calcining the impregnated catalyst, more extensive reaction may occur, leading in some cases to specific compound formation. A well known example of the latter occurs in the preparation of nickel catalysts supported on silica or alumina.

Data (e.g. refs 103, 104) indicate that for a number of oxide surfaces, we can expect the strength of cation adsorption to decrease in the order

$$\text{Fe(III)} > \text{Fe(II)} > \text{Cu(II)} > \text{Ni(II)} > \text{Co(II)} \qquad (4.20)$$

and some of the basic factors and methods have been discussed in Chapter 2. Under appropriate conditions the extent of cation adsorption may be substantial in relation to the metal loading which may be required in catalyst preparation. For instance, at 298 K the adsorption of Co(II) cations from an aqueous solution of concentration 10^{-3} mol dm^{-3} may reach about 6×10^{-6} and about 10^{-5} mol m^{-2} on silica and titania

---

footnotes continued from Table 4.4.
(i) support reagent grade magnesium oxide powder, 18 m^{2}g^{-1}: catalyst dried and calcined about 620 K.
(j) γ-alumina, 188 m^{2}g^{-1}: catalyst dried 380 K, H$_2$ reduced 670 K.
(k) γ-alumina, 188 m^{2}g^{-1}: catalyst dried 380 K, calcined in air 970 K 1 hour, H$_2$ reduced 670 K.
(l) γ-alumina, 188 m^{2}g^{-1}: catalyst dried 380 K, calcined in air 970 K 5 hours, H$_2$ reduced 670 K.
(m) γ-alumina, 188 m^{2}g^{-1}: catalyst dried 380 K, H$_2$ reduced 970 K 1 hour.

respectively.[104] If the oxide had a specific surface area of 200 m^{2}g^{-1}, this cationic uptake would lead to a metal loading of 7·2 and 12 wt. % respectively.

In general terms it seems probable that when the introduction of a heavy metal cation to a catalyst support does lead to extensive cation adsorption, other things being equal this will result in a better dispersion of the metal in the final catalyst than when only mechanical occlusion operates. However, at the same time adsorption may well lead to a greater likelihood of chemical reaction with the support.

The nature of the interaction between heavy metal cations and oxide support surface has been studied by various spectroscopic (visible, infrared) and magnetic (e.s.r., magnetobalance) means. The main ideas are common to all systems irrespective of whether they involve silica, alumina, zeolite or some other oxide. The treatment of silica gel with Ni(II), Co(II), Cu(II) and Cr(III) has been studied in some detail and is a good illustration of the sort of behaviour that can occur.[130,142,143,108,109]

When the metal-containing cation is exchanged onto a silica gel surface, it is immediately bound relatively strongly: it cannot be easily recovered by washing with water, although removal with dilute aqueous mineral acid is possible. Whatever else may occur, this resistance to removal by washing with water is a consequence of an unfavourable equilibrium for the necessary hydrolysis reaction (cf. reaction 2.6, Chapter 2). However, strong binding may also be due to ligand exchange between the incoming cation and a surface siloxyl group, and the importance of this will depend both on the conditions and on the nature of the cation. Thus with cations containing highly unreactive ligands such as [Cr(NH$_3$)$_6$]$^{3+}$, no ligand exchange with the surface occurs unless the exchanged surface is desiccated at room temperature or above.[174] On the other hand, with more reactive cations, some ligand exchange occurs with the initial introduction of the cation to the surface.

Direct impregnation of the unexchanged gel with a solution containing a salt (such as the nitrate or chloride) of the metal in question under conditions such that minimal initial ion exchange occurs, leads to a spectrum that is the same as for the bulk solution; thus at this stage there is no change in the ligand structure of the ions. As the impregnated samples are dried this degree of aqueous co-ordination of the isolated metal ions cannot be maintained, and there ensues a ligand exchange process with the entry of anions from solution and/or surface groups into the co-ordination sphere. The anions referred to here are those such as chloride or nitrate which may have been present in the original impregnation solution. Ligand exchange with surface groups always occurs to some extent, and if the drying temperature is limited to 370 K or so, it is likely that the surface ligands are the OH of silanol groups, and up to three may be used per metal ion to give adsorbed structures such as shown in Fig. 4.15. Octahedral (or distorted octahedral) symmetry is retained with Ni(II), Co(II) and Cr(III) (cf. Fig. 4.15a), but tetrahedral symmetry is likely for Cu(II) (cf. Fig. 4.15b). Again the reactivity for ligand exchange varies, and $[Cr(H_2O)_6]^{3+}$ is, for instance, relatively unreactive. At this stage, readsorption of water can remove the ions from the surface and regenerates the original aqueous complexes of the impregnation solution. However, if the impregnated and desiccated sample is heated to higher temperatures up to 770 K irreversible changes occur, and the metal ions become more strongly bound to the surface so that they cannot be removed by treatment with water. It is likely that the increased strength of binding to the surface is a consequence of the conversion of the surface ligands from silanol (Si–OH) to siloxyl (Si–O$^-$). At the same time, some or all of the salt anions (e.g. chloride or nitrate) are removed by the evolution of gaseous products such as hydrogen chloride or oxides of nitrogen and nitric acid, and those which come from the co-ordination sphere of the metal have their place taken by hydroxyl groups, the latter being provided by the dehydration and dehydroxylation of the rest of the silica surface.

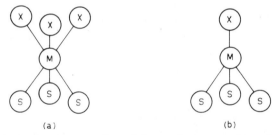

FIG. 4.15 Schematic representation of the environment of transition metal ions adsorbed by ligand exchange on a silica surface. The groups S are part of the surface and are oxygens of neutral or ionized silanol groups. The sites X can be occupied by anions or molecules.

Hydroxylated metal cations on the silica gel surface may react together with the elimination of water and the genesis of particles of metal oxide. The ease with which this occurs is dependent on the nature of the metal. For instance, it is relatively easy with cations of Cr(III) when the reaction occurs via the formation of oxide polymer.[174] Gaseous oxygen may also be involved if the metal can wholly or in part be converted to a higher oxidation number, and chromium is also an example of this with the generation of some Cr(V) and Cr(VI). Oxide formation is also dependent on the nature of the support, being easier, for instance, on silica than on alumina, and this difference is mainly a reflection of the different ways in which the cations are co-ordinated to the surface.

The uptake of heavy metal ions by γ-alumina has been studied on a number of occasions (e.g. refs 105, 144–146). Again as with silica, the metal cations in samples dried at temperatures up to 370 K or so retain the co-ordination symmetry of the aquocation; octahedral in the case of nickel and cobalt, for instance. Ion adsorption occurs with relative strengths as indicated by the sequence 4.20, and the ions are probably bound by co-ordination to surface ligand groups in much the same way as described for silica. The question then is that of the fate of the adsorbed cations on further thermal treatment. If we take the case of cobalt(II) by way of example, despite earlier magnetobalance data indicating octahedral site symmetry after calcination at 770 K,[145] it is now clear[105,144] that by such treatment the adsorbed cobalt ions move into positions of tetrahedral symmetry, that is, they move into vacant tetrahedral sites in the γ-alumina. At temperatures of calcination below about 670 K the ions remain predominantly at the γ-alumina surface and retain octahedral (or distorted octahedral) co-ordination. At temperatures in the range 370–670 K, the nature of the ligands probably changes in much the same way as we have seen to occur for cations adsorbed on silica. Similar processes to these also occur with adsorbed iron(III). With cobalt and iron, these processes only describe the behaviour of impregnated γ-alumina samples up to a metal loading of about 1–1·5 wt. %. Above this, some metal remains unadsorbed from solution occluded in the support pores, and on drying and calcining at 770 K this results in the formation of cobalt oxide (Co_3O_4) and iron oxide (probably Fe_2O_3), together with some aluminate reaction product. On calcining above 1070 K these metal oxides are readily removed completely by dissolution and reaction.

The general behaviour of nickel(II) on γ-alumina is different to that described above, since the suggestion[147] based on magnetobalance data that nickel ions which have entered the γ-alumina residue in octahedral sites agrees with the known structure of $NiAl_2O_4$ (about 80% inverted), while the residence of cobalt in tetrahedral sites agrees with the normal structure of $CoAl_2O_4$.

The incorporation of heavy metal ions is, of course, even more pronounced in co-precipitated samples, that is samples prepared by, for instance, the simultaneous precipitation of aluminium and nickel or cobalt hydroxide; with suitable stoichiometry the spinels are readily obtained after calcining.

The solution of heavy metal ions in other oxide supports is known, and an example with titania is the solubility of iron(III) studied by Selwood et al.[110]

There is an important distinction between silica and active alumina (e.g. γ-alumina) in terms of the solubility of heavy metal cations. The solubility in γ-alumina can be appreciable, but that in silica is quite low: for instance, the solubility of iron(III) in silica was found not to exceed 0·1 wt. %.[105] This difference is presumably a consequence of γ-alumina having a proportion of vacant lattice sites. However, this does not mean that some metal silicate is not formed during the preparation of silica-supported heavy metal catalysts. In fact it does occur and has its origin in two ways: a small amount may be formed by the direct precipitation of metal silicate during the impregnation stage, due to the finite solubility of silica particularly if the pH is alkaline; the second and generally more important source is direct reaction during calcining. Both of these sorts of processes also operate to generate metal aluminate with active alumina supports, and the evidence which we shall discuss subsequently indicates that aluminate is formed considerably more easily than silicate. The reactions are of the sort

$$\left.\begin{array}{l}2M^{2+} + 2O^{2-} + SiO_2 \rightarrow M_2SiO_4 \\ 2M^{2+} + 4OH^- + SiO_2 \rightarrow M_2SiO_4 + 2H_2O\end{array}\right\} \quad (4.21)$$

$$\left.\begin{array}{l}M^{2+} + O^{2-} + Al_2O_3 \rightarrow MAl_2O_4 \\ M^{2+} + 2OH^- + Al_2O_3 \rightarrow MAl_2O_4 + H_2O\end{array}\right\} \quad (4.22)$$

Detailed reports of the structure of nickel/silica catalysts have been published by Schuit and van Reijen[12] and by Coenen and Linsen[13] for workers in the Netherlands, while work by Morikawa and co-workers has also been summarized.[14]

The structure of the product and the extent to which it is formed are much dependent on the preparative details. In the unreduced materials, basic nickel silicates have frequently been identified by X-ray methods, particularly when the preparation involves precipitation from an aqueous nickel solution under alkaline conditions. Under these conditions nickel hydroxide is also present, and the basic nickel silicates are structurally related to the latter by partial replacement of OH layers by silicate layers. If the precipitating solution also contains anions such as sulphate, carbonate or chloride, these may also be incorporated into the precipitate either as impurities or as separate phases (e.g. basic nickel carbonates or sulphates), so the composition of the product may well not be uniquely defined. In any case, the precipitated nickel compounds are usually imperfectly crystalline, with considerable stacking disorder in the layer lattices. This degree of crystallinity is much reduced on calcining and hydrogen reduction, although there is no doubt that some form of nickel silicate is retained.

It is also worth noting that when nickel is introduced by impregnation and calcination, the proportion whose fate is some form of nickel silicate rather than nickel oxide decreases as the nickel loading increases. When the metal loading is very high it is not uncommon for the amount of nickel silicate to be negligible compared to the amount of nickel oxide: a typical example of this situation is quoted by Roman and Delmon[178] for a silica gel (334 m^{2}g^{-1}) impregnated with nickel nitrate to give a nickel loading of 18·9 wt. %, followed by calcining in air at 770 K.

Although catalysts are often prepared by impregnation or deposition with a nickel compound which would decompose thermally on calcining to give nickel oxide, these catalysts require hydrogen reduction at >770 K even though nickel oxide itself only requires 500–600 K for complete reduction. Schuit and van Reijen[12] concluded that impregnation of a silica support by, for instance, nickel nitrate solution resulted, on calcining, in the formation of particles of nickel oxide covered with a layer of nickel silicate, and it is the latter which impedes the reduction process. It is also possible that some nickel oxide particles become encapsulated within collapsed pores of the silica gel as a result of calcination. Nevertheless, some reduction to yield particles of metallic nickel clearly does occur, as is evident from magnetic and X-ray data (the latter are difficult to interpret because of line broadening due to the small particle size, and complexity from other phases). Catalysts prepared by nickel precipitation, and particularly those prepared by co-precipitation, are even more difficult to reduce than are impregnated ones. Typical compositions after reduction are given both by Coenen and Linsen and by Morikawa *et al.*: even extended hydrogen reduction at about 770 K may easily result in no more than about half the nickel being present as metallic nickel, the rest being present as nickel oxide and as some sort of nickel silicate. The range of composition is, however, highly variable, as may be judged from the composition ranges quoted by Coenen and Linson: crystalline metallic nickel, 25–81 mol %; nickel as nickel oxide, 3–29 mol %; nickel as a nickel silicate, 8–66 mol %. Clearly, the nickel silicate is highly resistant to hydrogen reduction or at the most, reduction to metallic nickel only occurs to a very limited extent in a surface layer of the silicate. The presence of residual sulphate is also important[13] since during hydrogen reduction this tends to be reduced to sulphide which collects on the nickel surface and is itself very resistant to removal by hydrogen reduction: the consequence is the presence of surface sulphide which can profoundly modify the surface behaviour of the nickel particles, for instance, by greatly reducing the extent of hydrogen chemisorption.

Schuit and van Reijen quoted a typical example of a 0·16 wt. % nickel catalyst, prepared by impregnation of a Davison 22 silica gel with a solution of nickel nitrate followed by calcining and reduction at 770 K: the average nickel particle diameter was about 4·4 nm. Reinen and Selwood[15] also found average nickel particle diameters in the region 3·0–4·5 nm for catalysts reduced at about 670 K. For a variety of different preparations, Coenen and Linsen quoted a range of average nickel particle diameters of about 3·0–20 nm. In all cases the metallic nickel was well crystalline, although with particles at the lower end of the size range (3·0–4·0 nm) there was X-ray evidence for an increase of some 0·34% in lattice parameter. Coenen and Linsen ascribed this behaviour to the epitaxial growth of the very small nickel particles on the surface of the nickel silicate. The model proposes that growth occurs with a (111) plane of the nickel in contact with an hexagonally arranged surface nickel layer of the nickel silicate. Since the Ni–Ni distance in the latter is rather larger than in nickel (111), some lattice dilation is detected in the epitaxed metal particles, provided they are small enough. There is a wide variety of shapes consistent with such an epitaxial model. Coenen and Linson assume a roughly hemispherical shape which exposes (111), (100) and (110) planes, and this is at least consistent with the very limited electron microscopic evidence. However, there are no data to make a valid selection between a number of possible crystal shapes which are very roughly circular in plan outline and on the surface of which low index planes predominate.

The structure of nickel/silica catalysts was also studied by van Hardeveld and van Montfoort using an "Aerosil" support.[8] The highest degree of dispersion was obtained by homogeneous hydrolysis (slow hydrolysis), the average nickel particle diameter after hydrogen reduction at 620–720 K being about 3·0 nm (for 11·1 wt.% nickel). An impregnation method (6·7 wt.% nickel) with reduction at 620 K gave an average nickel diameter of about 7·0 nm and calcining in air at 720 K for 16 h prior to reduction resulted in an average diameter of about 15·0 nm. Figure 4.16 shows the electron microscopic particle size distributions obtained by these workers for nickel/"Aerosil" catalysts, and Fig. 4.17 shows a typical micrograph of a sample prepared by impregnation (uncalcined). From these various results it emerges that smaller nickel particles are obtained at lower nickel contents, and under conditions where a high proportion of nickel is present as a nickel silicate. Although there is a trend for the growth of nickel particles during calcining and hydrogen reduction, this is less pronounced on silica gel than on "Aerosil", no doubt because the highly developed pore structure of silica gel tends to keep the nickel particles isolated. On the whole, it appears to be more difficult to produce a very high degree of metal dispersion with nickel than with platinum, and it is very difficult to obtain an average nickel particle size $< 2·0$ nm.

A recent account of the nature of nickel/alumina catalysts have been given

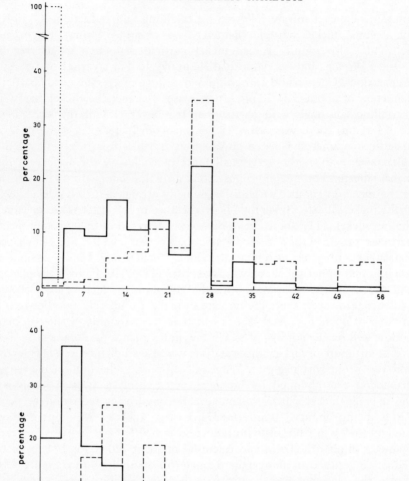

FIG. 4.16 Particle size distribution for nickel/"Aerosil" dispersions. Upper: dotted line, 11·1 wt.% nickel, prepared by homogeneous hydrolysis, hydrogen reduced at 620–720 K; full and broken lines, 6·7 wt.% nickel, prepared by impregnation, calcined in air at 720 K prior to hydrogen reduction at 620 K. Lower: 6·7 wt.% nickel, prepared by impregnation, hydrogen reduced at 620 K. The full lines and the dotted lines refer to number distributions, the broken lines to surface area distributions. Data from van Hardeveld, R. and van Montfoort, A. *Surface Sci.* **4**, 396 (1966).

by Morikawa et al.[14] Calcination of the catalyst prior to hydrogen reduction would be expected to yield nickel oxide, but the formation of nickel aluminate is competitive with this. The greater the extent of dilution of the nickel on the alumina, the greater is the chance of nickel aluminate formation: in the limit, preparation by co-precipitation or by the decomposition of a low concentration of nickel hydroxide on fresh alumina hydrogel, yields nickel aluminate exclusively. On the other hand, when, as in impregnation, larger particles of nickel compound are deposited, the calcination product is a mixture of nickel oxide and nickel aluminate. The proportion of nickel oxide increases when occlusion of the impregnation solution leads to a very non-uniform distribution.[30]

The nature of the nickel aluminate depends on the calcination temperature. Above 770 K, $NiAl_2O_4$ spinel is certainly formed, however at < 770 K the structure of the aluminate phase is indeterminate.

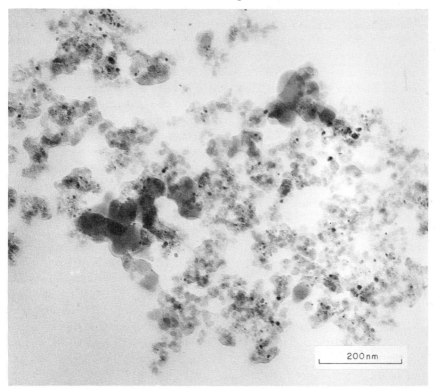

FIG 4.17 Electron micrograph of nickel/"Aerosil" dispersion, 6·7 wt.% nickel, prepared by impregnation, hydrogen reduced at about 620 K. Reproduced with permission from van Hardeveld, R. and van Montfoort, A. *Surface Sci.* **4**, 396 (1966). The small black dots are nickel particles.

As in the nickel/silica case, the rate of hydrogen reduction of nickel oxide when this is present on an alumina support is very slow compared to that for pure nickel oxide, requiring prolonged hydrogen treatment at 820 K, for instance. Presumably this is due to the presence of a thin skin of aluminate over the oxide particle. Nevertheless, reduction of oxide to discrete particles of metallic nickel certainly occurs. For instance, nickel particles have been observed directly with the electron microscope in a reduced catalyst examined by Shephard:[31] the extent of occurrence of metallic nickel has been estimated by acid leaching[14] and reaction with carbon monoxide to yield nickel carbonyl,[14,28] and nickel particle size has been estimated magnetically.[15]

On the whole, with impregnated catalysts, nickel/alumina is more difficult to reduce than nickel/silica, with nickel/silica-alumina occupying an intermediate position. With co-precipitated catalysts, both nickel/alumina and nickel/silica are very resistant to reduction.[32]

Morikawa et al. suggest that nickel aluminate itself undergoes hydrogen reduction only to a superficial extent, and then produces extremely small nickel particles as the reduction product. In this circumstance, the nickel particle size distribution in a reduced nickel/alumina catalyst will obviously depend on the preparative details which control the proportions of nickel oxide and nickel aluminate and the size of the particles in which these substances exist before reduction. This model does imply a bimodal nickel particle size distribution, and this was a conclusion reached by Brooks and Christopher from gas adsorption studies.[33] Directly determined (electron microscopic) nickel particle size distributions are not yet available. However, Morikawa et al. have pointed out that large nickel particles (from nickel oxide reduction) can be eliminated by first removing the nickel oxide by acid treatment. In Shephard's catalysts which were hydrogen reduced at 770 K,[31] the average nickel particle diameter was roughly constant at about $5 \cdot 0$ nm for calcining temperatures in the range 610–830 K, but the size increased by a factor > 2 as a result of calcining at 830–1430 K. Although the electron microscopic resolution of nickel particles in supported catalysts is relatively poor, Shephard concluded that some of his particles were of a multiply twinned structure. For comparison, Reinen and Selwood[15] obtained an average nickel diameter of $2 \cdot 5$–$5 \cdot 0$ nm by magnetic methods for a catalyst reduced at 720 K and Brooks and Christopher[33] obtained a value of about $5 \cdot 0$ nm for the average size of the large end of their bimodal distributions.

As would be expected, high temperature treatment results in nickel particle growth and loss of nickel surface area. This has been studied in some detail by Williams et al.[67] for a 75 wt.% nickel/alumina steam reforming catalyst. The fresh catalyst after hydrogen reduction at 720 K showed by electron microscopy a fairly even distribution of nickel particles which were < 10 nm in diameter. Treatment with steam at 770 K for 1000 h at 25 atm pressure resulted in the formation of particles in the 30–300 nm diameter range. At the same time, the mesopores in the γ-alumina support, which had an average diameter of about 7·0 nm were eliminated, although larger pores (200–1000 nm diameter) remained. Some α-alumina was also formed.

The average metal particle size was shown[36] to be considerably larger with nickel on silica–alumina supports than on silica, and the same is true for cobalt.[37] Again, a proportion of the metal undoubtedly remains incorporated into the support in a form highly resistant to hydrogen reduction (for instance, 75 % reduction to metallic nickel at 970 K determined by Richardson[39]). This appears to be the more important the higher the alumina content of the silica–alumina.

Unlike the noble metals such as platinum and palladium which do not form stable carbides, the existence of free metal on carbon supports with other metals which do form stable carbides is a matter which always needs to be approached with caution. It has been claimed that nickel crystallites do *not* exist in nickel/carbon catalysts.[47] This requires verification particularly since the carbide, if formed, should be reduced to the metal during hydrogen reduction, but it does draw attention to the fact that carbon is not inert towards many metals which can form carbides or intercalation compounds with graphite.

In general terms, the main structural characteristics of cobalt and iron catalysts supported on substances such as silica gel or high area alumina appear to resemble nickel, although the ease of reduction to the metal decreases in the order nickel, cobalt, iron. Ferric ions supported on alumina or silica in low concentrations (about 0·1 wt. %) can only be reduced to the ferrous state in hydrogen at 970 K:[182] presumably at this low concentration all of the iron ions are directly bound to the support surface. At higher iron loadings with occlusion, partial reduction to some particles of metallic iron becomes possible: for instance, a catalyst prepared by impregnation of microspheroidal silica with an aqueous solution of ferrous nitrate to give a metal loading of 10 wt. %, followed by drying at 380 K and hydrogen reduction at 820 K, was partly (but still only partly) reduced to give particles of metallic iron.[18] In some cases reduction beyond iron(II) by means of hydrogen is facilitated if the catalyst is co-impregnated with chloroplatinic acid.[182] Although Mössbauer spectroscopy indicates that the ultimate product of reduction is then bimetallic particles of iron and platinum, it seems reasonable to suppose that the chloroplatinic acid is first reduced to give very small metallic clusters of platinum which, being able readily to chemisorb hydrogen dissociatively, can provide hydrogen atoms by surface diffusion for reduction of adjacent iron(II) ions. However, the method was ineffective with an iron(II)-exchanged Y-zeolite.[224]

The position of copper(II) ions which had been exchanged into X- and Y-zeolites has been studied in some detail.[148-150] The copper enters the zeolite in the form of aquocations, with octahedral symmetry. After dehydration (670–770 K), the majority of the ions are situated close to hexagonal windows of the sodalite cage in sites $S_{I'}$ and $S_{II'}$ adjacent to S_I and S_{II}. In dehydrated X- and Y-zeolites exchanged with nickel(II) or cobalt(II), these ions occur at S_I (octahedral co-ordination) and S_{II} sites, with a modest preference for S_I.[225,230] In dehydrated A-zeolite exchanged with nickel(II), these ions occur in non-planar hexagonal sites[151,152] which resemble S_{II}: the co-ordination is possibly monoclinically distorted octahedral. These remarks also apply to iron(II) in X- and Y-zeolites,[230] but there is also Mössbauer evidence[183] which, while agreeing with some preference for S_I sites, suggests that the remaining iron(II) ions are tetrahedrally co-ordinated using three zeolitic oxygens and a hydroxyl ligand. Nickel(II) is known to favour octahedral co-ordination relative to tetrahedral more strongly than iron(II).

Pope and Kemball exchanged 13X zeolite with aqueous nickel nitrate, followed by vacuum drying and calcining at 740 K and by hydrogen reduction.[45] Their results (including e.s.r.) strongly suggest the formation of nickel crystallites at reduction temperatures >620 K. This agrees qualitatively with the result quoted by Yates that reduction at 670 K gave nickel particles of average size $24 \cdot 0$ nm.[46] Richardson[39] more recently used magnetic methods to estimate average nickel particle sizes for a series of Y zeolites exchanged with $2 \cdot 4$–$3 \cdot 1$ wt. % nickel, which had been hydrogen reduced at 670 K. Values in the range $12 \cdot 6$–$9 \cdot 5$ nm were obtained. This work also showed that the extent of reduction to metallic nickel was dependent on the nature of the exchangeable cation in the zeolite, the values being 100 % for Na^+; 80 % Li^+; 76 % Ca^{2+}; 45 % Mg^{2+}; 0 % NH_4^+. The reason is not known. It should be noted that the metal particle sizes quoted by Yates were obtained by X-ray line broadening and therefore very small crystallites would not have been detected, while Richardson's average sizes also do not rule out the possibility of a substantial proportion of nickel being present as very small particles.

Yates[46] also quoted $17 \cdot 0$ nm for the average size of silver particles formed by hydrogen reduction at 470 K of a silver-exchanged 13X zeolite, although particles a good deal larger than this have also been observed.[113] Very large metal particles, average diameter $54 \cdot 5$ nm, were also reported for a 1 wt. % copper catalyst on 13Y zeolite after hydrogen reduction at 810 K.[129] Silver-exchanged 13X zeolite collapses on reduction in hydrogen (but not in carbon monoxide).

Iron(II) ions exchanged into 13Y zeolite are completely resistant to hydrogen reduction at temperatures up to at least 800 K, and no metallic iron is produced.[183] However, the iron ions are readily interconverted between the iron(II) and iron(III) states by treatment at about 670 K with dry oxygen or hydrogen. Cobalt(II) ions exchanged into type A zeolite are largely resistant to hydrogen reduction at 520 K, although some slight evidence for incipient reduction has been observed.[184]

TABLE 4.5 Average metal particle sizes for various supported catalysts containing Ni, Co, Re, Cu, Ag

Catalyst*		Average metal particle diameter (nm)	Reference
10 wt. % Co/silica	(a)	12·0	37
10 wt. % Co/γ-alumina	(b)	8·6	37
10 wt. % Co/silica-alumina	(c)	40·0	37
10 wt. % Co/carbon	(e)	17·3	37
10 wt. % Re/silica	(d)	12·2	20
10 wt. % Ni/silica	(a)	5·0	36
5 wt. % Ni/silica	(a)	5·7	36
10 wt. % Ni/silica-alumina	(c)	9·9	36
5 wt. % Ni/silica-alumina	(c)	10·8	36
40 wt. % Cu/magnesia	(i)	11·4	129
31 wt. % Cu/magnesium silicate	(j)	9·5	129
10·5 wt. % Ag/γ-alumina	(f)	30–65	120
10 wt. % Ag/α-alumina	(g)	17·5–25	120
12 wt. % Ag/silica	(h)	30	120

(a) support Cabosil HS5, 300 m²g⁻¹: catalyst dried 380 K, H_2 reduced 640 K.
(b) support 295 m²g⁻¹: catalyst dried 380 K, H_2 reduced 640 K.
(c) support Davison DA-1, 87 mol % SiO_2, 450 m²g⁻¹: catalyst dried 380 K, H_2 reduced 640 K.
(d) support Cabosil HS5, 300 m²g⁻¹: catalyst dried 380 K, H_2 reduced 770 K.
(e) support Darco G-60 activated charcoal, 410 m²g⁻¹: catalyst dried 380 K, H_2 reduced 640 K.
(f) support Degussa P110, 110 m²g⁻¹: catalyst dried 420–470 K, H_2 reduced 420–470 K.
(g) support Alcoa A3, 16 m²g⁻¹: catalyst dried 470–520 K, H_2 reduced 470–520 K.
(h) support Degussa Aerosil, 100 m²g⁻¹: catalyst dried 420–440 K, H_2 reduced 420–440 K.
(i) support 140 m²g⁻¹: catalyst dried 470 K, H_2 reduced 470 K.
(j) support 276 m²g⁻¹: catalyst dried 420 K, H_2 reduced 420 K.

* Cobalt and nickel catalysts prepared by impregnation with aqueous solutions of metal nitrates; rhenium by impregnation with aqueous solution of perrhenic acid; silver by reactive precipitation onto the carrier from aqueous solution; copper catalysts were commercial samples.

Table 4.5 lists some typical examples of supported non-noble metal catalysts.

Ultrathin Evaporated Metal Films

For the present purpose, we take the term "ultrathin" to refer to an evaporated metal film where the concentration of metal on the support is low enough for the film to consist of small isolated metal crystals. If the average concentration of metal atoms on the support is of the order of a monolayer, the metal crystals are small enough for ultrathin films to serve as models for highly dispersed metal catalysts, but where surface cleanliness and catalyst structure can be better controlled.

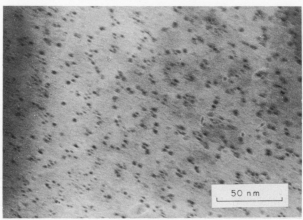

FIG. 4.18 Electron micrograph of ultrathin platinum film deposited on air cleaved mica at 545 K in UHV. Average film thickness $1 \cdot 2 \times 10^{-3}$ gm^{-2}.

The discontinuous nature of very thin metal films is well known (e.g. refs 43, 51) and arises from the growth of crystal nuclei on the substrate. One virtue of ultrathin film catalysts is that single crystal supports may be used, so that there is hope of forming metal crystals with a definite structure. In practice, ultrathin films can readily be prepared with average crystallite diameters in the range from several nm down to the practical limit of electron microscopic resolution of about $0 \cdot 5$ nm.

In actual catalytic experiments, ultrathin metal films have been used on mica, glass and silica supports.[52-55] It would also be possible to use supports such as cleaved or evaporated rocksalt or other crystalline materials, but this has not yet been done. An electron micrograph of a typical ultrathin platinum film, deposited on mica in UHV at 550 K is shown in Fig. 4.18. Here the average film thickness (obtained by chemical analysis after film deposition) is $1 \cdot 2 \times 10^{-3}$ g m^{-2}. For comparison, a close-packed monolayer of platinum atoms corresponds to $4 \cdot 9 \times 10^{-3}$ g m^{-2}. The individual platinum crystallites are clearly resolved, and the average diameter is about $2 \cdot 0$ nm. Furthermore, the particle density on the mica is $1 \cdot 3 \times 10^{16}$ m^{-2}, and it follows that, if one assumes a model crystallite geometry of a cylindrical prism standing end-on to the mica, the average height is about $1 \cdot 4$ nm. The somewhat non-uniform distribution of particles over the mica support that is evident in Fig. 4.18 has been frequently observed for various metals deposited under UHV conditions on air-cleaved mica. It is probably due to impurity adsorbed on the mica, since it does not occur with vacuum-cleaved mica. As the average film thickness grows, so does the particle size: a film of $5 \cdot 0 \times 10^{-3}$ g m^{-2} deposited under the same conditions as those described immediately above had, for instance, an average particle diameter of about $3 \cdot 5$ nm.

Films with specific weights in the range $8 \cdot 0$–$2 \cdot 0 \times 10^{-4}$ g m^{-2} have also been studied, and these all have average crystallite diameters in the region of $\leqslant 1 \cdot 5$ nm. Towards the lower end of this specific weight range, the apparent particle density is low, and the measurements are of poor accuracy due to inadequate electron microscopic resolution. In all cases, the average particle height appears to be about $1 \cdot 0$–$1 \cdot 5$ nm, although this estimate is again only very rough for films near the lower extreme of the specific weight range.

In all films there is a distribution of crystallite diameters. An example is shown in Fig. 4.19 for the film with a specific weight of $1 \cdot 2 \times 10^{-3}$ g m^{-2}. The smallest particles whose diameters can be measured in a micrograph

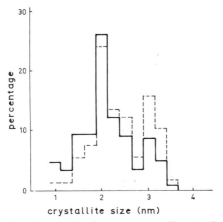

FIG. 4.19 Distribution of particle sizes corresponding to the specimen referred to in Fig. 4.18. The full line is the number distribution, the dotted line the surface area distribution.

(and then only very approximately) have diameters of about $1 \cdot 0$ nm, and this is the lower size limit used in Fig. 4.19. However, particles smaller than this can readily be observed in the micrograph, and there is no doubt that this type of film contains some crystallites down to the limit of microscopic resolution (about $0 \cdot 8$ nm in this case), and presumably beyond. However, their number appears to be relatively small. It is interesting to compare the specific film weight of these ultrathin platinum films with the amount of platinum per unit actual surface area of support for typical supported platinum catalysts. A typical supported catalyst would have 1 wt. % of platinum on a support which has a specific surface area of a few hundred m^2g^{-1}. Assuming all of the internal surface of the support is available (not always a valid assumption in practice) the concentration of platinum on the support surface is in the region of 10^{-6}–10^{-4} g m^{-2}, which is rather lower than the range for the ultra-thin films.

With ultrathin films of average particle diameter $< 5 \cdot 0$ nm or so, one is only able to infer the structure of the particles from observations made under conditions where larger particles occur (cf. Chapter 5). In fact, an examination of a dark-field micrograph of a field such as that of Fig. 4.18 confirms the presence of an appreciable proportion of pentagonal multiply twinned particles, and also shows that most of the particles are oriented with a (111) parallel to the substrate. For metals such as platinum, our conclusions about the likely structure of ultrathin films whose particles are too small for detailed structural study, are as follows:

(a) deposited at 550 K on mica; some tetrahedrally shaped crystallites (probably vertically truncated) together with more complex multiply-twinned particles of both well-defined and ill-defined shapes, and some untwinned particles of ill-defined shapes; crystallites of ill-defined shapes predominate.

(b) deposited at about 300 K on mica; many multiply-twinned and some untwinned particles, all of ill-defined shapes.

2. Unsupported Metal Catalysts

Metal Powders

Under this heading it is convenient to differentiate colloidal metallic dispersions from metal powders of larger particle size.

A summary of preparative methods for metal powders has been published:[50] see also, for instance, Schachter *et al.*,[116] Broadbent *et al.*,[117] McKee,[118] Kobayashi and Shirasaki,[119] Scholten *et al.*,[120] Best and Russell[136] and Hall and Emmett.[137] In many cases the metal powder has been prepared by hydrogen reduction of the oxide powder, the latter having been prepared via precipitation as the hydroxide, carbonate or basic carbonate, or sometimes by thermal decomposition of a metallic salt such as nitrate or oxalate. The particle size of the metal which is ultimately obtained depends on both the type of starting material and the conditions under which it is processed. In general, the lower the processing temperature, the more highly dispersed the metal. The metal particle sizes so obtained are usually quite large: specific surface areas are usually in the range $0 \cdot 1$–$1 \cdot 0$ m^2 g^{-1} with average particle diameters in the μm range. Iron, cobalt and nickel powders in the μm size range may also be prepared by decomposition of the carbonyls.

In the case of the platinum group metals, silver and gold, a variety of methods have been used to generate metallic dispersions. Of course, the route via the precipitated hydroxide indicated in the preceding paragraph is applicable and has a very long history for the preparation of platinum catalysts for laboratory purposes, the process having been used by Wöhler.

A metal oxide is the product from Adams' method (metal halide in fused alkali metal nitrate followed by leaching with water),[212,213] and this procedure has also been extensively used. The nature of the platinum oxide produced by Adams' method has recently been studied in some detail by Cahen and Ibers.[211] The material which had traditionally been formulated as $PtO_2 \cdot H_2O$ was shown to be a mixture of metallic platinum, α-PtO_2 (possibly hydrated) and a sodium platinum bronze $Na_xPt_3O_4$, the sodium originating from the sodium nitrate fusion process. If potassium nitrate is used no corresponding potassium platinum bronze is formed. The α-PtO_2 is reducible by hydrogen to metallic platinum, but the $Na_xPt_3O_4$ is only reduced to a minor extent.

A third method for the preparation of dispersed noble metal, particularly platinum, is by reductive precipitation from aqueous solution: in the traditional Willstatter process reduction is commonly effected with formaldehyde, but other reducing agents such as aldoses, hydrazine or sodium borohydride are possible.* In this case too the powdered material dispersed in the aqueous medium is probably an ill-defined mixture of platinum and platinum oxide. In all these cases it is the oxygen-containing powder initially produced which is used as a hydrogenation catalyst: reduction of the catalyst occurs *in situ* in the initial stages of the catalytic process. This is a matter of some importance since, in practice, it results in optimum activity. Attempts at rigorous purification of the surface of the

---

* Reduction of ruthenium salts with borohydride solution has been reported[220] to give a "metal powder" which, when dried, violently explodes when touched with water or with a spatula. A very unstable and poorly characterized "ruthenium hydride" is apparently formed. A safer alternative is to use aqueous solutions of hydrazine for the chemical reduction of ruthenium-containing solutions.

dispersed metal results in sintering with loss of surface area, and more importantly may result in the production of a surface which is subject to deactivation by self-poisoning.

The particle size obtained for noble metal dispersions extends over a wide range; from below colloidal dimensions to the μm range. Sermon[68] examined a range of palladium black sample by electron microscopy and by gas adsorption methods, and found average particle sizes ranging between about 100 nm and 5 nm. In all cases, these basic particles were consolidated into aggregates. These aggregates were dimensionally quite extensive and formed a rather porous macrostructure. Figure 4.20 shows

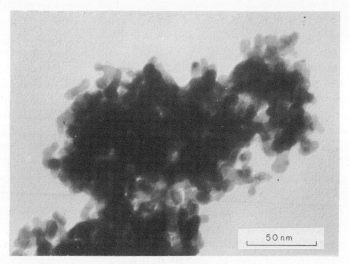

FIG. 4.20 Electron micrograph of palladium black, specific surface area 68 m^{2}g^{-1}, average particle diameter about 8 nm. Reproduced with permission from Sermon, P. A. *J. Catal.* **24**, 467 (1972).

a typical example. A platinum black prepared by Adams' method and hydrogen reduced at 420 K had an average particle diameter of 53 nm, and a ruthenium black similarly prepared gave 22 nm.[118] A number of ruthenium blacks prepared by Adams' or Willstatters' method and hydrogen reduced at 470–670 K all had average particle diameters in the region of 100 nm.[119] Fasman[208] has shown that if the reduction of the platinum oxide is carried out in an aqueous suspension, the presence of alkaline earth chloride in solution leads to a marked reduction in the ultimate platinum particle size. It is possible that non-noble transition metal chlorides have a similar effect, since these are known to augment the activity of platinum black for various hydrogenation reactions[209, 210] but in this case there is also the likelihood that the presence of the hetero-metal on the platinum surface greatly modifies its adsorption and catalytic behaviour.

The platinum metals, silver and gold can be readily obtained as colloidal dispersions. In this form they, particularly platinum and palladium, have a long history as laboratory catalysts. Usually an aqueous medium is used, but this is not necessarily so. The colloidal particles have diameters in the range 10^2–10^3 nm. When present without a stabilizer they form a hydrophobic colloid, and the particles which carry a negative charge are very susceptible to coagulation by the addition of traces of electrolyte, particularly those with highly charged cations. However, without a stabilizer the concentration of colloidal metal that can be produced is $<0\cdot1$ wt. %; for this reason and because of the high sensitivity to impurity, unstabilized metal colloids are seldom used as catalysts. Typical stabilizers are soluble protein (e.g. gelatin), polysaccharides or soluble synthetic polymers (e.g. polyvinyl alcohol). The adsorption of stabilizer onto the surface of the metal particles renders the colloid hydrophilic, stable to substantial concentrations of electrolytes and stable at concentrations that may reach 50 wt. %. In the use of a stabilized colloidal dispersion as a catalyst, the conditions are of course assumed to be such that the stabilizer is not itself decomposed. Furthermore, in practice it should be remembered that the colloidal metal cannot be removed from the reaction mixture by ordinary filtration methods.

Three general preparative methods have been used for metallic colloidal dispersions. Massive metal may be dispersed by striking an arc under water. Alternatively, the reductive precipitation process can be controlled to yield particles of colloidal dimensions. Thirdly, metal hydroxide particles of colloidal dimensions can be precipitated which are subsequently reduced: a stabilizer is often employed here too, and a typical example is the formation of colloidal platinum or palladium hydroxide in the presence of polyvinyl alochol.[214]

Unsupported metal powders pose a difficult problem for surface cleaning. Because the individual particles so readily come into direct contact, they are very susceptible to sintering and particle growth, and this will occur, to some extent at least, at any temperature above the preparation temperature and particularly when the powder is consolidated into a dry compact. Thus even the relatively mild temperature of 370 K used by Nace and Aston[69] for hydrogen reduction and subsequent outgassing of palladium black undoubtedly resulted in some sintering, while it seems very probable that similar treatment at 273 K as used for platinum black[70] and palladium black[68] would not have produced a completely clean surface.

An example of this strong dependence of surface area on sintering temperature is shown in Fig. 4.21 for a reduced platinum black: the surface area is here indicated by the relative rate of hydrogen peroxide decomposition.

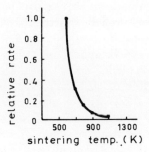

FIG. 4.21 Relative rate of hydrogen peroxide decomposition as a function of the temperature of vacuum sintering of platinum black catalyst. Sintering at each temperature carried out to constant reactivity ($\geqslant$ 2 hours). After Khassan, S. A., Emel'yanova, G. I., Lebedev, V. P. and Kobozev, N. I. *Russ. J. Phys. Chem.* **44**, 821 (1970).

Stabilized Porous Metals

In practice, catalysts of this type are largely confined to the metals iron, cobalt and nickel. Because of their considerable industrial importance they have been described in great detail in the literature (see refs 153–155).

Stabilized porous metal catalysts may be prepared by several alternative methods. Most of these proceed via the metal oxide, and they vary in the means by which the oxide is formed and in the way by which the stabilizer* and chemical promotor are distributed. The stabilizer, whose purpose is to confer porosity on the catalyst by preventing agglomeration of the metal particles, is usually present as a minor component relative to the amount of metal. If the proportion of stabilizer is increased, the catalyst type tends towards a conventional supported catalyst: there is no hard and fast demarcation line.

In one specific case the stabilizer (and usually the chemical promotor) is distributed by additions made to fused oxide. This case is magnetite (Fe_3O_4): the stabilizer is often alumina, magnesia or silica, while the chemical promotor is often potassium oxide (derived from the addition of potassium carbonate). After crushing to a suitable size, the catalyst is hydrogen reduced to give metallic iron. This is the traditional iron

---

* We use the term "stabilizer" to be synonymous with "structural promotor" which is often used in the literature.

synthetic ammonia catalyst, which has also been used as a Fischer-Tropsch catalyst.

The second method of preparation is co-precipitation, and this is applicable to iron, cobalt or nickel. Hydroxides of these metals and of the stabilizer are precipitated simultaneously from aqueous solution, and the catalyst is processed by washing, filtering, drying and hydrogen reduction. If a chemical promotor such as a water-soluble potassium salt is desired, it is added by impregnation of the catalyst before reduction. The stabilized catalyst is itself often distributed over a low or medium area support such as kieselguhr or granular low porosity silica or alumina. This is done to improve reactant accessibility. This support is introduced by slurrying with the solution before or during the co-precipitation stage. As an alternative to co-precipitation, oxides may be generated by thermal decomposition of metal nitrate, but this method is infrequently used.

The third method consists of making a porous compact from metal oxide powder, which is subsequently reduced with hydrogen. In practice this method has only been used to any extent with iron, starting with iron(III) oxide. Again, in order to minimize collapse of the porous structure during and after reduction, the iron oxide is sintered with a stabilizing agent such as alumina or borax which may also act as a binder.

Finally, a porous metal catalyst may be prepared starting with metal powder and sintering, sometimes with the use of an agent such as borax which acts to retain porosity. The metal powders would typically have particle sizes in the μm range, but it should be remembered that they may be pyrophoric in air, which makes handling difficult. Monolithic porous metal catalysts prepared in this way are used as electrocatalysts in fuel cell applications, some aspects of which have recently been summarized.[226] A commonly used hydrogen electrode for alkaline fuel cells consists of porous nickel, perhaps alloyed with other metals such as iron, molybdenum or titanium, and made more active as an electrocatalyst by depositing finely divided nickel, platinum or palladium on the nickel surface by the obvious method of impregnation and hydrogen reduction. In practice, a closely controlled pore structure is required to regulate liquid and gas transport.

We deal first with iron catalysts prepared by the fusion method. The way in which the stabilizer is distributed depends on its chemical nature. The oxides alumina, magnesia and titania certainly dissolve to some extent in magnetite. This is also true for the oxides CaO, Li_2O and Na_2O, but K_2O and BaO are insoluble (these alkali metal and alkaline earth metal oxides being relevant as chemical promotors).[221] However, with alumina and magnesia, if the proportion is >1 mol %, it is likely in practice that solution is incomplete[156] and some of it remains present as a separate phase and this is also borne out by a recent electron microprobe study.[157] Silica (and zirconia?) appear to be insoluble in magnetite, and when used as a stabilizer these materials are found in a layer separating the magnetite grains. However, silica does have the effect of inhibiting the solution of the more basic oxides (by compound formation) in the magnetite, and its presence thus makes more difficult the uniform distribution of alkali metal or alkaline earth metal oxides which may be intended as chemical promotors.[221]

When the magnetite is reduced, dissolved alumina or magnesia is precipitated and collects between the iron particles to augment the stabilizer material which may have been present as an intergranular component in the magnetite. This model has been confirmed electronmicroscopically.[162] However, the fate of the dissolved alumina depends on the reduction conditions.[215-218] The dissolved alumina is apparently present as $FeAl_2O_4$. Under severe reduction conditions (700 K in pure hydrogen at 1 atm pressure for 23 h),[218] this is all converted into iron and alumina, but under less severe conditions (e.g. 670 K in a nitrogen/hydrogen mixture),[215-217] some if not all of the $FeAl_2O_4$ is retained. It also appears likely from adsorption data[219] and from strain broadening of the X-ray diffraction lines of the α-iron, that a proportion of the alumina generated by reduction of the $FeAl_2O_4$ is present as small inclusions within the α-iron grains. In the cases studied (3 wt.% alumina) these inclusions appear to be about 3 nm in size within α-iron grains about 25 nm in size. Their presence should be easily confirmed by transmission electronmicroscopy, but this has not yet been done. The strain created by the alumina inclusions would be expected to shift the α-iron particle size toward smaller values.

The electron microprobe work of Chen and Anderson[157] also emphasizes the inhomogeneous structure of a typical fused magnetite catalyst prior to hydrogen reduction. The main stabilizer was 6·4 wt.% magnesia, and there was also 0·8 wt.% potassium oxide as chemical promotor (expressed in terms of the composition of the reduced catalyst). The major portion of the material consisted of magnetite grains which were homogeneous except for a few inclusions of potassium silicate impurity, and in which some magnesium was presumably dissolved. However, there was also a substantial minor component which consisted of alternate layers of magnetite and another phase which contained about 10 wt.% magnesium possibly as a mixed oxide of iron and magnesium. The inhomogeneous component was generally more readily reduced than the homogeneous one, probably as a result of the smaller size of the magnetite domains, and on this basis the inhomogeneous component also yields more highly dispersed iron particles after reduction. One expects the proportion of the inhomogeneous component to increase as the proportion of stabilizer is increased, with a corresponding increase in the surface area of the reduced catalyst. That this is the case is shown by the data in Table 4.6 for fused iron catalysts containing alumina stabilizer[158] and reduced at 670–770 K.

TABLE 4.6 Total surface areas of reduced iron synthetic ammonia catalysts as a function of alumina content

Alumina (wt. %)	Total surface area after hydrogen reduction at 670–770 K (m^2g^{-1})
0·15	~0·5
1·3	3·3
10·2	8·4

The porosity in this sort of catalyst is developed during the reduction process, so the total surface area and the reduced metal surface area (measured for instance by carbon monoxide chemisorption) progressively increase. This is demonstrated by the data in Table 4.7, due to Hall et al.,[159] for an iron catalyst prepared by fusion and containing 6·4 and 0·8 wt. % of magnesia and potassium oxide respectively (reduced catalyst basis). It will be seen that the average pore size is nearly constant but the total surface area increases linearly with the extent of reduction. The

TABLE 4.7 Surface areas and average pore diameters as a function of extent of reduction* for iron catalyst prepared by fusion

% reduction	Total surface area (m^2g^{-1})†	Carbon monoxide adsorption ($dm^3 kg^{-1}$ at STP)†	Average pore diameter‡ (nm)
0	0	0	0
20	2·1	0·16	34·3
40	4·2	0·29	33·3
60	6·3	0·43	33·0
80	8·4	0·57	33·8
100	10·1	1·00	35·2

* hydrogen reduction at 720 K.
† per g of unreduced catalyst.
‡ estimated from surface area and pore volume.

surface area available for carbon monoxide adsorption does not, however, increase in a like manner, but there is a more rapid increase when the reduction is nearing completion. This is probably due to the presence of chemisorbed oxygen and water vapour which inhibit carbon monoxide adsorption, and which are removed from the iron surface to an increasing extent at later stages of the reduction process.

As the reduction temperature is increased, the surface area of the reduced catalyst is decreased and the average pore size is increased. For instance, if the catalyst specified in Table 4.7 was reduced at 820 K

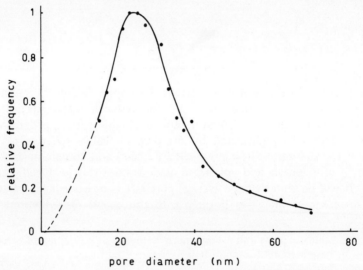

FIG. 4.22 Relative pore size distribution for porous iron catalyst (synthetic ammonia catalyst prepared from fused magnetite), containing 3 wt.% alumina and 1 wt.% potassium oxide, after hydrogen reduction at 720 K. After Zweitering, P. and Koks, H. L. T. *Nature* **173**, 683 (1954).

instead of 720 K, the total surface area at 100% reduction decreased by a factor of about 2.5, while the average pore diameter increased by about the same factor. At the same time, this increase in reduction temperature caused the fraction of the total surface which could chemisorb carbon monoxide to decrease by some 8%, presumably as a result of some agglomeration of iron particles. The pores in a reduced catalyst are, of course, distributed over a range of diameters, and Fig. 4.22 gives an illustration of this for a catalyst containing 3 wt.% alumina and 1 wt.% potassium oxide, after hydrogen reduction at 720 K.[159] The data were obtained from a mercury porosimeter. The average pore diameter was

35·4 nm, that is, quite close to the values recorded in Table 4.7 for a reduced magnesia stabilized catalyst.

The average iron particle size has been estimated by X-ray line broadening[160] for a typical reduced catalyst which had a total specific surface are of 8 m² g⁻¹, and a value of 36 nm was obtained. This sample contained calcium oxide and potassium oxide as chemical promoters, together with alumina. This value is also quite close to that obtained[161] for a catalyst containing about 1 wt.% alumina and silica (present as impurity in the original magnetite), prepared by fusion, crushing and hydrogen reduction at 650 K, namely about 32 nm. Use of the catalyst in a Fischer-Tropsch synthesis reaction resulted in some degree of iron particle growth. If the original magnetite was reduced without prior fusion, the iron particles were extremely large; presumably in the original magnetite the alumina and silica were present as large particles and/or inclusions, and fusion was necessary to distribute them throughout the magnetite.

The concept that the stabilizer and chemical promotor exist between the iron particles implies that these components should occupy a portion of the iron surface. Potassium oxide is extremely potent in this respect, and as the proportion of potassium oxide in a catalyst increased from 0·09 to 1–1·5 wt.%, the proportion of the iron surface covered by potassium oxide increased from 16 to 60–70%. On the other hand, with alumina contents in the range 0·4–10·2 wt.%, the proportion of the iron surface covered by alumina varied in the range 14–55%. There is, however, some interaction between these components in the sense that a higher content of potassium oxide tends to depress the proportion of the surface covered by alumina.[154]

The iron surfaces in a reduced catalyst probably expose (111) planes preferentially (but not exclusively).[154, 163]

We turn now to co-precipitated catalysts. The surface area of this type of catalysis has been compared with the type prepared by magnetite fusion, by Hall *et al.*[164] The surface areas of the precipitated catalysts are high before reduction (>100 m²g⁻¹), but after reduction the catalysts had much the same surface areas and pore structures as the reduced fused catalysts, and it is reasonable to conclude that in the reduced form, their general morphologies are similar.

Stabilized porous nickel and cobalt catalysts are, like iron, readily prepared by co-precipitation. In many of these (intended as Fischer-Tropsch catalysts), thoria or a combination of thoria and magnesia is the stabilizer, and they are formed by hydroxide precipitation along with the heavy metal hydroxide. After drying, hydrogen reduction is typically effected at 620–720 K. A wide variety of formulations have been recorded, and Table 4.8 summarizes a few of the cobalt catalysts listed by Storch

TABLE 4.8 Surface area and average pore diameter data for some reduced porous cobalt catalysts

Catalyst*	Surface area (m^2g^{-1})†	Average pore diameter (nm)§
cobalt/magnesia 100/12	60·4	—
cobalt/thoria 100/6	23·8	—
cobalt/thoria/magnesia 100/6/12	84·1	38
cobalt/thoria/kieselguhr 100/18/100	42·3	69
cobalt/thoria/magnesia/kieselguhr 100/6/12/200	47–99‡	22–80‡
cobalt/kieselguhr 100/200	22·8	406

* hydrogen reduction at 670 K.
† per g of reduced catalyst.
‡ depending on kieselguhr type.
§ estimated from surface area and pore volume.

et al.[153] and by Anderson.[155] The more limited data for nickel are of the same order as for cobalt.[155] In this type of catalyst, reduction is incomplete at 620–670 K. For instance, with cobalt the maximum reduced proportion is typically 60–80%. Unreduced metal presumably remains combined in various substances formed by reaction with the stabilizer or the support. This situation is reminiscent of that found with some conventional supported nickel catalysts. On the other hand, with fused magnetite reduction appears to be essentially complete at 720 K if sufficient time is allowed.

With these porous metal catalysts very many empirical variations on composition and form have been described, and many of these variations influence the catalytic properties in ways that are at present not understood. At the experimental level an empirical approach remains unavoidable.

Skeletal Metals

Raney catalysts are prepared by leaching out with aqueous alkali the aluminium from certain alloys. The original and the best known is Raney nickel,[72–74] but a number of other metals have been prepared in this form including Raney cobalt,[75-77] iron,[78-79] copper,[80] silver (cf. ref. 120), and rhenium.[121] Most work has been done with Raney nickel, and the subsequent discussion will mainly refer to this substance.

The composition of Raney nickel is somewhat variable depending on the preparative conditions. Some residual aluminium and hydrated aluminium oxide (mostly bayerite in the undehydrated catalyst) are always present. Even when strenuous efforts are made (e.g. ref. 81) to minimize these residues they are still appreciable, for instance, metallic aluminium 3·5 wt. % and aluminium oxide 0·03 wt. %.[81] With more conventional preparative conditions these residues are at considerably greater levels, and Kokes and Emmett[82] quote analytical data for a Davison commercial preparation giving 8 wt. % metallic aluminium and 1 wt. % aluminium oxide, and for a specimen of very high catalytic activity prepared according to methods given in the literature,[74, 83] metallic aluminium 3 wt. %, aluminium oxide 20 wt. %.

The fact that Raney nickel is not pure nickel is not a demerit from the catalytic point of view; indeed some of the desirable features of these catalysts have their origin here. The aluminium oxide has a similar function to the stabilizer in the stabilized metal catalysts which were discussed in the previous section, in that it has a dramatic inhibiting effect on the ease of thermal sintering, and Raney catalysts have much greater thermal stability with respect to particle and surface area integrity than do pure nickel catalysts such as may be prepared, for instance, by the reduction of nickel oxide powder with hydrogen. Thus, Kokes and Emmett[82] have reported that their catalyst which contained 20 wt. % aluminium oxide only suffered a 20 % reduction in surface area on heating to 770 K, while their Davison catalyst which contained about 1 wt. % aluminium oxide was without appreciable area reduction up to 520 K and still only underwent an area reduction by a factor of about two on being heated to 750 K.

The alloy from which Raney nickel is prepared usually contains 40–50 mol % nickel, and the main phases present are $NiAl_3$ and Ni_2Al_3 in roughly comparable proportions, together with a lesser proportion (2–25 mol %) of a eutectic. Freel et al.[84] have studied how the morphology of the catalyst develops during alkali attack on the alloy. For any of the phases present, the attack advances as a front yielding a sharp gradient between alloy and catalyst. The eutectic phase does not yield catalyst but appears to disintegrate to generate voids in the final structure. The Ni_2Al_3 phase is the least reactive although this difference is only apparent when using concentrated alkali solution.

A detailed study of the microstructure of Raney nickel has been made by Anderson and his collaborators[84, 85, 87, 88] and by Fouilloux et al.,[81] using a variety of techniques including electron microscopy, X-ray diffraction and gas adsorption methods. Studies with a scanning electron microscope showed that the bayerite crystallites occur over most of the nickel surface, and this is clearly the reason for their inhibiting action on thermal sintering. The fraction of the total nickel surface not covered by bayerite varies in the range 85–55 % for conventional preparations, and the amount of bayerite is greater the more dilute the alkali solution used for the leaching process. The amount of residual bayerite can also be reduced by continued extraction with fresh alkali solution. Nevertheless, the change in the available nickel surface does not correlate well with catalytic activity, and, in practice, there is little point in trying to minimize the amount of residual bayerite. The main nickel particles are fairly large (>100 nm), but these are made up from smaller nickel crystallites which are individually 2·5–15 nm in size, and which are loosely packed without any preferred orientation, to give a porous structure. The total surface areas vary somewhat with the conditions of preparation. Low temperature ($\leqslant 320$ K) alkali leaching favours the retention of a high area, typically 80–100 m^2g^{-1}, and also the retention of smaller pores. Average pore diameters for various preparations are in the range 2·6–12·8 nm, and there is some evidence for a bimodel distribution of pore size.[88] The bayerite tends to function to block some of the pores.

When Raney nickel is heated, considerable hydrogen is evolved, and evolutions in the range 50–100 STP dm^3kg^{-1} are common on heating to 770 K. This is considerably more than could be accounted for by adsorption and normal solution processes ($\leqslant 10$ STP dm^3 kg^{-1}). There has in the past been a good deal of speculation (cf. ref. 82) that this apparently abnormal amount of hydrogen is present substitutionally in the nickel catalyst and that this unique structure accounts for the high catalytic activity of the material. It now seems quite certain that this suggestion is not correct; rather the extra hydrogen which is evolved on heating originates from a chemical reaction between residual metallic aluminium and water bound in the bayerite.[86]

Pearce and Lewis[132] used X-ray diffraction line broadening to determine microstrains and stacking faults in Raney nickel and Raney copper catalysts. In general, these Raney catalysts were appreciably defected, but the defect concentration decreased with increasing time and temperature of digestion in the preparative process. Furthermore, ageing of Raney nickel under ethanol resulted in a decreased defect concentration, and it is also known that ageing under similar conditions for extended periods (about six months) resulted in a decreased surface area due to grain growth.[133]

When dry, Raney nickel is pyrophoric and is preserved out of contact with oxygen often under water, dilute aqueous alkali or ethanol.

3. Dispersed Multimetallic Catalysts

At the present time only very sparse information is available about the structure of supported multimetallic catalysts, and information about the nature of the metallic particles is particularly rudimentary and difficult to obtain.

If one of the component metals is ferromagnetic in the pure massive state while the other is not, the magnetization is strongly dependent on composition and is easily measured. One can readily compare the values observed for the dispersed catalyst with those known for massive alloy, so

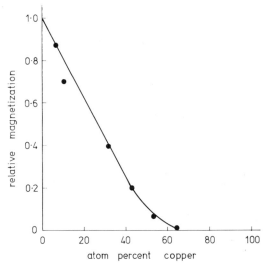

FIG. 4.23 Relative magnetization of nickel–copper alloys as a function of composition. The points are values for dispersed unsupported catalysts obtained at 77 K: the curve is derived from data at 0 K due to Ahern, Martin and Sucksmith (ref. 175) for bulk alloys. Reproduced with permission from Sinfelt, J. H., Carter, J. L. and Yates, D. J. C. *J. Catal.* **24**, 283 (1972).

that compositions can also be compared. This assumes that the particles are not sufficiently small for collective paramagnetic behaviour to occur (cf. pp. 370–375), i.e. for the magnetization to become particle size dependent. An example of this approach is shown in Fig. 4.23 which gives magnetization *vs* composition data for some dispersed unsupported nickel–copper catalysts,[174] compared with data for bulk alloys prepared from molten metals.[175] The latter data give the ratio of the magnetization of an alloy to that of pure nickel at 0 K. The data on the dispersed catalysts were derived from magnetization values at 77 K determined at field strengths of $1 \cdot 2 \times 10^5$–$5 \cdot 2 \times 10^5$ A m^{-1} (about 1500–6500 Oe). Under these conditions the magnetization was virtually independent of field strength. The comparison indicates that the dispersed particles consist of alloy of the nominal composition. Some comments on methods of magnetic measurement are to be found in Chapter 6.

X-ray diffraction data can also give information about particle composition. The method depends on the variation in lattice spacing and thus reflection angle with composition. Within a single phase composition range, it is not uncommon for the lattice parameter to change by $0 \cdot 02$–$0 \cdot 05\%$ for each 1 atom % change in composition: Vegard's linear relationship between lattice parameter and composition is often, but not always obeyed. For comparison, it may be noted that under good conditions, the Debye-Scherrer method can fairly readily yield a precision in the estimation of lattice parameter of about $0 \cdot 01\%$, and with care this can be improved by up to a factor of four. Pearson has provided a useful summary.[176] However, the precision of the method is severely affected by particle size effects which need to be minimal if acceptable accuracy in composition estimation is to be achieved. Again, a comparison may be made between data for the dispersed catalyst and data for bulk alloys of known composition. An example is provided by Sinfelt et al. for dispersed unsupported nickel–copper catalysts.[174]

Gray et al. prepared a range of catalysts by impregnation of η-alumina with chlorides of the following mixtures ($0 \cdot 01$–$0 \cdot 5$ wt. % metal):[60] platinum–rhodium, palladium–rhodium, platinum–iridium, platinum–ruthenium, platinum–osmium. Although alloy formation was assumed after reduction at 720–770 K, no detailed structural information was given. Sokolskii and his collaborators prepared a range of bimetallic supported catalysts using group VIII metals and mainly using a high area alumina support (350 m^{2}g^{-1}), but also occasionally using silica gel or barium sulphate supports: the systems include palladium–ruthenium,[203-207] platinum–ruthenium, platinum–rhodium, palladium–rhodium, platinum–palladium.[206, 207] The total metal loading was $0 \cdot 5$–1 wt. %. Details of the structure of the dispersed metal particles were not obtained, although there is some evidence[206, 207] from hydrogen adsorption measurement that the surface area of the dispersed metal does not necessarily vary monotonically with composition at a fixed total metal loading.

Bond and Webster[188] prepared a number of bimetallic powder catalysts in the form of "blacks" by the use of Adam's method. Among the systems studied were platinum–ruthenium, platinum–iridium, platinum–rhodium, platinum–palladium, platinum–iron, platinum–cobalt, platinum–nickel, palladium–ruthenium, palladium–rhodium, palladium–cobalt, palladium–nickel. Unequivocal evidence for particle composition and structure was not obtained in most cases. However, in some cases (e.g. platinum–ruthenium) electron diffraction or X-ray diffraction data suggest complete solid solution formation between the two metals, while in other cases (e.g. palladium–ruthenium, platinum–nickel) there was evidence for incomplete component mixing.

A wide range of Raney nickel catalysts have been prepared by Fasman and Sokolskii and their collaborators, containing one and sometimes two extra transition metal components. Among the additional metals studied are molybdenum,[190,191] iron,[190,192] manganese,[193] chromium,[194] ruthenium,[195] copper,[192] rhodium,[196] platinum,[197] rhenium,[198] vanadium,[199] zirconium,[200] tantalum,[200] niobium,[200] palladium;[201] titanium and manganese, titanium and vanadium, titanium and molybdenum, vanadium and molybdenum.[202] The catalysts were prepared by the usual Raney method, that is the production of an aluminium alloy (about 50 wt. % aluminium) followed by alkali leaching. The structure of the resulting Raney catalyst has only been elucidated in any detail in a few cases. In most cases alloying with an extra metal results in a reduction in grain size. The nature of the metallic phases which are present after leaching is much dependent on the solubility relations between the metals and the tendency for compound formation to occur. However, when the added metal was present in low enough concentration, cases were shown where, after leaching, most if not all of the added metal was present in solid solution in the nickel (e.g. < 3 wt. % rhenium; < 7 wt. % palladium; < 10 wt. % platinum; < 7 wt. % rhodium; in each case the composition being expressed in terms of the amount of added metal and nickel). When compound formation occurs the behaviour becomes quite complicated and is poorly understood. If the added metal enters a compound with aluminium, subsequent leaching with alkali may result in the generation of a dispersion of particles of the added metal. The leached product may also contain particles of compounds formed between nickel and the added metal, either generated initially or by the removal of aluminium from a ternary compound.

Catalysts containing both nickel and copper on silica–alumina have been studied by X-ray methods by Swift et al.,[61] who showed that provided the molar ratio Ni/Cu was in the range 1/1–2/1, the metallic component was inhomogeneous.

A number of platinum-containing systems have been studied. Platinum–rhenium catalysts on alumina have assumed considerable importance.[62–64] Both metals are present in roughly comparable amounts with the total metal content in the vicinity of 0·5–2·0 wt. %. After calcining and hydrogen reduction at about 620–970 K, X-ray and electron diffraction studies indicated that a platinum–rhenium alloy was present,[62] and no separate rhenium phase could be detected provided the proportion of rhenium was less than about 40 mol % of the metallic constituent, in agreement with the bulk phase diagram. Anderson et al.[65] have studied the structure of silica-supported platinum–copper catalysts (0·55 and 5 wt. % metal on Davison 950 and 62 silica gels respectively, prepared by impregnation and hydrogen reduction at 750 K). X-ray diffraction, electron microscopy and hydrogen adsorption measurements were made. In general, the alloy catalysts had about the same particle size as did catalysts with pure platinum prepared under similar conditions. It was concluded that the structure of the metallic particles depended on their diameter. For diameters increasing above 3·0 nm, there was an increasing enrichment of the surface with copper, reaching an amount equivalent to about 2 monolayers of copper for 8·0 nm diameter particles. Below 3·0 nm diameter, no evidence for surface enrichment was encountered. Except for this tendency for copper segregation at the surface, the alloy was present as a solid solution. Bartholomew and Boudart[131] have recently used Mössbauer spectroscopy to examine the state of platinum–iron catalysts supported on Graphon carbon (87 m^{2}g^{-1}). The catalysts were prepared by impregnation using a solution of chloroplatinic acid and ferric nitrate in a solvent consisting of a 4 : 1 mixture of benzene and ethanol, followed by drying at 370–420 K and hydrogen reduction at 670–770 K. Platinum–iron particles, free of platinum and iron phases, and having average diameters in the range 1·5–3·0 nm, were obtained. The surface composition for freshly reduced catalysts which had then been exposed to oxygen at 298 K depended on the bulk composition. For bulk compositions in the range 25–50 mol % iron, the surface composition was the same as the bulk; for a bulk composition 7·9 mol % iron, the surface composition was 13·5 mol % iron.

With silica-supported rhodium–silver catalysts (5 wt. % metal on Davison 62 silica gel, prepared by impregnation),[66] the phase diagram suggests very little solid solubility of the metals, and the catalyst does in fact appear to consist of discret and separate crystallites of rhodium-rich and silver-rich phases distributed over the support. As would be expected from the known behaviour of the individual metals, the average diameter of the rhodium-rich particles (5·0–7·0 nm) was much less than that of the silver-rich (40–70 nm). There was evidence for a layer of silver, perhaps up to a few tenths of a nanometre thick on the surface of each rhodium crystallite.

Unsupported powders of palladium–platinum, palladium–rhodium and platinum–ruthenium have been described by McKee and Norton.[134,135] They were prepared by sodium borohydride reduction of solutions of the chloride mixtures, followed by drying of the precipitated metal powder at 370–390 K and hydrogen reduction at 570 K. The alloy particle diameters lay in the range 10–50 nm. X-ray examination showed the palladium–platinum and palladium–rhodium catalysts to be homogeneous and in solid solution over the entire composition range studied (approximately 10/90 mol % to 90/10 mol %). However, in the case of platinum–ruthenium, solid solutions only existed up to about 50 mol % ruthenium: above this, there was some free ruthenium present. This is in approximate agreement with the bulk phase diagram.

Alloy powders have also been prepared via the mixed oxides by hydrogen reduction, and the general preparative techniques resemble those used for single metals (see, for instance, refs 136–140). A recent examination of nickel–copper alloy powder formed in this way has produced evidence that the equilibrium structure can be reached, corresponding to that expected from the thermodynamic data of Fig. 3.33 and the particle structure represented in Fig. 3.34 for equilibrated nickel–copper alloy films.[17] This was achieved by (a) initially reducing the mixed oxides at the minimum possible temperature (373–410 K) so as to maximize the production of defects which would promote diffusion and phase separation; (b) extended subsequent thermal treatment—12 h hydrogen reduction at 620 K followed by vacuum annealing for 192 h at 620 K, and finally slow cooling to room temperature over a 48 h period. Hydrogen adsorption was used to monitor the particle structure, and Fig. 4.24 compares the observed data with what is expected for equilibrated particles. The full line represents the behaviour expected if phase equilibrium is reached and the particles have the structure of Fig. 3.34, while

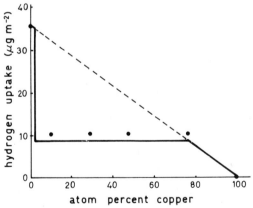

FIG. 4.24 Hydrogen adsorption on nickel–copper alloy powder at 77 K. The uptakes were measured for a hydrogen pressure of about 0·27 Pa (about 2×10^{-3} Torr). Reproduced with permission from Cadenhead, D. A. and Wagner, N. J. *J. Catal.* **27**, 475 (1972).

the broken line is what would be expected for a continuous series of solid solutions: these expectations also include the assumption that the extent of hydrogen adsorption is proportional to the amount of nickel in the surface.

Nevertheless, at the moment there is too little known about the behaviour of dispersed multimetallic systems, particularly the influence of particle size on the ease of equilibration, to enable predictive statements to be made with much confidence. Such general comments as can be currently offered are discussed in Chapter 5. For the moment it suffices to add that surface enrichment with the component of lowest surface energy is always likely, although even this behaviour will be extensively modified by the presence of a chemisorbable gas which reacts preferentially with one of the component and so leads to surface enrichment with that particular component; this situation is thus quite analogous to that discussed in more detail in Chapter 3 for massive metal catalysts.

References

1. Innes, W. B. *In* "Catalysis" (P. H. Emmett, ed.), Vol. 1, Reinhold, New York (1954) p. 245.
2. Gil'debrand, E. I. *Internat. Chem. Eng.* **6**, 449 (1966).
3. Moss, R. L. *Platinum Metals Rev.* **11**, 1 (1967).
4. Dorling, T. A., Eastlake, M. J. and Moss, R. L. *J. Catal.* **14**, 23 (1969).
5. Cormack, D. and Moss, R. L. *J. Catal.* **13**, 1 (1969).
6. Dorling, T. A. and Moss, R. L. *J. Catal.* **7**, 378 (1967).
7. Dorling, T. A., Lynch, B. W. J. and Moss, R. L. *J. Catal.* **20**, 190 (1971).
8. van Hardeveld, R. and van Montfoort, A. *Surface Sci.* **4**, 396 (1966).
9. Benesi, H. A., Curtis, R. M. and Studer, H. P. *J. Catal.* **10**, 328 (1968).
10. Adams, C. R., Benesi, H. A., Curtis, R. M. and Meisenheimer, R. G. *J. Catal.* **1**, 336 (1962).
11. Poltorak, O. M. and Boronin, V. S. *Russ. J. Phys. Chem.* **40**, 1436 (1966).
12. Schuit, G. C. A. and van Reijen, L. L. *Advances in Catalysis* **10**, 242 (1958).
13. Coenen, J. W. E. and Linsen, B. C. *In* "Physical and Chemical Aspects of Adsorbents and Catalysts" (B. G. Linsen, ed.), Academic Press, London (1970), p. 471.
14. Morikawa, K., Shirasaki, T. and Okada, M. *Advances in Catalysis* **20**, 98 (1969).
15. Reinen, D. and Selwood, P. W. *J. Catal.* **2**, 109 (1963).
16. Webb, A. N. *Ind. Eng. Chem.* **49**, 261 (1957).
17. Sinfelt, J. H. and Yates, D. J. C. *J. Catal.*, **8**, 82 (1967).
18. Sinfelt, J. H. and Yates, D. J. C. *J. Catal.*, **10**, 362 (1968).
19. Yates, D. J. C. and Sinfelt, J. H. *J. Catal.* **8**, 348 (1967).
20. Yates, D. J. C. and Sinfelt, J. H. *J. Catal.* **14**, 182 (1969).
21. Aben, P. C. *J. Catal.* **10**, 224 (1968).

22. Avery, N. R. and Sanders, J. V. *J. Catal.* **18**, 129 (1970).
23. Wilson, G. R. and Hall, W. K. *J. Catal.* **17**, 190 (1970).
24. Spenadel, L. and Boudart, M. *J. Phys. Chem.* **64**, 204 (1960).
25. Adler, S. F. and Keavney, J. J. *J. Phys. Chem.* **64**, 208 (1960).
26. Gruber, H. L. *J. Phys. Chem.* **66**, 48 (1962).
27. Cusumano, J. A., Dembinski, G. W. and Sinfelt, J. H. *J. Catal.* **5**, 471 (1966).
28. Swift, H. E., Lutinski, F. E. and Tobin, H. H. *J. Catal.* **5**, 285 (1966).
29. Scholten, J. J. F. and van Montfoort, A. *J. Catal.* **1**, 85 (1962).
30. Hill, H. F. and Selwood, P. W. *J. Amer. Chem. Soc.* **71**, 2522 (1949).
31. Shephard, F. E. *J. Catal.* **14**, 148 (1969).
32. Holm, V. C. F. and Clark, A. *J. Catal.* **11**, 305 (1968).
33. Brooks, C. S. and Christopher, G. L. M. *J. Catal.* **10**, 211 (1968).
34. Figueras, F., Mencier, B., de Mourgeses, L., Naccache, C. and Trambonze, Y. *J. Catal.* **19**, 315 (1970).
35. Cusumano, J. A., Dembinski, G. W. and Sinfelt, J. H. *J. Catal.* **5**, 471 (1966).
36. Taylor, W. F., Sinfelt, J. H. and Yates, D. J. C. *J. Phys. Chem.* **69**, 3857 (1965).
37. Yates, D. J. C., Sinfelt, J. H. and Taylor, W. F. *Trans. Faraday Soc.* **61**, 2044 (1965).
38. Carter, J. L., Cusumano, J. A. and Sinfelt, J. H. *J. Phys. Chem.* **70**, 2257 (1966).
39. Richardson, J. T. *J. Catal.* **21**, 122 (1971).
40. Firth, J. G. and Holland, H. B. *Trans. Faraday Soc.* **65**, 1891 (1969).
41. Lewis, P. H. *J. Catal.* **11**, 162 (1968).
42. Rabo, J. A., Schomaker, V. and Pickert, P. E. *In* "Proceedings 3rd International Congress on Catalysis" (W. M. H. Sachtler, G. C. A. Schuit and P. Zwietering, eds), North-Holland, Amsterdam (1965), p. 1264.
43. Avery, N. R. *J. Catal.* **19**, 15 (1970).
44. McHenry, K. W., Bertolacini, R. J., Brennan, H. M., Wilson, J. L. and Seelig, H. S. *In* "Actes du Deuxieme Congres de Catalyse", Editions Technip Paris (1961), p. 2295.
45. Pope, C. G. and Kemball, C. *Trans. Faraday Soc.* **65**, 619 (1969).
46. Yates, D. J. C. *J. Phys. Chem.* **69**, 1676 (1965).
47. Nesterov, O. V. and Evdokimou, V. B. *Zhur. Fiz. Khim.* **35**, 376 (1961).
48. Brownlie, I. C., Fryer, J. R. and Webb, G. *J. Catal.* **14**, 263 (1969).
49. Trimm, D. L. and Cooper, B. J. *Chem. Comm.* **1970**, 477.
50. Ciapetta, F. G. and Plank, C. J. *In* "Catalysis" (P. H. Emmett, ed.), Vol. 1, Reinhold, New York (1954), p. 315.
51. "Single Crystal Films" (M. H. Francombe and H. Sato, eds), Pergamon Press, London (1964).
52. Macdonald, R. J. Ph.D. Thesis, Flinders University, Adelaide, Australia (1970).
53. Anderson, J. R. and Macdonald, R. J. *J. Catal.* **19**, 227 (1970).

54. Anderson, J. R., Macdonald, R. J. and Shimoyama, Y. *J. Catal.* **20**, 147 (1971).
55. Anderson, J. R. and Shimoyama, Y. Unpublished results, Flinders University, (1972).
56. Rossington, D. R. *In* "Chemisorption and Reactions on Metallic Films" (J. R. Anderson, ed.), Vol. 2, Academic Press, London (1971), p. 211.
57. Moss, R. L., Thomas, D. H. and Whalley, L. *Thin Solid Films*, **5**, R19 (1970).
58. Moss, R. L., Gibbens, H. R. and Thomas, D. H. *J. Catal.* **16**, 117 (1970).
59. Sachtler, W. M. H. and Jongepier, R. *J. Catal.* **4**, 665 (1965).
60. Gray, T. J., Masse, N. G. and Oswin, H. G. *In* "Actes du Deuxieme Congres de Catalyse", Editions Technip, Paris (1961), p. 1697.
61. Swift, H. E., Lutinski, F. E. and Kehl, W. L. *J. Phys. Chem.*, **69**, 3268 (1965).
62. Kluksdahl, H. E. *U.S. Patent* No. 3415737, (1968).
63. Jacobson, R. L., Kluksdahl, H. E. and Spurlock, B. *U.S. Patent* No. 3434960, (1969).
64. Edeleanu, A. G., Blue, E. M. and McCoy, C. S. *Erdol u. Kohle* **23**, 17 (1970).
65. Anderson, J. H., Conn, P. J. and Brandenberger, S. G. *J. Catal.*, **16**, 326 (1970).
66. Anderson, J. H., Conn, P. J. and Brandenberger, S. G. *J. Catal.*, **16**, 404 (1970).
67. Williams, A., Butler, G. A. and Hammonds, J. *J. Catal.* **24**, 352 (1972).
68. Sermon, P. A. *J. Catal.* **24**, 467 (1972).
69. Nace, D. M. and Aston, J. G. *J. Amer. Chem. Soc.* **79**, 3619, 3623, 3627 (1957).
70. McKee, D. W. *J. Phys. Chem.* **67**, 841 (1963).
71. Pope, D., Smith, W. L., Eastlake, M. J. and Moss, R. L. *J. Catal.* **22**, 72 (1971).
72. Raney, M. *U.S. Patents* 1563787, (1925); 1628191, (1927); 1915473, (1933).
73. Adkins, H. "Reactions of Hydrogen", Wisconsin University Press, Madison, Wis., 1937.
74. Adkins, H. and Billica, H. R. *J. Amer. Chem. Soc.* **70**, 695 (1948).
75. Dupont, G. and Piganiol, P. *Bull. Soc. Chim. France* **6**, 322 (1939).
76. Faucounau, L. *Bull. Soc. Chim. France* **4**, 63 (1937).
77. Fischer, F. *Ber.* **67**, 253 (1934).
78. Paul, R. and Hilly, G. *Bull. Soc. Chim. France* **6**, 218 (1939).
79. Paul, R. and Hilly, G. *Comptes Rendus* **206**, 608 (1938).
80. Faucounau, L. *Bull. Soc. Chim. France* **4**, 58 (1937).
81. Fouilloux, P., Martin, G. A., Renonprez, A. J., Moraweck, B., Imelik, B. and Prettre, M. *J. Catal.* **25**, 212 (1972).
82. Kokes, R. J. and Emmett, P. H. *J. Amer. Chem. Soc.* **81**, 5032 (1959).
83. Smith, H. A., Chadwell, H. J. and Kirdis, S. S. *J. Phys. Chem.* **59**, 820 (1955).
84. Freel, J., Pieters, W. J. M. and Anderson, R. B. *J. Catal.* **16**, 281 (1970).

85. Freel, J., Robertson, S. D. and Anderson, R. B. *J. Catal.* **18**, 243 (1970).
86. Mars, P., Scholten, J. J. F. and Zwietering, P. *In* "Actes du Deuxieme Congres de Catalyse", Editions Technip, Paris (1961), p. 1245.
87. Robertson, S. D., Freel, J. and Anderson, R. B. *J. Catal.* **24**, 130 (1972).
88. Freel, J., Pieters, W. J. M., and Anderson, R. B. *J. Catal.* **14**, 247 (1969).
89. Dalla Betta, R. A. and Boudart, M. *In* "Proceedings 5th International Congress on Catalysis" (J. W. Hightower, ed.), North-Holland, Amsterdam, (1973), p. 1329.
90. Kubo, T., Arai, H., Tominaga, H. and Kunugi, T. *Bull. Soc. Chem. Japan* **45**, 607 (1972).
91. Yenanzi, L. M. *In* "Platinum Group Metals and Compounds", *Advances in Chemistry Series* **98**, 66 (1971).
92. Poltorak, O. M. and Boronin, V. S. *Russ. J. Phys. Chem.* **39**, 1476 (1965).
93. Hoekstra, H. R., Siegel, S. and Gallagher, F. X. *In* "Platinum Group Metals and Compounds", *Advances in Chemistry Series* **98**, 39 (1971).
94. Poltorak, O. M. and Boronin, V. S. *Russ. J. Phys. Chem.* **37**, 1174 (1963).
95. Pozpelova, A. P., Kobozev, N. I., and Eremin, E. N. *Russ. J. Phys. Chem.* **35**, 143 (1961).
96. Mal'tsev, A. N., Kobozev, N. I., Agronomev, A. E. and Voronova, L. V. *Russ. J. Phys. Chem.* **37**, 322 (1963).
97. Sancier, K. M. *J. Catal.* **20**, 106 (1971).
98. Kluksdahl, H. E. and Houston, R. J. *J. Phys. Chem.* **65**, 1469 (1961).
99. Cha, D. Y. and Parravano, G. *J. Catal.* **18**, 200 (1970).
100. Matsumoto, H., Saito, Y. and Yoneda, Y. *J. Catal.* **22**, 182 (1971).
101. Kikuchi, E., Tsurumi, M. and Morita, Y. *J. Catal.* **22**, 226 (1971).
102. Brooks, C. S. *J. Coll. Interface Sci.* **34**, 419 (1970).
103. Vesely, V. and Pekarek, V. *Tantala* **19**, 219 (1972).
104. James, R. O., Ph.D. Thesis, University of Melbourne, Australia (1971).
105. Pott, G. T. and McNicol, B. D. *Discussions Faraday Soc.* **52**, 121 (1971).
106. Ermakova, S. I. and Zaidman, N. M. *Kinetics and Catal.* **10**, 1158 (1969).
107. Zaidman, N. M., Dzis'ko, V. A., Karnaukhov, A. P., Kefeli, L. M., Krasilenko, N. P., Koroleva, N. G. and Ratner, I. D. *Kinetics and Catal.* **10**, 313 (1969). Zaidman, N. M., Dzis'ko, V. A., Karnaukhov, A. P., Krasilenko, N. P. and Koroleva, N. G. *Kinetics and Catal.* **10**, 534 (1969).
108. Cornet, D and Burwell, R. L. *J. Amer. Chem. Soc.* **90**, 2489 (1968).
109. Burnell, R. L., Pearson, R. G., Haller, G. L., Tjok, P. B. and Chock, S. P. *Inorg. Chem.* **4**, 1123 (1965).
110. Selwood, P. W., Ellis, M. and Wethington, K. *J. Amer. Chem. Soc.* **71**, 2181 (1949).
111. Roth, J. F. and Reichard, T. E. *J. Res. Inst. Catal. Hokkaido Univ.* **20**, 85 (1972).
112. Contour, J. P. and Pannetier, G. *Bull. Soc. Chim. France* **1968**, 3591.
113. Jutasi, E., Beyer, H. and Czaran, E. *Acta Chim. Acad. Sci. Hung.* **58**, 427 (1968).
114. Ravi, A. and Sheppard, N. *J. Catal.* **22**, 389 (1971).

115. Guerra, C. R. and Schulman, J. H. *Surface Sci.* **7**, 229 (1967).
116. Schachter, K. and Tetenyi, P. *Acta Chim. Hung.* **46**, 229 (1965).
117. Broadbent, H. S., Campbell, G. C., Bartley, W. I. and Johnson, J. H. *J. Org. Chem.* **24**, 1847 (1959).
118. McKee, D. W. *J. Catal.* **8**, 240 (1967).
119. Kobayashi, M. and Shirasaki, T. *J. Catal.* **28**, 289 (1973).
120. Scholten, J. J. F., Konvalinka, J. A. and Beekman, F. W. *J. Catal.* **28**, 209 (1973).
121. Balandin, A. A., Karpeiskaya, E. I. and Polkovnikov, B. D. *Doklady Akad. Nauk. SSSR* **139**, 1101 (1961).
122. Khassan, S. A., Emel'yanova, G. I., Lebedev, V. P. and Kobozev, N. I. *Russ. J. Phys. Chem.* **44**, 821 (1970).
123. Maatman, R. W. and Prater, C. D. *Ind. Eng. Chem.* **49**, 253 (1957).
124. Maatman, R. W. *Ind. Eng. Chem.* **51**, 913 (1959).
125. Harriott, P. *J. Catal.* **14**, 43 (1969).
126. Weisz, P. B. *Trans. Faraday Soc.* **63**, 1801 (1967).
127. Weisz, P. B. and Hicks, J. S. *Trans. Faraday Soc.* **63**, 1807 (1967).
128. Weisz, P. B. and Zollinger, H. *Trans. Faraday Soc.* **63**, 1815 (1967).
129. Scholten, J. J. F. and Konvalinka, J. A. *Trans. Faraday Soc.* **65**, 2465 (1969).
130. Anderson, J. H. *J. Catal.* **26**, 277 (1972).
131. Bartholomew, C. H. and Boudart, M. *J. Catal.* **25**, 173 (1972).
132. Pearce, C. E. and Lewis, D. *J. Catal.* **26**, 318 (1972).
133. Schwab, G. M. and Markenthal, H. *In* "Proceedings 2nd International Congress on Surface Activity" (J. H. Schulman, ed.), Butterworths, London (1957), p. 64.
134. McKee, D. W. and Norton, F. J. *J. Catal.* **3**, 252 (1964).
135. McKee, D. W. and Norton, F. J. *J. Phys. Chem.* **68**, 481 (1964).
136. Best, R. J. and Russell, W. W. *J. Amer. Chem. Soc.* **76**, 838 (1954).
137. Hall, W. K. and Emmett, P. H. *J. Phys. Chem.* **62**, 816 (1958).
138. Emmett, P. H. and Skau, N. *J. Amer. Chem. Soc.* **65**, 1029 (1943).
139. Long, J. H., Frazer, J. C. W. and Oh, E. *J. Amer. Chem. Soc.* **56**, 1101 (1934).
140. Cadenhead, D. A. and Masse, N. G. *J. Phys. Chem.* **70**, 3559 (1966).
141. Nikolajenko, V., Bosacek, V. and Danes, V. *J. Catal.* **2**, 127 (1963).
142. Anderson, J. H. *J. Catal.* **28**, 76 (1973).
143. Dugger, D. L., Stanton, J. H., Irby, B. N., McConnell, B. L., Cummings, W. W. and Maatman, R. W. *J. Phys. Chem.* **68**, 757 (1964).
144. Ashley, J. H. and Mitchell, P. C. H. *J. Chem. Soc.* (*A*) **1969**, 2730.
145. Tomlinson, J. R., Keeling, R. O., Rymer, G. T. and Bridges, J. M. *In* "Actes du Deuxième Congres de Catalyse", Editions Technip, Paris (1961), p. 1831.
146. Sacconi, L. *Discussions Faraday Soc.* **7**, 173 (1949).
147. Rymer, G. T., Bridges, J. M. and Tomlinson, J. R. *J. Phys. Chem.* **65**, 2152 (1961).
148. Leith, I. R. and Leach, H. F. *Proc. Roy. Soc.* **A330**, 247 (1972).

149. Mikheikin, I. D., Shvets, V. A. and Kazanskii, V. B. *Kinetics and Catal.* **11**, 609 (1972).
150. Gallezot, P., Ben Taarit, Y. and Imelik, B. *Comptes Rendus* **C272**, 261 (1971).
151. Polak, R. and Cerny, V. *J. Phys. Chem. Solids* **29**, 945 (1968).
152. Klier, K. and Ralek, M. *J. Phys. Chem. Solids* **29**, 951 (1968).
153. Storch, H. H., Golumbic, N. and Anderson, R. B. "The Fischer-Tropsch and Related Syntheses", Wiley, New York (1951).
154. Bokhoven, C., van Heerden, C., Westrik, R. and Zwietering, P. *In* "Catalysis" (P. H. Emmett, ed.), Vol. 3, Reinhold, New York (1955), p. 265.
155. Anderson, R. B. *In* "Catalysis" (P. H. Emmett, ed.), Vol. 4, Reinhold, New York (1956), p. 29.
156. Maxwell, L. R., Smart, J. S. and Brunauer, S. *J. Chem. Phys.* **19**, 303 (1951).
157. Chen, H.-C. and Anderson, R. B. *J. Coll. and Interface Sci.* **38**, 535 (1972).
158. Emmett, P. H. and Brunauer, S. *J. Amer. Chem. Soc.* **59**, 310, 1553 (1937).
159. Zwietering, P. and Koks, H. L. T. *Nature* **173**, 683 (1954).
160. Nielsen, A. and Bohlbro, H. *J. Amer. Chem. Soc.* **74**, 963 (1952).
161. Herbstein, F. H. and Smuts, J. *J. Catal.* **2**, 69 (1963).
162. Schafer, K. *Z. Electrochem.* **64**, 1190, 1194 (1960).
163. Westrik, R. and Zwietering, P. *Proc. Konin. Nederlandse Akad.* **B56**, 492 (1953).
164. Hall, K. W., Tarn, W. H. and Anderson, R. B. *J. Amer. Chem. Soc.* **72**, 5436 (1950).
165. Garner, W. E. *J. Chem. Soc.* **1947**, 1239.
166. Anderson, J. S. *Discussions Faraday Soc.* **4**, 163 (1948).
167. Pease, R. N. and Taylor, H. S. *J. Amer. Chem. Soc.* **43**, 2179 (1921).
168. Delmon, B. *Bull. Soc. Chim. France* **1961**, 590.
169. Bandrowski, J., Bickling, C. R., Yang, K. H. and Hougen, O. A. *Chem. Eng. Sci.* **17**, 379 (1962).
170. Charcosset, H., Frety, R., Trambonze, Y. and Prettre, M. *In* "Reactivity of Solids: Proceedings 6th International Symposium on the Reactivity of Solids" (J. W. Mitchell, R. C. De Vries, R. W. Roberts, and P. Cannon, eds), Wiley-Interscience, New York (1969), p. 171.
171. Cadenhead, D. A. and Wagner, N. J. *J. Catal.* **27**, 475 (1972).
172. Corolleur, C., Gault, F. G., Juttard, D., Maire, G. and Muller, J. M. *J. Catal.* **27**, 466 (1972).
173. Fox, P. G., Ehretsmann, J. and Brown, C. E. *J. Catal.* **20**, 67 (1971).
174. Sinfelt, J. H., Carter, J. L. and Yates, D. J. C. *J. Catal.* **24**, 283 (1972).
175. Ahern, S. A., Martin, M. J. C. and Sucksmith, W. *Proc. Roy. Soc.* **A248**, 145 (1958).
176. Pearson, W. B. "A Handbook of Lattice Spacings and Structures of Metals and Alloys," Pergamon, Oxford (1958).
177. Delmon, B. and Pouchot, M. T. *Bull. Soc. Chim. France* **1966**, 2677.
178. Roman, A. and Delmon, B. *J. Catal.* **30**, 333 (1973).
179. Newkirk, A. E. and McKee, D. W. *J. Catal.* **11**, 370 (1968).

180. Schmahl, N. G. and Minzl, E. *Z. Phys. Chem.* (N. F.) **41**, 78 (1964).
181. Dollimore, D., Gilland, R. D. and McKenzie, E. D. *J. Chem. Soc.* **1965**, 4479.
182. Garten, R. L. and Ollis, D. F. *J. Catal.* **35**, 232 (1974).
183. Garten, R. L., Delgass, W. N. and Boudart, M. *J. Catal.* **18**, 90 (1970).
184. Klier, K. *In* "Molecular Sieve Zeolites—I", *Advances in Chemistry Series* **101**, 480 (1971).
185. Massoth, F. E. *J. Catal.* **30**, 204 (1973).
186. Biloen, P. and Pott, G. T. *J. Catal.* **30**, 169 (1973).
187. Seshadri, K. S. and Petrakis, L. *J. Catal.* **30**, 195 (1973).
188. Bond, G. C. and Webster, D. E. *Ann. New York Acad. Sci.* **158**, 540 (1969).
189. Bozorth, R. M. "Ferromagnetism", Van Nostrand, New York (1951).
190. Nalibaev, T. N., Fasman, A. B. and Inayatov, N.Sh. *Russ. J. Phys. Chem.* **45**, 211 (1971).
191. Fasman, A. B., Kabiev, T., Sokolskii, D. V., Molyukova, N. I., Batkov, A. A., Kirilyus, I. V. and Chernousova, K. T. *Russ. J. Phys. Chem.* **40**, 56 (1966).
192. Molyukova, N. I., Fasman, A. B. and Khizhnyak, I. V. *Russ. J. Phys. Chem.* **42**, 876 (1968).
193. Petrov. B. F., Fasman, A. B. and Sokolskii, D. V. *Russ. J. Phys. Chem.* **44**, 1736 (1970).
194. Fasman, A. B., Molyukova, N. I., Kabiev, T., Sokolskii, D. V. and Chernousova, K. T. *Russ. J. Phys. Chem.* **40**, 948 (1966).
195. Fasman, A. B., Isabekov, A. and Almashev, B. K. *Russ. J. Phys. Chem.* **42**, 470 (1968).
196. Pushkareva, G. A., Fasman, A. B. Klyuchnikov, Yu. F. and Sapukov, I. A. *Russ. J. Phys. Chem.* **46**, 843 (1972).
197. Fasman, A. B., Isabekov, A., Sokolskii, D. V., Presnyakov, A. A. and Chernousova, K. T. *Russ. J. Phys. Chem.* **40**, 1125 (1966).
198. Fasman, A. B., Sokolskii, D. V., Kabiev, T., Almashev, B. and Chernousova, K. T. *Russ. J. Phys. Chem.* **40**, 1190 (1966).
199. Molyukova, N. I., Petrov, B. F., Fasman, A. B. and Sokolskii, D. V. *Russ. J. Phys. Chem.* **41**, 748 (1967).
200. Fasman, A. B., Kabiev, T. and Yagudeev, T. A. *Russ. J. Phys. Chem.* **41**, 1511 (1967).
201. Isabekov, A., Fasman, A. B. and Almashev, B. K. *Russ. J. Phys. Chem.* **41**, 1013 (1967).
202. Fasman, A. B., Almashev, B. K., Klyuchnikov, Yu.F. and Sapukov, I. A. *Russ. J. Phys. Chem.* **46**, 1468 (1972).
203. Sokolskii, D. V., Dukhovnaya, T. M. and Dzhardamelieva, K. K. *Russ. J. Phys. Chem.* **44**, 1614 (1970).
204. Sokolskii, D. V., Gildebrand, E. I., Dukhovnaya, T. M. and Dzhardamelieva, K. K. *Russ. J. Phys. Chem.* **45**, 1039 (1971).
205. Sokolskii, D. V., Dzhardamelieva, K. K. and Dukhovnaya, T. M. *Russ. J. Phys. Chem.* **43**, 275 (1969).

206. Popov, N. I., Sokolskii, D. V., Bizhanov, F. B. and Pechenkina, B. F. *Russ. J. Phys.* **45**, 1493 (1971).
207. Popov, N. I., Sokolskii, D. V., Bizhanov, F. B. and Akchalov, Zh. G. *Russ. J. Phys. Chem.* **46**, 363 (1972).
208. Fasman, A. B., Gorokhov, A. P., Sokolskii, D. V. Klyuchnikov, Yu. F. and Sapukov, I. A. *Russ. J. Phys. Chem.* **46**, 1153 (1972).
209. Carothers, W. H. and Adams, R. *J. Amer. Chem. Soc.* **47**, 1047 (1925).
210. Maxted, E. B. and Akhtar, S. *J. Chem. Soc.* **1959**, 3130.
211. Cahen, D. and Ibers, I. A. *J. Catal.* **31**, 369 (1973).
212. Adams, R. and Voorhees, V. *J. Amer. Soc.* **44**, 1683 (1922).
213. Adams, R., Voorhees, V. and Shriner, R. L. *Org. Syn.* **1**, 452 (1932).
214. Rampino, L. D. and Nord, F. F. *J. Amer. Chem. Soc.* **63**, 2745, 3268 (1941).
215. Hosemann, R., Preisinger, A. and Vogel, W. *Ber. Bunsenges.* **70**, 796 (1966).
216. Hosemann, R., Lemm, K., Schonfeld, A. and Wilke, W. *Kolloid Z.* **216–217**, 103 (1967).
217. Hosemann, R. *Chem. Ing. Tech.* **42**, 1252, 1325 (1970).
218. Topsoe, H., Dumesic, J. A. and Boudart, M. *J. Catal.* **28**, 477 (1973).
219. Solbakken, V., Solbakken, A. and Emmett, P. H. *J. Catal.* **15**, 90 (1969).
220. Cusumano, J. A. *Nature* **247**, 456 (1974).
221. Dry, M. E. and Ferreira, L. C. *J. Catal.* **7**, 352 (1967).
222. Zeliger, H. I. *J. Catal.* **7**, 198 (1967).
223. Taylor, K. C. Proceedings Third North American Meeting of the Catalysis Society, San Francisco (1974).
224. Huang, Y. Y., and Anderson, J. R., unpublished work, CSIRO Division of Tribophysics, (1974).
225. Olson, D. H. *J. Phys. Chem.* **72**, 4366 (1968).
226. Bacon, F. T. and Fry, T. M. *Proc. Roy. Soc.* **A334**, 427 (1973).
227. Koberstein, E. *German Patent* No. 1086106, (Appln. 1957).
228. Giraitis, A. P. and Whaley, T. P. *U.S. Patent* No. 3162606, (1964).
229. Smith, R. M. *U.S. Patent* No. 3003973, (Appln. 1959).
230. Angell, C. L. and Schaffer, P. C. *J. Phys. Chem.* **70**, 1413 (1966).

CHAPTER 5

Structure and Properties of Small Metal Particles

	page
1. IDEAL CRYSTALLOGRAPHIC PARTICLES	246
2. OBSERVED EQUILIBRIUM CRYSTALLITE STRUCTURE	253
3. SURFACE ATOMS OF LOW CO-ORDINATION	260
4. BIMETALLIC PARTICLES.	263
5. PROPERTIES OF SMALL METAL PARTICLES	266
6. METAL CLUSTER COMPOUNDS	270
7. INTERACTION OF METAL PARTICLES WITH A NON-METALLIC SUPPORT	275
Metal–Support Bonding	275
Electrical Phenomena at the Interface	277
8. PARTICLE GROWTH	280
Particle Migration and Coalescence	280
Interparticle Transport	282
Relation to Observed Sintering Behaviour	282

In the early stages of growth of evaporated metal films, individual three-dimensional metal particles are formed via very small nuclei consisting of only a few atoms. Whether these crystal nuclei, or atom aggregates, are two-dimensional or three-dimensional will depend both on the number of metal atoms in an aggregate and their interaction with the support. If the energy of interaction of a metal atom with the support is small compared to the metal–metal interaction energy, the aggregate geometry would be such as to maximize the number of nearest neighbour metal–metal interactions. If one assumes constancy and additivity of the nearest neighbour interaction energy (x), the dissociation energy for M_2 would be x; linear M_3, 2x; equilateral triangular M_3, 3x; linear M_4, 3x; square M_4, 4x; tetrahedral M_4, 6x. On this basis, an equilateral triangle and a tetrahedron would be the most stable arrangements for M_3 and M_4 respectively. Nevertheless, the influence of the support on the aggregate geometry

5. STRUCTURE AND PROPERTIES OF SMALL METAL PARTICLES 245

cannot always be ignored, and a knowledge that some f.c.c. metals can be grown epitaxially as evaporated films on an ionic support such as rocksalt, with metal (100) parallel to rocksalt (100) would suggest that M_4 with a square arrangement can also be stable if the interaction with the support is strong enough. Supports which are less well defined crystallographically are presumably of lesser influence. On the whole, however, predictions about the likely geometry of small aggregates need to be approached with some caution because the theoretical models upon which they are based are still relatively rudimentary. Baetzold[18] has, for instance, recently concluded from the use of molecular orbital calculations that in aggregates containing up to 30 atoms, linear chains are the most stable with silver, but normal three-dimensional geometry is the most stable with palladium. Some direct evidence for two-dimensional aggregates has recently been provided by an electron microscopic study of a 1 wt.% rhodium/Cabosil HS5 catalysts after hydrogen reduction at 670 K.[58] The visible aggregates appear to contain about six atoms.

An immediate question is the extent to which isolated (unaggregated) metal atoms in an oxidation state zero, M(0), may exist in dispersed metal catalysts. One needs to consider both thermodynamic and kinetic factors.

In the first place we should recall (cf. p. 170) that the generation of M(0) by ordinary chemical reduction processes (e.g. hydrogen reduction) is in most cases thermodynamically unfavourable if one starts with a precursor with about the same free energy of formation as the metal oxide. Nevertheless, it is instructive to consider the fate of an isolated M(0) on a support surface.

Values for the equilibrium energy of attraction between the atoms in an M_2 pair are, in the gas phase, Al_2, 193; Cu_2, 191; Ag_2, 158; Au_2, 216; Pd_2, 96·5 kJ mol^{-1}. The low value for palladium is probably connected with the filled level structure of the palladium atom, $4d^{10}5s^0$. Since atomic platinum is $5d^9 6s^1$, one would expect Pt_2 to be considerably more stable than Pd_2, but there no experimental data to confirm this. The addition of further metal atoms to yield larger aggregates is, of course, also thermodynamically favourable.

These favourable energies for the formation of small atomic aggregates show that single metal atoms will only be thermodynamically stable on a support with respect to aggregation if the interaction energy of a metal atom with the support is considerabley greater as an isolated atom than as an atom in an aggregate. This difference in interaction energy would need to be at least as large as the energy of dissociation per atom of the aggregate, that is, the energy of interaction of an isolated atom with the support would need to be at least 40–80 kJ mol^{-1} greater. The nature of the interaction between a metal crystallite and an oxide support is discussed in a subsequent section, and the conclusion emerges that, in general and provided one avoids oxidizing conditions which result in the formation of metal oxide at the interface, the interaction forces are essentially van der Waals in nature. The extra binding energy of an isolated metal atom can conceivably result from the presence of adsorption sites which are available to isolated atoms and which offer a high binding energy, but which for steric reasons are unavailable to atoms in an aggregate. Clearly for this to be so, the interaction energy of the isolated metal atom with the support would need to involve more than ordinary van der Waals forces, and presumably some degree of charge transfer between the metal atom and the nearest neighbour support atoms would be required. This apparently occurs with silver on silver sulphide[21] where silver atoms exist in appreciable concentration: however, this is probably a fairly rare situation since interaction energies in the region of 20–40 kJ mol^{-1}, that is of the magnitude of van der Waals interaction energies, have been found for a number of systems of metal atoms adsorbed on oxide or alkali halide surfaces.[41] The interaction energy of metal atoms with a carbon (graphite) surface is apparently considerably larger than for oxide surfaces.[69]

The removal of M(0) by aggregation would require the surface diffusion of M(0) through a sufficient distance. A simple example is indicative of what might be expected. If M(0) interacts physically with the support surface, the activation energy for surface diffusion of M(0) would be expected to be perhaps 12 kJ mol^{-1}. If one then assumes a normal value for the pre-exponential factor in the equation relating the diffusion coefficient to temperature, one obtains an estimate for the diffusion coefficient of M(0) at, say 700 K, of about 10^{-7} m^{2}s^{-1}. From this it would follow that the RMS migration distance in, say 10 min at 700 K would be many times the value required for complete aggregation to occur.

1. Ideal Crystallographic Particles

Although crystallites observed in practice are seldom geometrically perfect, it is convenient to be able to refer to the shapes of ideal crystals and to the statistics of surface atoms and sites. We restrict ourselves to crystal shapes which expose only low energy planes, e.g. (111) and (100) for f.c.c., (110) and (100) for b.c.c., (001) and (101) for hexagonal.

This subject has been treated thoroughly by van Hardeveld and Hartog,[1] to whom reference may be made for further details and on whose work the present discussion is largely based. Figures 5.1a–d, 5.2a, b and 5.3a, b show the ideal shapes and surface atom arrangements. Surface atoms differ from atoms in the bulk in that they have an incomplete set

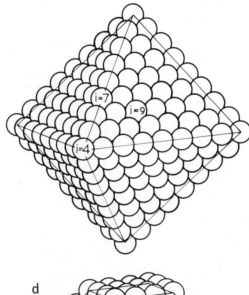
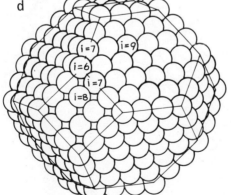

Fig. 5.1 f.c.c. (a) tetrahedron; (b) cube; (c) octahedron; (d) cubo-octahedron. Values for i specify number of nearest neighbours.

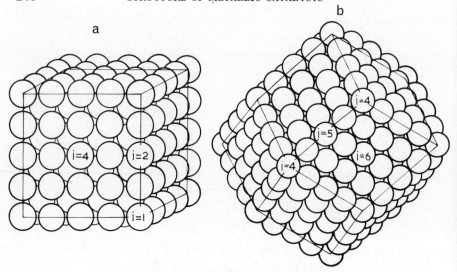

Fig. 5.2 b.c.c. (a) cube; (b) rhombic dodecahedron. Values for i specify number of nearest neighbours.

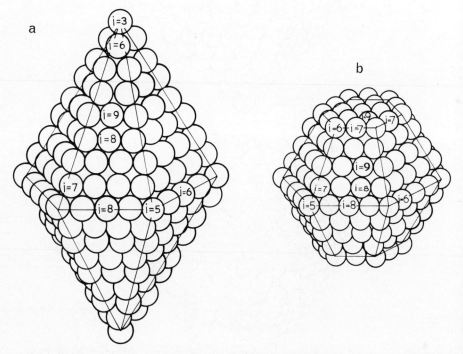

Fig. 5.3 h.c.p. (a) hexagonal bipyramid; (b) truncated bipyramid. Values for i specify number of nearest neighbours.

of neighbours. We define by A_i an atom in the crystal having i nearest neighbours, and by F_i the fraction of all surface atoms of the type A_i. In fact, A_i and F_i can often be subsclassified, depending on the detailed geometric environment of the surface atom. We shall ignore this. Figures 5.4a–d, 5.5a, b and 5.6a, b show F_i as a function of crystallite size, the latter being expressed as the relative diameter, d_{rel}, which is defined as the ratio of the diameter of a sphere with a volume equal to $N_{(T)}$ times the volume occupied by an atom in the unit cell, to the atom diameter; $N_{(T)}$ is the total number of atoms in the crystallite in question. For

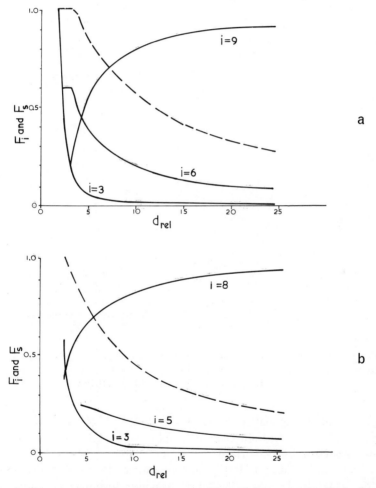

FIG. 5.4 Surface atom statistics for f.c.c. crystallites. Full curves give F_i, broken curves, F_s. The dotted curves in (c) and (d) give F_s for a pentagonal bipyramid and icosahedron, respectively. (a) tetrahedron; (b) cube; [*continued overleaf*

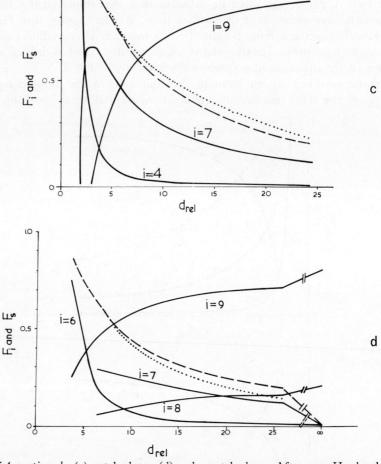

FIG. 5.4 *continued* (c) octahedron; (d) cubo-octahedron. After van Hardeveld, R. and Hartog, F. *Surface Sci.* **15**, 189 (1969) and Allpress J. G. and Sanders, J. V. *Austral. J. Phys.* **23**, 23 (1970).

f.c.c. and hexagonal crystals $d_{rel} = 1 \cdot 105\ N_{(T)}^{1/3}$; for b.c.c. crystals $d_{rel} = 1 \cdot 137\ N_{(T)}^{1/3}$. Figures 5.4–5.6 also show F_S, the fraction of total atoms present in the surface.

Van Hardeveld and Hartog[1] also investigated surface statistics for ideal f.c.c. crystals. In particular, they provided formulae for computing the total number of various types of adsorption sites as a function of crystal size. The sites considered are (i) S_2, a pair of nearest neighbour metal atoms; (ii) S_3, three metal atoms situated at the vertices of an equiliateral triangle with sides equal to the atomic diameter; (iii) S_4, four metal atoms situated at the corners of a square with sides equal to the atomic diameter; (iv) S_5, terrace-edge sites (see below). In general, the ratio of the total number of each kind of site to the total number of surface atoms is almost independent of d_{rel}. The S_5 sites are not present on any of the perfect crystallite shapes we have considered here; however, they are formed, for instance, on f.c.c. (111) and (100) faces which are partly covered by an

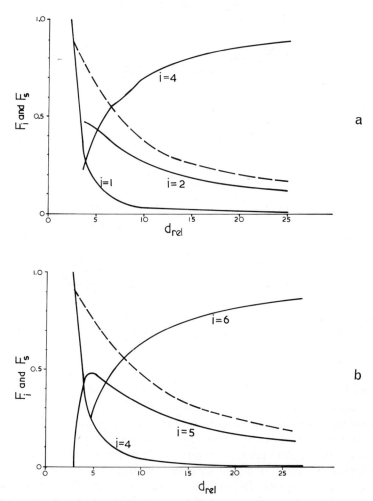

FIG. 5.5 Surface atom statistics for b.c.c. crystallites. Full curves give F_i, broken curves, F_s. (a) cube; (b) rhombic dodecahedron. After van Hardeveld, R. and Hartog, F. *Surface Sci.* **15**, 189 (1969).

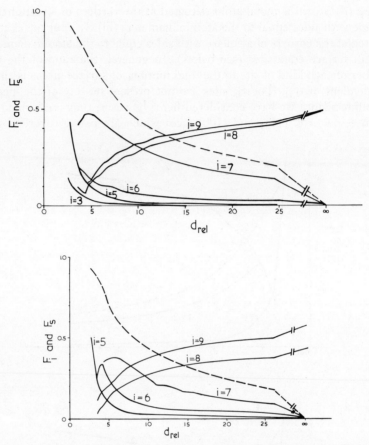

FIG. 5.6 Surface atom statistics for h.c.p. crystallites. Full curves give F_i, broken curves, F_s. (a) hexagonal bipyramid; (b) truncated bipyramid. After van Hardeveld, R. and Hartog, F. *Surface Sci.* **15**, 189 (1969).

extra layer of atoms and which therefore have a step at the edge of the incomplete layer. In terms of nearest neighbour geometry, two sorts of S_5 sites are possible, depending on the exact arrangement of the five atoms: one type has f.c.c. (110) geometry, the other (311) geometry (Fig. 5.7). Because incomplete planes are quite likely in practice, it would be unwise to ignore S_5 sites. As would be generally anticipated, the concentrations of S_2, S_3 and S_4 sites fall in that decreasing order. The concentration of S_5 sites depends on the extent of the incomplete atomic planes; but if the incomplete planes are as extensive as possible, the number of S_5 sites is maximized, and for a cubo-octahedral crystal, for instance, this number is then approximately equal to the number of S_4 sites. More recently and as an extension to this, Schlosser[2] has evaluated the way in which the proportion of S_5 sites varies as the surface of a regular cubo-octahedron becomes progressively more faceted towards a spherical geometry.

The surface energy of very small metallic crystallites have been calculated by Romanowski,[70] taking only nearest neighbour interactions into account. The forms with minimum surface energies are: f.c.c. cubo-octahedron; b.c.c. rhombic dodecahedron; h.c.p truncated hexagonal bipyramid.

FIG. 5.7 Alternative S_5 site geometries formed at terrace edges on (111) and (100) f.c.c. Left, S_5 with (110) geometry; right, S_5 with (311) geometry. After van Hardveld, R. and van Montfoort, A. *Surface Sci.* **4**, 396 (1966).

2. Observed Equilibrium Crystallite Structure

The variation of surface energy with crystallographic orientation determines the equilibrium crystallite shape. With an increasing degree of anisotropy of surface energy, the expected equilibrium shape changes from spherical to polyhedral, and the Wulff construction (cf. Herring, refs 44, 45) allows the shape to be computed for specified degrees of anisotropy. For the degree of anisotropy indicated by theoretical calculations,[46] the expected equilibrium shape lies between the above extremes; that is, the expected shape is of the form of a polyhedron with markedly rounded corners.

The actual equilibrium shapes of f.c.c. metal crystallites has been determined by Sundquist,[10,11] Winterbottom[12] and by Pilliar and Nutting.[13] The particles were generated on ceramic substrates, using the fact that if a continuous film is heated it may break up into islands which transform into particles of diameter about 1 μm, the shapes of which can readily be examined by electron microscopy. Supports of beryllia, magnesia and α-alumina were used, and these were outgassed under vacuum at temperatures up to 2273 K prior to metal deposition. The point about this work is that a deliberate attempt was made to ensure particle equilibrium: thus, films were typically heated in vacuum or in hydrogen at 970–1270 K for about 100 h under conditions of metal–vapour equilibrium. Gold, silver, copper, nickel and γ-iron particles were examined.

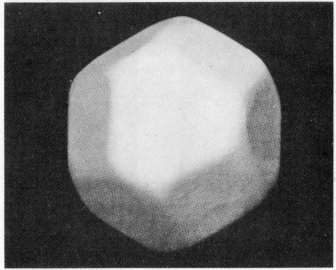

FIG. 5.8 Model representation of crystallite shape formed by break-up of a continuous film of f.c.c. metal on a ceramic substrate due to heating. Reproduced with permission from Sundquist, B. E. *Acta. Met.* **12**, 67 (1964).

The general conclusion is that the crystallites take the form of polyhedra with somewhat rounded corners: Fig. 5.8, due to Sundquist, gives a model representation of the typical crystal shape, this shape being dictated by the anisotropy of surface energy. Winterbottom's observation that separate small crystallites are much more difficult to produce by disintegration of a continuous silver film in UHV than in HV is probably because silver has a particularly low surface energy anisotropy, and because a continuous film prepared in UHV will consist of much larger crystals than when prepared in HV.

5. STRUCTURE AND PROPERTIES OF SMALL METAL PARTICLES

On a mechanical model one would expect the average lattice parameter of a small metal particle to be reduced below the value for massive metal, owing to the surface stress which places the particle in compression. Experimental verification of this is difficult because diffraction data are strongly influenced by line broadening. However, a change in average lattice parameter in the expected direction has recently been observed by electron diffraction measurements on UHV-deposited metal films. With gold[47] and silver,[48] the decrease in average lattice parameter was some 0·26% and 0·70% respectively, and these changes were parametized in terms of surface stresses of 1·175 and 1·415 N m^{-1}, respectively.

Evidence has been obtained on the shape and structure of ultrathin films of platinum,[8,14,15] nickel, palladium, gold and silver deposited in UHV on vacuum-cleaved mica at temperatures in the range 370–770 K under conditions where relatively large crystallites (5–20 nm diameter) are formed.

With nickel, palladium and gold,[8] the films are found to consist of well-

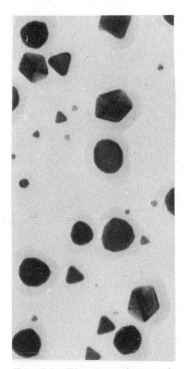
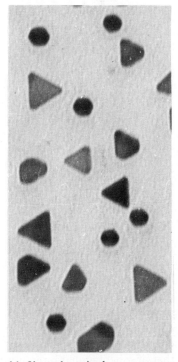

FIG. 5.9 Electron micrographs of ultrathin gold films deposited on vacuum-cleaved mica at about 520 K in UHV. Reproduced with permission from Allpress, J. G. and Sander, J. V. *Surface Sci.* **7**, 1 (1967).

formed crystallites, mostly of regular geometric shapes with a (111) metal plane parallel to the support. Examples are shown in Fig. 5.9 for gold and Fig. 5.10 for palladium. Many crystals with a triangular plane shape are seen, and these are undoubtedly tetrahedral in three dimensions, probably truncated in a vertical direction: they expose, of course, only (111) facets. Other important plane shapes to be seen are pentagonal (Fig. 5.9) and hexagonal (Figs 5.9 and 5.10). Both of the latter non-ideal structures can be considered as multiple twins on (111) tetrahedral faces. Other twinning arrangements are occasionally observed.[8] Evidence for multiple twinning may be obtained from both dark-field micrograph

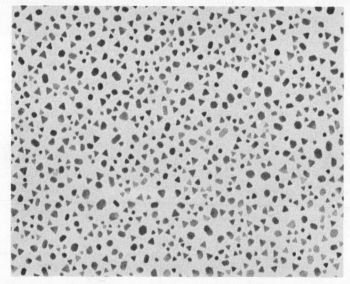

FIG. 5.10 Electron micrograph of ultrathin palladium film deposited on vacuum-cleaved mica at 730 K in UHV. Reproduced with permission from Allpress, J. G. and Sanders, J. V. *Surface Sci.* **7**, 1 (1967).

images and from relative intensities in diffraction.[8] It is particularly obvious with a pentagonal crystallite that it cannot have an ideal crystallographic structure with cubic symmetry. In fact, very small pentagonal metal particles have been previously observed in a number of systems, including silver smoke,[3] gold deposited on gold,[4] gold precipitated from aqueous solution,[5] gold deposited on rocksalt,[6,7] as well as the systems cited previously.[8] The structure of the pentagonal particles is, in fact, that of a pentagonal bipyramid (Fig. 5.11a). It can be imagined as formed by multiple twinning of five tetrahedra on their (111) faces: this would give Fig. 5.11b.[6,3,8] The electron microscope images show no evidence

5. STRUCTURE AND PROPERTIES OF SMALL METAL PARTICLES

of strains, dislocations or misfits, corresponding to the gap in Fig. 5.11b. Apparently the actual crystallites relax structurally to avoid dislocations between the twins. A structure with hexagonal plan symmetry which has the experimentally required diffraction properties is first obtained by further twinning to a total of 16 tetrahedra; however, the irregular nature of one face of this structure makes it unlikely that it has significant frequency of occurrence in practice. Further twinning to a total of 20 tetrahedra gives a particle with the general three-dimensional shape of an icosahedron (Fig. 5.11c), which also has a hexagonal plan outline and the required diffraction properties. In practice, crystals with hexagonal plan outline are almost certainly icosahedra.

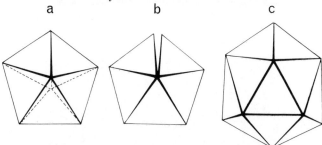

FIG. 5.11 Multiply-twinned particles of an f.c.c. metal, (a) ideal pentagonal bipyramid; (b) structure formed by twinning of five tetrahedra on (111) faces and giving (a) on relaxation; (c) twinning of twenty tetrahedra on (111) gives on relaxation an icosahedron which also has a hexagonal plan outline.

The relative energies of various crystallographically ideal and non-ideal multiply twinned particles have been computed using a pair-wise interaction model.[9, 16, 43] These computations confirm that, at least in some instances, regular f.c.c. atom clusters can be expected to have a greater lattice energy than non-ideal structures. For instance, a 55-atom cubo-octahedron is less stable than an icosahedron (Fig. 5.11c) by an energy difference of about 8%. The nearest neighbour distance between surface atoms in a non-ideal structure is close to the ideal value. On the other hand, the computations showed that an octahedron (19-atom) and a rhombicubo-octahedron (43-atom) are stable ideal f.c.c. structures.

For the sake of reference, we have included plots for F_s for the pentagonal bipyramid and the icosahedron in Figs 5.4c and d, respectively.

The structure of these non-ideal particles has its origin in the very early stages of growth when the nuclei contain only a few atoms.[9]

Generally, for a given film, the untwinned tetrahedral crystallites are always the largest, pentagonal crystallites somewhat smaller, and hexagonal crystallites smallest. The proportion of multiply-twinned crystallites

FIG. 5.12 Electron micrograph of ultrathin platinum film deposited on vacuum-cleaved mica in UHV. (a) deposited at 770 K, (b) deposited at 600 K. Courtesy J. V. Sanders.

with these metals is generally substantial, and is not sensitive to support temperature during deposition in the range 370–770 K. However, at least with gold, multiple twinning disappears with very thin films at very low deposition rates ($<2 \times 10^{-4}$ nm s^{-1}).[8] These films always contain a proportion, usually minor, of crystallites of indefinite shape, often approaching a circular plan shape. These presumably are multiply faceted, but with individual facets not large enough to be resolved. Crystallites of this type become dominant at deposition temperatures approaching room temperature.

Figures 5.12a and b show deposits of platinum prepared on mica at 770 K, and 600 K respectively.[15] The crystallites in Fig. 5.12a show the essential features described above for the other metals, although in general the crystallites are not quite so well formed. At the lower support temperature of Fig. 5.12b, this tendency is more pronounced because the particles, although generally showing marked angularity, are somewhat more irregular than at 770 K. The average particle diameters are about 12 and 10 nm for Figs 5.12a and b, respectively. In all cases the crystallites have a strongly preferred orientation with a (111) parallel to the mica.

In contrast to platinum, palladium, nickel and gold, with silver only crystallites with an approximately circular plan shape are seen[14] and there is no worthwhile evidence of angularity. A typical example is shown in Fig. 5.13, where the average crystallite diameter is 8·1 nm. Crystallites are also essentially circular for diameters down towards the limit of observation. Some evidence for the three-dimensional shape of these

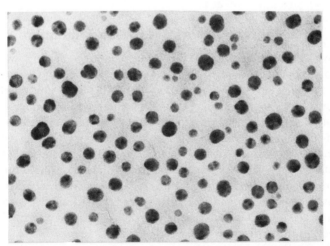

FIG. 5.13   Electron micrograph of ultrathin silver film deposited on vacuum-cleaved mica at 300 K in UHV. Courtesy J. V. Sanders and H. Jaeger.

crystallites may be obtained from the contrast in micrographs due to stacking faults,[14,15] from which it follows that the likely shape when sectioned in a plane normal to the support is that of a flattish, curved dome. With these silver films, particle orientation with (111) parallel to the support is only extensive if the support is >670 K during deposition. At lower temperatures the orientation becomes increasingly random, and is completely so for films deposited at 300 K. In all cases, the proportion of multiply twinned particles is substantial.

On an atomic scale, the actual surfaces of small metal particles are subject to the same sort of departures from ideality as have already been discussed for the surfaces of massive metal specimens, and reference for this may be made to Chapter 3.

The structure of the metal particles dispersed on a silica powder support ("Aerosil" 380, 7 nm average silica particle size) has been studied by Avery and Sanders using electron microscopy in both bright and dark field,[74] to determine the extent to which the metal particles were multiply twinned or of ideal structure. Platinum, palladium and gold were examined. These catalysts were prepared by impregnation using an aqueous solution of metal chloride derivatives, were dried at 370–420 K and were hydrogen reduced at 670 K. The proportion of metal was in the range 5–15 wt. %. The range of electron microscopically visible metal particle sizes were platinum $1 \cdot 0$–$8 \cdot 0$ nm, palladium $2 \cdot 0$–$10 \cdot 0$ nm, gold $10 \cdot 0$–$35 \cdot 0$ nm. In all cases metal particles were randomly oriented on the substrate and the proportion of multiply twinned particles was low, certainly not exceeding 2%.

3. Surface Atoms of Low Co-ordination

We have seen in section 2 of this chapter, dealing with ideal crystallographic particles, that there are present in the surface various sorts of atoms which are differentiated in terms of the number of nearest neighbours. For any particular crystal geometry, the number of nearest neighbours will increase for surface atoms in corner, edge and face positions, in that order; and this is also the order in which the proportions of these classes of atoms increase in the surface. It is reasonable to believe that surface atoms with different nearest neighbour co-ordination numbers should in some way have this fact reflected in terms of their reactivity, and this should be particularly so for corner atoms for which the co-ordination number is lowest. The idea that corner atoms may act as catalytic sites with special properties has been alluded to in Chapter 1.

If the metal particles in a dispersed catalyst are of crystallographically ideal shape, the proportion of corner atoms in the surface will decrease

5. STRUCTURE AND PROPERTIES OF SMALL METAL PARTICLES   261

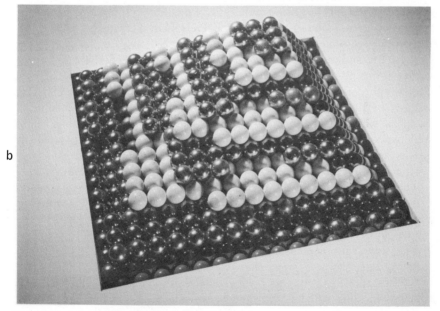

FIG. 5.14 Pyramidal-shaped crystal with (a) terraces only, (b) terraces and kinks.

rapidly with increasing crystallite size (cf. Figs 5.4–5.6); and this has led to attempts to interpret particle size effects in metallic catalysis in these terms (e.g. refs 59–61). Nevertheless, it needs to be recognized that one can seldom expect the metal particles in an actual dispersed catalyst to be of ideal shape. One must consider the influence of roughness and imperfections in the particle surface. For the present we shall confine the discussion to some simple examples. Inasmuch as the details of surface topography at this level are not experimentally accessible, this discussion must remain essentially speculative.

Consider, for example, planes of atoms stacked on an f.c.c. (111) plane as base, to yield the general shape of a pyramid on a triangular base: if the crystallite were geometrically ideal it would, of course, be a tetrahedron exposing four equivalent (111) faces. If, however, the planes of atoms which are stacked are incomplete with respect to the size required to generate a tetrahedral crystallite, one could form an irregularly shaped pyramid with (say) monatomic terraces (Fig. 5.14a), and these terraces may have kinks (Fig. 5.14b). It is obvious that an irregular crystallite of this sort has a much higher proportion of surface atoms in corner positions than does an ideal crystallite. The number of surface atoms ($N_{(S)}$) in an ideal tetrahedrally shaped crystal varies *roughly* as l^2, where l is a linear dimension of the crystal (this is approximate because a term in l^2 is merely the leading term in a quadratic expression for $N_{(S)}(l)$). Since the number of corner atoms ($N_{(C)}$) in an ideal tetrahedral crystal is constant (4), the proportion of corner atoms in the surface ($N_{(C)}/N_{(S)}$) varies roughly as l^{-2}. In the situation envisaged in Fig. 5.14a, in which there are three corner atoms on each terrace, $N_{(C)}$ will vary roughly as l, so $N_{(C)}/N_{(S)}$ will vary roughly as l^{-1}, and this will also apply if each terrace had a constant number of kinks in it, although in this latter situation the actual value of $N_{(C)}/N_{(S)}$ would be greater for a given crystallite size. However, one might well expect that the number of kinks in a terrace could be greater the greater the terrace length: if one assumed that their number was proportional to terrace length, one would expect $N_{(C)}/N_{(S)}$ to vary as l^0.

There is a different sort of behaviour for a crystal with a curved surface. Curvature is provided by terraces and kinks in the manner illustrated in Fig. 5.15. The smaller the radius of curvature (R), the greater will be the terrace and kink density, and $N_{(C)}/N_{(S)}$ then varies roughly as R^{-2}.

The important conclusion to be drawn is that in practice, one can certainly not assume *a priori* that $N_{(C)}/N_{(S)}$ will decrease with increasing crystalline size in the manner expected for ideally shaped crystals. The actual behaviour will be very much dependent on the detailed surface topography; and an added complication is that the surface roughness itself may vary with crystal size due to the details of the specimen prepara-

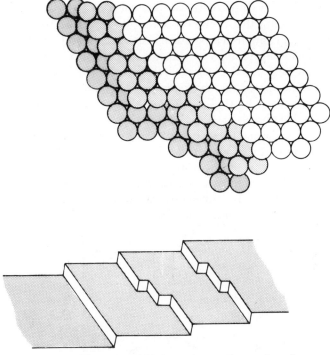

FIG. 5.15   Terrace and kink structure of curved surface.

tion. It will be clear that the above discussion, which has been couched in terms of individual crystallites, applies equally to the surface structure of (say) pyramidal excrescences on an otherwise flat surace.

4. Bimetallic Particles

When considering the likely structure of dispersed bimetallic catalysts, the question arises whether the bulk phase diagram is likely to be a reasonable guide to the behaviour of small particles. This is a problem which has been discussed by Ollis[76] and Hoffman,[77] but the problem has not yet been completely resolved.

Consider a model temperature-composition (T, x) phase diagram (Fig. 5.16). The two points of immediate concern are: (a) at equilibrium does the position of the line separating the solid solution field from the field in which two solid phases are in equilibrium, depend on particle size (an equilibrium particle size effect) and, (b) is the ease of attaining this equilibrium affected by particle size (a kinetic particle size effect). We shall first deal with the latter of these factors.

There is no doubt that if one cools a solid solution to below the equilibrium phase separation temperature, substantial undercooling is often observed and equilibrium is very difficult to achieve. This is because if the phases are to be formed by a nucleation/diffusion mechanism, the low diffusion coefficients in the solid place a very severe limitation on both nucleation and growth. If the solid solution is present as a very small particle, the likelihood of phase separation will be much lower still, because nucleus formation is in any case a relatively infrequent event, particularly for particles where the defect concentration is low. In this situation, nucleation would be restricted to a very small proportion of the particles near the equilibrium transformation temperature. Eventually, as the temperature falls sufficiently far below the equilibrium transformation temperature, some other process may occur.

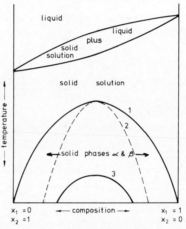

Fig. 5.16 Model phase diagram.

Decomposition of the solid solution into the two equilibrium phases may occur as a spinodal decomposition. Alternatively, if the solid solution exhibits equilibrium polymorphism, that is phases of different basic crystal structure type at constant composition (a feature not represented in Fig. 5.16), the transformation of one form to another with changing temperature may occur by nucleation and growth at the equilibrium transformation temperature, but again in solids and with very small particles nucleation may be strongly suppressed, and in the supercooled state transformation may eventually occur by a martensitic or massive process.

In general, the T, x region for spinodal decomposition lies within the equilibrium two phase field. Two spinodal lines can be distinguished; the chemical (or incoherent) and the coherent. Both are indicated in a qualitative sense in Fig. 5.16. In the field beneath the chemical spinodal but above the coherent spinodal, incoherent nuclei of the phases α and β can be formed by a spinodal mechanism, that is by a gradual transformation of a relatively large volume of material in which regions of inhomogeneity develop and which become more intense as the temperature falls. However, once an incoherent nucleus has been formed, it has to grow by a diffusion process, and in the solid this remains extremely slow. The point about the coherent spinodal is that the rearranged structure remains coherent with the matrix, the misfit being accommodated by elastic strain, so growth of the new phase is diffusionless and rapid. In many practical situations diffusion is so slow and the cooling rate is such that phase separation is mainly determined by the accessibility of the coherent spinodal, and this would be particularly so for small particles.

The extent to which the coherent spinodal is depressed below the equilibrium phase separation line can be very substantial. An important parameter is the lattice distortion coefficient η, which is the rate of change of lattice parameter, a, with changing solute concentration (dlna/dx). The depression of the coherent spinodal temperature increases with increasing η: in the case of a regular solution approximation it varies as η^2.[78] For example, with the nickel–gold system which displays a solid state miscibility gap with a critical solution temperature at about 1090 K, $\eta \approx 0 \cdot 13$ per atom fraction, and the top of the coherent spinodal has been estimated at about 270 K. On the other hand with the aluminium–zinc system for which $\eta \approx 0 \cdot 027$ per atom fraction, the coherent spinodal is only depressed by some 35 K below the chemical spinodal.

It remains to assess the extent to which these considerations are relevant to dispersed bimetallic catalysts as they are normally prepared. The point here is that the preparative conditions are such that the particles are generally assembled atom by atom at relatively low temperatures compared to the melting point of the alloy. The temperature would seldom exceed 800 K at any point of the catalyst processing. Therefore the fate of a relatively defect-free solid solution particle when it is cooled into the two phase field of the phase diagram is probably not of much relevance, since the particle will have been generated via a highly disordered state and at a temperature that will often be below the equilibrium phase separation temperature. In this circumstance we believe it likely that more often than not a particle so formed will be in its equilibrium state with regard to phase separation simply because this is easily achieved during particle genesis.

We turn now to the equilibrium particle size effect. An equilibrium particle size effect occurs via the influence which the phase interface energy in a biphasic particle may have on the total free energy. For specimens of macroscopic dimensions, this will be an effect of totally negligible proportions. However, its relative importance would be expected to increase the smaller the specimen size. A quantitative analysis[76,77] is only available in the approximation which assumes regular solution theory throughout the phase diagram. The difficulty is that the magnitude of the effect is dependent on the magnitude of the phase interface energy, and this is necessarily small for a system which departs from ideal behaviour only to the extent of forming a regular solution. Within this approximation the critical temperature for phase separation only departs from the bulk value by 20% when the particle diameter has fallen to 1·0 nm, and the effect is insignificant ($< 2\%$ difference) for particle diameters above about 10 nm.

However, there is no doubt that in many bimetallic systems the phase interface energy may be considerably larger than the value expected for a regular solution since more extensive deviation from ideal behaviour than a regular solution would be the rule rather than the exception. Thus, on the basis that the phase interface energy is of the same order as the high angle grain boundary energy for a one component system, values in the range $0·6$–$1·2$ J m^{-2} are likely for the group VIII and refractory metals. Although no evaluation of the effect of the phase interface energy on the equilibrium phase separation temperature is available for departures from non-ideality beyond a regular solution, it is reasonable to adopt the point of view of Ollis[76] and obtain rough guidance by using the regular solution treatment, but retaining the phase interface energy as a disposable parameter to which values may be assigned outside the limits allowed for regular solutions. As would be expected, a substantial reduction in the equilibrium phase separation temperature can result. For instance, with the phase interface energy equal to $0·6$ J m^{-2}, and with a particle diameter of 10 nm, the equilibrium phase separation temperature can be reduced to two-thirds of its bulk value in K. The situation is badly in need of clarification from reliable experimental data since the theory is currently so inadequate. Nevertheless we consider there are good grounds for believing that an equilibrium particle size effect can be of considerable importance in stabilizing small bimetallic particles (say <10 nm) into a solid solution.

5. Properties of Small Metal Particles

One may approach the problem of the properties of small metal particles from various points of view, depending for practical purposes on what sort

5. STRUCTURE AND PROPERTIES OF SMALL METAL PARTICLES

of properties happen to be important. The chemist is mainly interested in understanding the propensity for the surface atoms to enter into chemical reactions and to form chemical bonds. The proportion of atoms present in the particle surface increases as the particle size decreases, and the geometric models for small particles which we have discussed in a previous section have indicated how, for idealized cases, there are surface atoms in various crystallographic environments which would be expected to offer varying propensities for chemical reaction. However, in addition we have to be concerned with how the electronically controlled chemical properties of the surface atoms are dependent on particle size. In other words, it is useful to distinguish the effect of particle size on surface geometry *per se* (for instance, the proportion of surface atoms of various co-ordination) from the effect of particle size on the electronic properties of the particle.

That electronic properties must become dependent on particle size at some size range is obvious if one reflects on the different properties displayed by the two limiting cases of massive metal on the one hand, and a single atom on the other. It is not difficult to judge that the properties of the massive metal will be retained down to a very small particle size, and a change in electronic properties will only start to be apparent for particles less than about 2 nm in diameter.

For a metal crystal of macroscopic size, the electronic energy levels are, of course, very closely spaced. The spacing between the levels varies as l^{-3} where l is a dimension of the specimen, and for small particles the average spacing between adjacent levels (δ) is approximately[17]

$$\delta \simeq 4\varepsilon_F/3N \qquad (5.1)$$

where ε_F is the Fermi energy and N is the number of atoms in the particle. For example, with ε_F of the order of 10 eV and with $N \approx 10^3$ for a particle of about 2·8 nm diameter, $\delta \approx 14 \times 10^{-3}$ eV. If we take as a criterion for departure from normal behaviour $\delta \approx kT$, it follows as a very rough and ready sort of estimate with only order of magnitude accuracy, that a particle with $N \approx 500$ (diameter about 2 nm) should show some deviations at room temperature. This simple model probably gives an upper limit to the estimate of particle size for which modified electronic properties may occur. Other data indicate a smaller estimate. It may be compared, for instance, with recent conclusions by Ross *et al.*[89] based on X-ray photoelectron spectroscopy. This work showed that with very small platinum particles (each on the average of not more than about ten atoms or so in size) and dispersed on a silica support, the most significant difference compared to bulk platinum is to be found in the valence band region of the spectrum, where a strong new peak is to be seen centred at about

8 eV below the Fermi level. By reference to known data[90] this peak cannot be indexed in terms of free-atom d electron states which, on this energy scale, would be expected in the region of 4–6 eV. For reasons that will be apparent from our discussion of metal/support interactions, we are also disinclined to attribute this peak to bonding valence states arising from platinum–silica bonding: rather it seems most likely to be a true reflection of the intrinsic electronic properties of platinum particles of this size.

For massive metal, the ionization potential and the electron affinity are, of course, of equal magnitude and equal to the magnitude of the workfunction. For single metal atoms on the other hand, the ionization potential is numerically greater than the electron affinity. Molecular orbital calculations by Baetzold[18,79] show that this difference between ionization potential and electron affinity is retained for very small aggregates of metal atoms, although these parameters converge as the aggregate size increases: in the cases of silver and palladium, the convergence points appear to be for aggregates containing about 20 atoms and 4 atoms respectively. The absolute value of these convergence points do not command much confidence because of the limitations to the method of calculation, but the important point is that convergence does occur at a very small aggregate size. It also appears likely from these calculations that when the ionization potential and the electron affinity first converge to a common magnitude, this is itself in excess of the work function of the massive metal, but the latter is approached as the aggregate size increases further. These calculations also showed the way in which the metallic band structure emerges as the aggregate size increases. For instance, in the case of palladium, as more metallic atoms are added to the aggregate the $5s$ band spreads and eventually overlaps the narrower $4d$ band until the band structure for massive palladium is generated. This band overlap results in the transfer of some electrons from the $4d$ to the $5s$ band, and in massive palladium there is, on the average, about 0·4–0·6 electrons in the $5s$ band and 0·4–0·6 holes in the $4d$ band per atom: it turns out that a figure of 0·6 is reached by the time the palladium aggregate has grown to about 10 atoms, and the same is true for nickel.[79] A band structure with the main features resembling those for massive metal is developed with two dimensional as well as three dimensional aggregates.

One aspect of the collective electron approach to small metallic particles has a very long history, that is the optical absorption caused by a dispersion of small metal particles. The theory of this absorption was first developed by Mie[23] using classical electromagnetic theory. This absorption results from a collective, or plasma, oscillation of the metallic electrons. On this model, and assuming the particles are small compared to the

5. STRUCTURE AND PROPERTIES OF SMALL METAL PARTICLES   269

wavelength of the light (λ), the optical absorption coefficient is proportional to

$$\frac{n\, n_0^3 \varepsilon_2}{\lambda[(\varepsilon_1 + 2n_0^2)^2 + \varepsilon_2^2]} \qquad (5.2)$$

where n is the number of particles per unit volume, n_0 is the refractive index of the medium in which the particles are dispersed, and ε_1 and ε_2 are, respectively, the real and imaginary parts of the dielectric constant of the metal. Data have been reviewed by Doremus.[24] On the whole, there is good agreement between the observed and calculated optical absorption maxima for particles of the alkali metals. These are, of course, the metals for which a free electron model would be expected to be most satisfactory. On the other hand, for metals of group 2 of the periodic classification there is only order of magnitude agreement. For transition metals one would expect even worse agreement.

According to the theory of Gorkov and Eliashberg,[36] the electric polarizability of very small metal particles should be greatly enhanced compared to the massive metal. However, in a recent experimental study of glass samples containing large concentrations of small silver particles of average diameter <10 nm, no enhancement of the dielectric constant was observed.[39]

The magnetic properties of small metal particles have also been studied. The collective paramagnetism of small particles of ferromagnetic metals is discussed in Chapter 6. There has been some recent work on the paramagnetic behaviour of finely dispersed transition metals such as palladium and platinum which, in the massive form, exhibit Pauli temperature-dependent paramagnetism. Even at the experimental level the situation is confused by contradictory results, and the reason almost certainly lies with the use of specimens that are very poorly defined in a chemical sense. Although most workers have been at some pains to eliminate metallic impurity, particularly ferromagnetic metals, other impurity, particularly adsorbed oxygen, was often not well controlled.

Instances where it was found that the magnetic susceptibility increased with an increasing degree of metal dispersion include platinum on silica gel and carbon,[25, 28] palladium on silica gel,[29] and palladium blacks,[33] while a contrary variation has been reported with palladium blacks,[30] and supported palladium.[31, 32] Disagreement is confined to the palladium data, while that for platinum is at least semi-quantitatively consistent. The platinum specimens were subject to thorough hydrogen reduction followed by outgassing in vacuum, but no hydrogen reduction was used with palladium. Although we do not believe that the surface of dispersed platinum was rendered completely clean by this treatment, there can be no doubt that the impurity level must have been much higher for palladium than for platinum. On the whole, we consider that the trend to higher susceptibilities with increasing dispersion observed for platinum is genuine, but we remain sceptical about the reliability of any of the palladium data. As has been suggested,[29, 33] the observed trend probably arises from the population of surface states with unpaired electrons. The paramagnetic properties of small metal particles have recently been studied theoretically,[17, 34-36] but without yet providing illumination for the observed behaviour of experimental systems because of inadequacies in knowledge of actual quantum states, including surface states.

Electron spin resonance of the conduction electrons (within about kT of the Fermi level) appears to be observed more readily with small metal particles than with massive metal: that is, the natural line width is narrower with small particles. Examples are measurements with small particles of gold,[37] lithium,[38] and platinum.[42] It has been suggested that the effect is due to a reduction in electron–phonon interaction in small particles. Taupin[38] also observed an n.m.r. absorption with lithium particles for which the Knight shift was absent. Dispersion of small metal particles merit a good deal more work by magnetic resonance techniques, since this should offer some hope for quantifying their electronic properties.

6. Metal Cluster Compounds

There is evidence of a strictly chemical nature for the existence of very small aggregates of metal atoms, and we refer here to the so-called cluster compounds. Sources to the compounds and data discussed in the following paragraphs are summarized in references 22, 40 and 71. Almost all cluster compounds with no more than four metal atoms have an 18-electron rare gas configuration for each metal atom. The electronic configuration of octahedral clusters is less well understood. The co-ordination number of the metal atoms in clusters is often similar to that found in other derivatives of the same metal atom in the same oxidation state. However, in a few cases the metal co-ordination number is increased to an unusually high value. An example is $(C_5H_5Fe\ CO)_4$. This tendency may be correlated with the relatively small solid angle of the co-ordination sphere of the metal atom in the cluster which is occupied by metal–metal bonds, since this leaves a relatively large portion of the co-ordination sphere available for bonding to other ligands. The possible analogy for the behaviour of corner atoms in small metallic crystallites is obvious.

The strength of the metal–metal bonds in cluster compounds increases

5. STRUCTURE AND PROPERTIES OF SMALL METAL PARTICLES

upon descending a column of the periodic table. Thus the iron triangle in $Fe_3(CO)_{12}$ is ruptured much more easily than the ruthenium triangle in $Ru_3(CO)_{12}$ upon reaction with reagents such as tertiary phosphines or cyclo-octatetrene. This trend, of course, parallels the cohesive energy of the massive metals.

Hexameric cluster compounds are known for a number of metals which include niobium, tantalum, molybdenum, tungsten, rhodium and ruthenium, and some typical examples are $[Nb_6Cl_{12}]^{2+}$, $[Ta_6Cl_{12}]^{2+}$, Ta_6Cl_{14}, $[Mo_6Cl_8]^{4+}$, $[Mo_6Cl_{14}]^{2-}$, $[W_6Cl_8]^{4+}$, $Rh_6(CO)_{16}$, $[Co_6(CO)_{14}]^{4-}$, $Ru_6(CO)_{18}H_2$. In all cases the metal atoms are arranged in an octahedral configuration: in some cases the octahedron is regular, in others it is somewhat distorted. Figure 5.17 shows the geometry for the

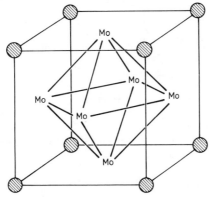

FIG. 5.17 Geometry of cluster compound cation $[Mo_6Cl_8]^{4+}$. Shaded circles = chlorine.

cations $[M_6Cl_8]^{4+}$ where $M \equiv Mo$, W (the octahedra are slightly distorted in these cases). All octahedral carbonyls such as $M_6(CO)_{16}$ are diamagnetic and would require 11 metal–metal bonds to provide the favoured 18-electron rare gas configuration for each metal atom. How this is accommodated into a structure that by virtue of the geometry has 12 nearest neighbour metal–metal interactions has not yet been elucidated. Mixed octahedral clusters containing tantalum and molybdenum are known.

Polymorphs of palladium(II) chloride and platinum(II) chloride are known which are hexameric with all the metal atoms octahedrally arranged. However in these cases the metal–metal distances are relatively long (0·33–0·34 nm in the case of platinum) and so the metal–metal bonds must be quite weak.

Cluster compounds containing more than six metal atoms are very rare: the best known is a derivative containing the ligand 1,1-dicyanoethylene-2,2-dithiol, $Cu_8[S_2CC(CN)_2]_6^{4+}$ in which the eight copper atoms are located at the corners of a cube.

There are tetrameric cluster compounds, for instance involving cobalt, rhodium, iridium, osmium, ruthenium, iron, rhenium, nickel and manganese. Some examples are $M_4(CO)_{12}$, $M \equiv Rh$, Ir; $[M_4(CO)_{13}]^{2-}$, $M \equiv Os$, Ru, Fe; $Re_4(CO)_{12}H_4$. These tetrameric compounds have the metal atoms disposed at the corners of a regular or slightly distorted tetrahedron. There exists an azulene compound containing ruthenium, the probable structure of which is shown in Fig. 5.18, which

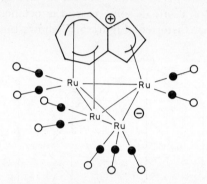

FIG. 5.18 Geometry of cluster compound $Ru_4(CO)_9$(azulene). Full circles = carbon; open circles = oxygen.

shows the π-co-ordination of three separate parts of an organic ligand to three separate metal atoms, and this is at least illustrative of what should be possible on the surface of massive metal with the appropriate geometry.

A substantial number of trimeric cluster compounds are known, probably the most thoroughly studied being $[Re_3Cl_{12}]^{3-}$ and $M_3(CO)_{12}$, $M \equiv Fe$, Ru, Os. Figure 5.19 shows the structure of $[Re_3Cl_{12}]^{3-}$, and this structure is based upon an equilateral triangle of rhenium atoms. The rhenium-rhenium distance is relatively short at $0 \cdot 248$ nm and this probably results from double bonding between the metal atoms. Such double bonding also provides for the rhenium atoms to have the 18-electron rare gas configuration. Trimeric cluster compounds are known for a substantial number of metals including rhenium, molybdenum, iron, ruthenium, osmium, manganese, cobalt, rhodium, nickel, palladium, and platinum.

A number of cluster compounds are known in which the cluster contains more than one type of metal atom. With trimeric clusters, examples are $FeOs_2(CO)_{12}$, $Fe_2Os(CO)_{12}$ and analogous compounds

5. STRUCTURE AND PROPERTIES OF SMALL METAL PARTICLES

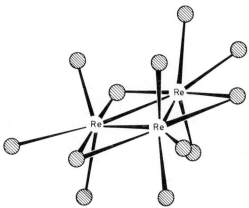

FIG. 5.19 Geometry of cluster compound cation $[Re_3Cl_{12}]^{3-}$. Shaded circles = chlorine.

with ruthenium in place of osmium; $Co_2Fe(CO)_9S$; $Co_2Os(CO)_{11}$; $Fe_2Rh(CO)_9(C_5H_5)$; $FeRh_2(CO)_9(C_5H_5)_2$; while in tetrameric clusters examples are $Fe_2Rh_2(CO)_8(C_5H_5)_2$; $FeCo_3(CO)_{12}H$ and analogous compounds with ruthenium or osmium in place of iron. The structure of $Fe_2Rh_2(CO)_8(C_5H_5)_2$ is known in detail and is shown in Fig. 5.20. It probably has a zwitterion structure as shown. It is tempting to regard these mixed clusters as analogues of alloys on the macroscopic scale. A zwitterion structure may provide a site at which heterolytic fission of a reactant is favourable.

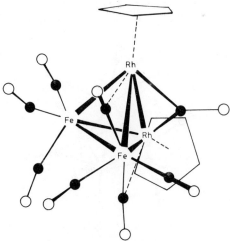

FIG. 5.20 Geometry of cluster compound $Rh_2Fe_2(\pi\text{-}C_5H_5)_2(CO)_8$. Full circles = carbon; open circles = oxygen.

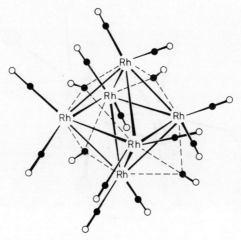

FIG. 5.21 Geometry of cluster compound $Rh_6(CO)_{16}$. Full circles = carbon; open circles = oxygen.

Cluster compounds show how very small metal aggregates can form more or less normal sorts of chemical bonds at their periphery. π-bonded ligands such as are shown in Figs 5.18 and 5.20 can occur, and this is reminiscent of the type of π-adsorbed species which is often postulated as an adsorbed reaction intermediate in some catalytic processes such as the hydrogenation of aromatic hydrocarbons or in dehydrocyclization reactions. The structures of cluster compounds with carbon monoxide ligands show examples of both linear and bridged CO groups and both of these are well known forms of carbon monoxide adsorbed on

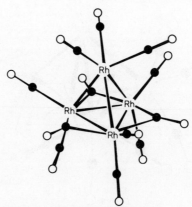

FIG. 5.22 Geometry of cluster compound $Rh_4(CO)_{12}$. Full circles = carbon; open circles = oxygen.

metal surfaces.[44] Examples are seen in Figs 5.21 and 5.22 which show the structures of $Rh_6(CO)_{16}$ and $Rh_4(CO)_{12}$. A hydrogen atom may also exist in a bridging situation, examples being in $M_3(CO)_{12}$ H where M ≡ Ru, Os, in which the hydrogen atom bridges one edge of the triangle, and $Re_3(CO)_{12}H_3$ with a hydrogen atom bridging each edge of the triangle. On the other hand, there are a number of compounds known in which hydrogen atoms have a simple monodentate function, e.g. $RuCo_3(CO)_{12}H$.

Easily decomposable metallic molecular cluster compounds may be used as precursors for the generation of dispersed supported metallic catalysts. An example is the use of $Rh_2Co_2(CO)_{12}$ for the preparation of a highly dispersed rhodium–cobalt bimetallic catalyst on a high area silica support.[88] The attraction of this approach is that it offers some hope for better control over particle structure than is available with conventional preparative methods. Collman et al.[91] have described the binding of $Rh_4(CO)_{12}$ or $Rh_6(CO)_{16}$ to a phosphenated polystyrene support.

7. Interaction of Metal Particles with a Non-Metallic Substrate

Metal–Support Bonding

The first question to be considered is the nature of the bonding between a metal crystallite and a non-metallic support upon which it resides. Most of the worthwhile information on this point has come from a study of the contact angle at the metal–support interface,[49-51] from a direct measurement by a mechanical method of the adhesion of a metal film on a support,[52,53] or from an electron microscopic study of the interface itself.[75] Results from various sources are not entirely consistent, and much depends on temperature and gaseous environment.

The interpretation of the contact angle in terms of the work of adhesion requires, of course, equilibration at a high temperature, and this has been done either above (e.g. ref. 49), or below (e.g. ref. 51) the melting point of the metal. Some of the data are summarized in Table 5.1, where it will be seen that, under a hydrogen atmosphere at high temperature, the work of adhesion lies in the range $0 \cdot 2$–$0 \cdot 9$ J m^{-2} for a variety of metals on ceramic-type supports. If one assumes that there are $1 \cdot 8 \times 10^{19}$ atom m^{-2} of metal in the interface, we find that these values for the work of adhesion correspond to the range 8–30 kJ mol^{-1}. Clearly, these values for the work of adhesion imply no more than a physical, van der Waals interaction between the metal atoms and the support. However, the magnitude of the work of adhesion depends very much on the environment in which the metal–support interface is established, particularly on oxygen or other reactive gases. The influence of oxygen is made clear from the data of

TABLE 5.1   Work of adhesion between metals and oxide support

System*	Work of adhesion ($J\ m^{-2}$)	Reference
Solid Ag on α-Al_2O_3	0·435	51
Solid Cu on α-Al_2O_3	0·475	51
Solid Au on α-Al_2O_3	0·530	51
Solid Ni on α-Al_2O_3	0·645	51
Solid γ-Fe on α-Al_2O_3	0·800	51
Molten Ni on BeO	0·187	49
Molten Fe on ThO_2	0·900	49

* In all cases a hydrogen atmosphere was used.

Benjamin and Weaver.[52,53] For a variety of metal films deposited on a glass support, they showed that if the interface was oxidized either by the presence of oxygen during film deposition, or by treatment with oxygen after deposition, the work of adhesion was markedly increased. On the other hand, if the support was an alkali halide, this effect of oxygen was not evident. Apparently, oxidation of the metal generates metal oxide which can interdiffuse with an oxide support, but this does not occur with alkali halide support.

The ease with which oxidation of the interface occurs is, of course, dependent on the nature of the metal. Nevertheless, even a gold film deposited on silica has its adherence increased by heating in oxygen,[54] while the sputtering of gold in an oxygen atmosphere also produces strongly adherent films,[55] so the phenomenon ought to be regarded as a general one.

However, there is also evidence which indicates that a metal can chemically react with the oxide support in the absence of added oxygen. One case is the reaction between nickel or iron and a titania support.[49] However, it is now known from observations in an electron microscope that at temperatures $\geqslant 1200$ K even metals such as platinum or palladium can undergo an interfacial reaction with supports such as alumina or magnesia.[75] When observed in the electron microscope this reaction occurs, of course, in a vacuum and it appears that a separate reaction product phase is produced in the interfacial region.

Conventional methods for the preparation of dispersed metal catalysts on oxide supports would often generate an oxidized metal–support interface. Much of this will presumably be removed during hydrogen reduction, but total removal will be quite uncertain. The final result will certainly influence the ease of particle mobility on the support, and would thus influence the ease of catalyst sintering. On the whole, it is probable that the temperatures used for catalyst processing are not sufficiently high to produce the type of interfacial reaction mentioned above for platinum and alumina or magnesia.

The interaction energy between metals and carbon supports is probably rather greater than with oxide-type supports. For instance, data obtained by the measurement of the temperature coefficient of nucleation of silver crystallites on graphite, indicate an interaction energy of some 92 kJ mol^{-1},[72] which is of a magnitude to suggest a chemical interaction. This system, as well as palladium on graphite has also been investigated by Baetzold[69, 79] using molecular orbital theory, and the conclusion emerged that the interaction energy is substantial. Moreover, it appears that the interaction energy per metal atom is largest for a single atom, and falls as the size of the metal aggregate increases. This interaction with carbon also has the result that a two dimensional aggregate of six palladium atoms is predicted not to have a d band when the aggregate is situated on a carbon support.

Electrical Phenomena at the Interface

The problem we are concerned with in this section refers to the possibility that electron transfer may occur between metal and support to such an extent that the gross electronic properties of the metal particle are modified, so that a modification of the metallic catalytic properties results. This question has often been approached by considering the metal–support interface as a metal–semiconductor contact according to conventional space-charge theory.[56, 57, 62] According to this, electrons will migrate towards the metal or the semiconductor according to which has the higher work-function, establishing a potential difference equal to the work-function difference. A space-charge of appropriate sign then exists in the semiconductor, the charge density decreasing with increasing distance into the semiconductor from the interface; and a charge of equal and opposite sign is induced at the metal surface. However, a quantitative examination of this theory raises some very serious problems, and it appears as a consequence that it is most unlikely to be an adequate description of reality so far as the properties of the metal are concerned. This may be illustrated by reference to the discussion due to Baddour and Deibert[62] for the behaviour of thin films of nickel evaporated on to germanium substrates which had various degrees of n- and p-type doping: the catalysts were used for formic acid dehydrogenation. The charge transferred is proportional to $(nV)^{1/2}$, where n is the charge carrier concentration in the semiconductor, and V is the potential difference which becomes established across the interface. The most important variable here is n, which can range over many orders of magnitude depending on the nature of the semiconductor. It turns out that even in the case of germanium as a substrate for which n can be varied widely over the range 10^9–10^{17} mm^{-3}, the charge transferred, σ, is relatively very small. Thus, with nickel/ germanium, σ lies in the range 10^{15}–10^{16} electrons m^{-2}. This means that even if the nickel layer is very thin, the actual change in the electron concentration in the metal is negligibly small. In the case of Baddour and Deibert's work, the nickel layer was up to about 20 atomic layers in thickness, and thus the increase in electron concentration would only amount to about 10^{-4} electron per nickel atom even in the most favourable circumstances. Even if it is argued that the transferred charge is entirely held at the free surface of the nickel, the density would still only be of the order of 10^{-3} electron per surface nickel atom. If one turns to oxide supports, the situation will in most cases be more unfavourable still, because the carrier concentration, n, will be considerably lower, and this is particularly so for many of the commonly used support materials which are at best only very weakly semiconducting.

When one comes to the behaviour of very small aggregates of just a few metal atoms, the situation is rather different from that described above. Certainly the simple space-charge theory for the metal–semiconductor contact would be inapplicable, and the situation ought to be described by an atomistic model. If an aggregate had (say) only three atoms, the loss of only one electron to the support would be expected to have a major effect on the properties of the aggregate. Various estimates have been made for the ionization potentials of small aggregates of silver and palladium.[18,21] For instance, for Ag_4 (tetrahedral) the ionization potential appears to lie in the region of 4·7–6·0 eV, so this represents the minimum electron affinity of the support required if Ag_4^+ is to be formed from Ag_4: for Pd_4 (tetrahedral), we would estimate the value for the ionization potential to be in the region of 5·5–8·0 eV basing our estimate upon a scaling factor using single atom ionization potentials as well as Baetzold's calculations.[18] Some insulating oxides possess surface electron acceptor sites, and the identification of the positive ions formed from adsorbed polycylic hydrocarbons (e.g. perylene) has been extensively used as a diagnostic tool for this type of site. The ionization potentials for the aromatic hydrocarbons which have been used in this way lie in the region 6·4–8·4 eV. It follows that it is quite reasonable to expect electron transfer from small metal aggregates to the same sort of electron acceptor site as is known to be effective for positive ion formation from aromatic hydrocarbons. Supports which are known to possess surface acceptor sites of this sort include silica–alumina and chlorinated alumina.[64,65]

Alumina does not possess strong acceptor sites even in the presence of oxygen unless dehydrated above about 820 K.[66] Pure silica is inactive.

Charge transfer between small silver aggregates and a graphite support has been investigated theoretically by Baetzold.[69] The interaction involves a transfer of charge from the silver to the graphite, and the average charge on an aggregate atom is shown in Fig. 5.23 as a function of aggregate size. Again, the extent of charge transfer is only appreciable when the aggregate contains just a few atoms. In the case of very small palladium aggregates, similar calculations indicate partial electron transfer in the opposite direction, that is, from the graphite to the palladium. E.s.r. studies with a platinum catalyst dispersed on a carbon support[42] also indicate some sort of electron transfer between the metal and the support, probably from the platinum to the carbon.[73]

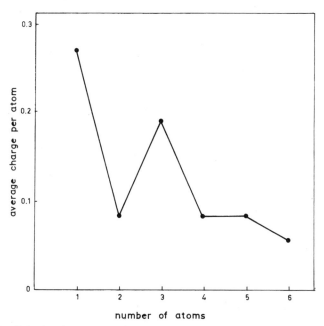

FIG. 5.23 Calculated average charge per silver atom transferred from silver aggregates to the graphite support, as a function of aggregate size. After Baetzold, R. C. *Surface Sci.* **36**, 123 (1972).

The reverse effect to that which has been discussed above, that is the alteration of the carrier concentration in a semiconductor by contact with a metal, can be quite appreciable, and Schwab has shown that this can lead to alteration of the catalytic activity of the semiconductor.[67,68] This effect would be expected to be considerably more pronounced than that involving a modification in the electron concentration in the metal, since the relative change in the charge concentration in the semiconductor will be much greater.

The nature of the metal–support interface must also be considered in relation to the possible presence of sites of special catalytic activity at the interface between small metal particles and the support in a dispersed catalyst. If the interaction were purely physical, it would be reasonable to assume that the metal atoms in contact with the support suffer only minor perturbation to their normal character, and special interface sites will not exist. On the other hand, if a chemical interaction exists, the metal atoms in the interface will be chemically modified, and may have distinct catalytic properties. However, even in this case, the proportion of metal atoms which are in the interface and which are accessible to reactant gas will be very small. Without having much firm evidence as guidance, we incline to the view that in conventional, well reduced, dispersed noble metal catalysts, the interface will be well reduced, and the metal–support interaction essentially physical in nature.

8. Particle Growth

We are concerned in this section with the processes that occur during the thermal treatment of dispersed catalysts, particularly particle growth. Particle growth will result in a reduction of the total free energy of the system. Growth can occur by two distinct mechanisms: (i) particle migration and coalescence; and (ii) the transfer of metal atoms individually from one particle and their deposition on another, and in this case we can envisage transport via surface diffusion across the support, or via vapour transport. These processes have been the subject of detailed discussion by Wynblatt and Gjostein [80, 81] and by Ruckenstein and Pulvermacher. [82, 87]

Particle Migration and Coalescence

There is more than one mechanism for particle migration. If the particles are more than just a few atoms in size, for instance, migration can occur by virtue of random metal atom movements over the metal particle surface: that is, if the migrating metal atoms happen to accumulate at some moment on one side of the particle, the net effect will be the migration of the particle as a whole in that direction. However, if the particle is only a small cluster with, for instance, four atoms in a tetrahedron, or six atoms in an octahedron, this migration model is not applicable, and the cluster probably translates across the support surface as an unperturbed unit, the process being dominated by the geometric relationships between the metal atoms and the support atoms which are in contact. In the following discussion we are concerned with the larger particles to which the discussion in references 80, 81, 82 and 87 applies. On the basis of the proposed model one can expect to be able to relate the particle diffusivity, D_p, to the diffusivity, D_a, of metal atoms on the particle surface. Two models have been used. The first assumes spherical particles with no surface energy anisotropy, and using the result obtained to describe the surface diffusion controlled migration of voids in solids[83, 84] we have

$$D_p = 4 \cdot 816 D_a (a/d)^4 \quad (5.3)$$

where a is the atom diameter and d is the particle diameter. Using data for platinum at 873 K,[80] $a = 0 \cdot 277$ nm and $D_a = 5 \cdot 0 \times 10^{-14}$ m^2 s^{-1}, we obtain the dependence of D_p on d shown by the full line in Fig. 5.24.

Alternatively, if one assumes a faceted particle so that the process is controlled by the nucleation of new monatomic layers on the facets, an application of nucleation theory gives

$$D_p = (D_a d/2l) \exp\left[-(\pi g \gamma_e / kT) d\right] \quad (5.4)$$

where l is the atom jump distance, g is the ratio of the facet diameter to the particle radius, and γ_e is the edge energy of a two-dimensional cluster of metal atoms. Using parameter values from reference 80, that is,

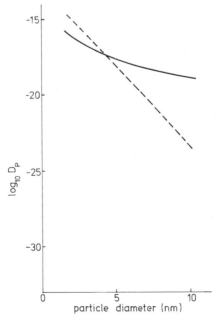

FIG. 5.24 Variation of particle diffusivity, D_p with particle diameter, according to equation 5.3 (full line), and equation 5.4 (broken line). The calculation is for platinum using parameters given in the text. D_p in m^2 s^{-1}.

$\gamma_e = 2 \times 10^{-11}$ J m^{-1}, g = 0·5, and assuming l = 0·277 nm, we obtain the dependence D_p on d shown by the broken line in Fig. 5.24.

The quantitative results from the two models are only in order of magnitude agreement for particles in the diameter range < about 6 nm. For the present purpose it suffices to say that for platinum particles with d = 2 nm at 873 K D_p is in the region 10^{-15}–10^{-16} m^2 s^{-1}, while with d = 6 nm. D_p is in the region of 10^{-18}–10^{-19} m^2 s^{-1}.

Interparticle Transport

Depending on whether atom diffusion across the support, or the transport of atoms across the particle edge is rate controlling, two different growth laws are obtained.[80] Both have the form

$$(\bar{d})^r \propto t \tag{5.5}$$

where r = 4 for diffusion control, and r = 3 for edge control.

Control of the concentration of diffusing atoms is dominated by the metal heat of atomization (ΔH_m) less the atom–support interaction energy (ΔH_s). If ΔH_s is about equal to a physical adsorption energy, the atom concentration will be negligible at normal sintering temperatures, and would only become significant if ΔH_s were at least half ΔH_m: under reducing conditions this is improbable, at least on clean oxide supports. Under oxidizing conditions ΔH_s will be increased, but with noble metals a mobile molecule of metal oxide becomes dominantly favourable, and interparticle transport then becomes quantitatively reasonable either via surface or vapour. For vapour transport, equation 5.5 also holds[80] but with r = 2.

Relation to Observed Sintering Behaviour

Experimental data are commonly fitted to some sort of rate–power law: using exposed metal surface area as the variable

$$\frac{dS}{dt} = -K_s S^n \tag{5.6}$$

or
$$1/S^{n-1} - 1/S_0^{n-1} = K_s' t \tag{5.7}$$

while using the mean particle diameter, $\bar{d}$, as the variable

$$\frac{d(\bar{d})}{dt} = K_d(\bar{d})^{-m} \tag{5.8}$$

or
$$(\bar{d})^{m+1} - (\bar{d}_0)^{m+1} = K_d' t \tag{5.9}$$

where the zero subscript indicates a value at zero time. These are connected via the relation

$$\bar{d} = \frac{6V}{S} \tag{5.10}$$

5. STRUCTURE AND PROPERTIES OF SMALL METAL PARTICLES

assuming spherical particles, where V is the total volume of particulate metal, and thus 5·6–5·9 are all equivalent, and

$$m = n-2 \tag{5.11}$$

Especially when n or m is large, it should not be imagined that fitting data to a single value of n or m necessarily implies that this value is characteristic of a single process. Indeed, the value assigned in this way in most cases represents some sort of empirical average, the entire sintering regime consisting of a sequence of processes with changing rate–power exponent.

The complexities of the physical situation are such that a quantitative calculation of the sintering rate for comparison with experiment is currently of limited utility, certainly if the comparison is intended to validate some proposed model. The unknown factors include: the effect of impurity on the metal surface on D_a, and depending on the nature of the impurity D_a may be either augmented or depressed; the extent to which the support surface contains topographic irregularities or trapping sites which depress the rate of particle or atom transport; and the extent to which the metal particles wet the support (also dependent on surface composition), since this influences the particle shape and hence D_p. Nevertheless, it is worth observing that, using a very crude model, particle migration can predict a sintering rate that is somewhere near the correct magnitude compared with experimental values. Assume one has a 1 wt.% platinum catalyst with a support of specific surface area 200 m² g⁻¹, and that all of the particles are initially of the same diameter, 2 nm. The particle density on the support is $5·55 \times 10^{14}$ m⁻², and the RMS particle velocity, calculated by taking $D_p = 5 \times 10^{-16}$ m² s⁻¹ at 873 K, is $3·33 \times 10^{-10}$ m s⁻¹. If we evaluate the *initial* sintering rate by merely calculating the binary collision rate, assuming the particles behave like a two-dimensional gas, one obtains for the relative rate of area reduction ($\Delta S/S$), 2×10^{-3}% s⁻¹, which is within an order of magnitude of the initial rates of sintering of some dispersed platinum catalysts under reducing conditions.[80] The only point to be made from this comparison is to validate the proposal that particle migration is a possible mechanism.

If particles migrate and coalesce, either migration or coalescence could be rate controlling. The kinetics of coalescence have been treated,[85, 86] and the quantitative conclusion[80] is that for particles with d < 50 nm, coalescence and attainment of ultimate shape are very fast compared to migration, so the latter must be rate controlling. Provided this comparison is made under similar conditions of surface composition, this conclusion is expected to be valid irrespective of the presence or absence of surface impurity, since although this (including chemisorbed gas) can reduce D_a, both coalescence and migration rates are proportional to D_a.

In assessing the significance of the particle diffusion mechanism, it is also worth comparing calculated average particle migration distances with the average interparticle distance in a typical supported catalyst. A 1 wt.% platinum catalyst in which the platinum particles of average diameter 6 nm are distributed over a support with a specific surface area of 200 m² g⁻¹ would have the particles spaced, on the average, about 200 nm apart. One would then take the view that in any situation where the average particle migration distance was substantially less than the average interparticle spacing, particle migration would not be an important process for particle growth. The previous discussion in this chapter would certainly lead to the conclusion that platinum particles are faceted, so the results for the faceted model in Fig. 5.24 are probably the more appropriate. Allowing for the possibility that equation 5.4 somewhat underestimates D_p, it is still reasonable to take $D_p \approx 3 \times 10^{-19}$ m² s⁻¹ for platinum particles at 873 K with d = 6 nm. The RMS platinum particle migration distance, x_p in time t may be evaluated from D_p, from the usual equation for two dimensional diffusion

$$x_p = 2\sqrt{D_p t} \tag{5.12}$$

Taking t equal to 5 h and a D_p value of 3×10^{-19} m² s⁻¹ leads to a value of x_p of 147 nm. In general then it seems reasonable to conclude that a diameter of 10 nm is a conservative upper limit beyond which particle migration can make no significant contribution to particle growth.

We summarize the situation as follows. For very small metal particles (d less than about 10 nm for platinum) growth occurs predominantly by particle migration with particle diffusion rate controlling. For larger particles, growth occurs predominantly by interparticle transport. Assessment of the likely significance of vapour transport is in conflict (cf. refs 80, 82), and the answer will no doubt vary a great deal from metal to metal, and depending on the way in which the surface chemistry can provide volatile species. However, Wynblatt and Gjostein's[80] electron microscopic observation of the gradual disappearance of platinum particles, initially of d̄ = 14 nm, in air at 973–1273 K clearly shows that this mechanism is possible provided the temperature and oxygen pressure are high enough.

Ruckenstein and Pulvermacher[82, 87] have considered in detail how a binary collision model based on particle migration leads to a growth law. They showed that a growth law of the form of equation 5.6 is produced, and the exponent n is related to the exponent q of the term $(a/d)^q$ in the relation between D_p and D_a, assumed to be of the form

$$D_p = AD_a(a/d)^q \tag{5.13}$$

where A is a constant. The relation is

$$n = q + 4 \tag{5.14}$$

In the case of spherical particles, equation 5.13 becomes 5.3 with $q = 4$, so $n = 8$. However, for the migration of faceted particles, the effective value of q varies with particle size and, for instance, when equation 5.13 is combined with the D_p vs d dependence given by equation 5.4, the value of q rises rapidly above 5 as the particle diameter rises above 2 nm. Ruckenstein and Pulvermacher consider the possibility of $q<4$, a situation which has been observed experimentally in terms of a sintering law with $n<8$.

Since the forms of equations 5.6–5.9 are all interconvertible, it is convenient to summarize the values of the rate–power law exponents, n and m, associated with the various sintering processes we have previously discussed. This is done in Table 5.2.

TABLE 5.2 Rate–power law exponents for various sintering processes

Mechanism	Power law exponents	
	n in equations 5·6 and 5·7	m in equations 5·8 and 5·9
spherical particle migration, diffusion control	8	6
faceted particle migration, diffusion control	>8	>6
interparticle transport, diffusion control	5	3
interparticle transport, edge control	4	2
interparticle transport, vapour control	2	1

Growth curves representing the entire range of exponents quoted in Table 5.2 have been observed in various experimental studies, and data are summarized in reference 80.

On the whole, we expect sintering by any of the first four mechanisms to be slower the greater the cohesive energy of the metal, so in this regime the increasing order of stability for the metals of interest as supported dispersed catalysts is

$$Ag<Cu<Au<Pd<Fe<Ni<Co<Pt<Rh<Ru<Ir<Os<Re$$

On the other hand, when the dominant mechanism is vapour transport using an oxygen-containing species, the equilibrium pressure of the latter is the important factor, and a different sequence is to be expected.[80]

$$Os < Ru < Ir < Pt < Pd \approx Rh$$

The factors which were mentioned previously as complicating any attempt at a quantitative calculation of sintering rates, also are of relevance to dispersion stability. In particular, stability will be enhanced by the use of a support containing positions at which metal particles or atoms are held more securely: holes in the support surface which have a radius of curvature comparable to that of the metal particle form sites with this property. Or again, stability will be improved by partial wetting of the support by the metal particle, as recently discussed.[82] Both the addition of dissolved or adsorbed material to the metal, and a modification to the chemical nature of the support surface may be contemplated, but the former carries the hazard of having undesirable consequences for the metal's catalytic activity.

References

1. van Hardeveld, R., and Hartog, F. *Surface Sci.* **15**, 189 (1969).
2. Schlosser, E. G. *Ber. Bunsen Ges.* **73**, 358 (1969).
3. Kimoto, K. and Nishida, I. *J. Phys. Soc. Japan* **22**, 940 (1967).
4. Schwoebel, R. L. *J. Appl. Phys.* **37**, 2515 (1966).
5. Bottoms, W. R. and Morriss, R. H. *In* "25th Annual Meeting of Electron Microscopy Society of America (1967)".
6. Ino, S. *J. Phys. Soc. Japan* **21**, 346 (1966).
7. Ino, S. and Ogawa, S. *J. Phys. Soc. Japan* **22**, 1369 (1967).
8. Allpress, J. G. and Sanders, J. V. *Surface Sci.* **7**, 1 (1967).
9. Allpress, J. G. and Sanders, J. V. *Austral. J. Phys.* **23**, 23 (1970).
10. Sundquist, B. E. *Acta. Met.* **12**, 67 (1964).
11. Sundquist, B. E. *Acta. Met.* **12**, 585 (1964).
12. Winterbottom, W. L. *Acta. Met.* **15**, 303 (1967).
13. Pilliar, R. M. and Nutting, J. *Phil. Mag.* **16**, 181 (1967).
14. Jaeger, H., Mercer, P. D. and Sherwood, R. G. *Surface Sci.* **11**, 265 (1968).
15. Sanders, J. V., private communication, CSIRO Division of Tribophysics, (1973).
16. Hoare, M. R. and Pal, P. *Nature* **236**, 35 (1972).
17. Kubo, R. *J. Phys. Soc. Japan* **17**, 975 (1962).
18. Baetzold, R. C. *J. Chem. Phys.* **55**, 4363 (1971). *Comments on Solid State Physics* **4**, 62 (1972).
19. Schissel, P. *J. Chem. Phys.* **26**, 1276 (1957).
20. Cotton, F. A. and Haas, T. E. *Inorg. Chem.* **3**, 10 (1964).
21. Mitchell, J. W., private communication, Dept. of Physics, Univ. of Virginia (1972).

22. Canterford, J. H. and Colten, R. "Halides of the Second and Third Row Transition Metals", Wiley, New York (1968).
23. Mie, G. *Ann. Phys.* **25**, 377 (1908).
24. Doremus, R. H. *J. Appl. Phys.* **35**, 3546 (1964).
25. Evdokimov, V. B., Ozeretokovskii, N. I. and Kobozev, N. I. *Zhur. Fiz. Khim.* **26**, 135 (1952).
26. Kobozev, N. I., Evdokimov, V. B., Zubovich, I. A. and Mal'tsev, A. N. *Zhur. Fiz. Khim.* **26**, 1349 (1952).
27. Evdokimov, V. B. and Kobozev, N. I. *Zhur. Fiz. Khim.* **28**, 362 (1954).
28. Bylina, E. A., Evdokimov, V. B. and Kobozev, N. I. *Russ. J. Phys. Chem.* **36**, 1392 (1962).
29. Reyerson, L. H., Solbakken, A. and Zuehlke, R. W. *J. Phys. Chem.* **65**, 1471 (1961).
30. Kubicka, H. *J. Catal.* **5**, 39 (1966).
31. Trzebiatowski, W., Kubicka, H. and Silva, A. *Roczniki Chem.* **31**, 497 (1957).
32. Trzebiatowski, W. and Kubicka, H. *Z. Chem.* **3**, 262 (1963).
33. Zuehlke, R. W. *J. Chem. Phys.* **45**, 411 (1966).
34. Czerwonko, J. *Phys. Status. Solidi* **30**, 723 (1968).
35. Denton, R., Muhlschlegel, B. and Scalapino, D. J. *Phys. Rev. Letters* **26**, 707 (1971).
36. Gor'kov, L. P. and Eliashberg, G. M. *Zhur. Eksp. Teor. Fiz.* **48**, 1407 (1965).
37. Dupree, R., Forwood, C. T. and Smith, M. J. A. *Phys. Status Solidi* **24**, 525 (1967).
38. Taupin, C. *J. Phys. Chem. Solids* **28**, 41 (1967).
39. Dupree, R. and Smithard, M. A. *J. Phys. C.* **5**, 408 (1972).
40. Mingos, D. M. P. *Nature, Phys. Sci.* **236**, 99 (1972).
41. Geus, J. W. *In* "Chemisorption and Reactions on Metallic Films" (J. R. Anderson, ed.) Academic Press, London, Vol. 1 (1971), p. 129.
42. Nicolau, C., Tom, G. and Pobichka, E. *Trans. Faraday Soc.* **55**, 1430 (1959).
43. Hoare, M. R. and Pal. P. *Adv. Phys.* **20**, 161 (1971).
44. Herring, C. *Phys. Rev.* **82**, 87 (1951).
45. Herring, C. *In* "Structure and Properties of Solid Surfaces" (R. Gomer and C. S. Smith, eds), University of Chicago Press, Chicago (1953), p. 5.
46. Nicholas, J. F. *Austral. J. Phys.* **21**, 21 (1968).
47. Mays, C. S., Vermaak, J. S. and Kuhlmann-Wilsdorf, D. *Surface Sci.* **12**, 134 (1968).
48. Wassermann, H. J. and Vermaak, J. S. *Surface Sci.* **22**, 164 (1970).
49. Humenik, M. and Kingery, W. D. *J. Amer. Ceram. Soc.* **37**, 18 (1954).
50. Kingery, W. D. *J. Amer. Ceram. Soc.* **37**, 42 (1954).
51. Pilliar, R. M. and Nutting, J. *Phil. Mag.* **16**, 181 (1967).
52. Benjamin, P. and Weaver, C. *Proc. Roy. Soc.* **A254**, 163 (1960); **A261**, 516 (1961); **A274**, 267 (1963).
53. Weaver, C. *Chem. and Ind.* **1965**, 370.
54. Moore, D. C. and Thornton, H. R. *J. Res. Natl. Bur. Std.* **62**, 127 (1959).
55. Mattox, D. M. *J. Appl. Phys.* **37**, 3613 (1966).

56. Schwab, G. M., Block, J. and Schultze, D. *Angew. Chem.* **71**, 101 (1959).
57. Solymosi, F. *Catal. Rev.* **1**, 233 (1968).
58. Prestridge, E. B. and Yates, D. J. C. *Nature* **234**, 345 (1971).
59. Anderson, J. R., Macdonald, R. J., and Shimoyama, Y. *J. Catal.* **20**, 147 (1971).
60. Poltorak, O. M. and Boronin, V. S. *Russ. J. Phys. Chem.* **40**, 1436 (1966).
61. Anderson, J. R. and Shimoyama, Y. *In* "Fifth International Congress on Catalysis" (J. W. Hightower, ed.) North-Holland, Amsterdam (1973), p. 695.
62. Baddour, R. F. and Deibert, M. C. *J. Phys. Chem.* **70**, 2173 (1966).
63. Figueras, F., Mencier, B., de Mourgues, L., Naccache, C. and Trambouze, Y. *J. Catal.* **19**, 315 (1970).
64. Flockhart, B. D., Scott, J. A. N. and Pink, R. C. *Trans. Faraday Soc.* **62**, 730 (1966).
65. Basset, J., Naccache, C., Mathieu, M. V. and Prettre, M. *J. Chim. Phys.* **66**, 1522 (1969).
66. Flockhart, B. D., Leith, I. R. and Pink, R. C. *Trans. Faraday Soc.* **65**, 542 (1969).
67. Schwab, G. M. *Surface Sci.* **13**, 198 (1969).
68. Schwab, G. M. and Zettler, H. *Chimia* **23**, 489 (1969).
69. Baetzold, R. C. *Surface Sci.* **36**, 123 (1972).
70. Romanowski, W. *Surface Sci.* **18**, 373 (1969).
71. King, R. B. *Prog. in Inorg. Chem.* **15**, 287 (1972).
72. Lewis, B. *Surface Sci.* **21**, 273, 289 (1970).
73. Nicolau, C. S. and Thom, H. G. *Z. Anorg. Allgem. Chem.* **303**, 133 (1960).
74. Avery, N. R. and Sanders, J. V. *J. Catal.* **18**, 129 (1970).
75. de Bruin, H. J., Moodie, A. F., and Warble, C. E. *J. Materials Sci.* **7**, 909 (1972).
76. Ollis, D. F. *J. Catal.* **23**, 131 (1971).
77. Hoffman, D. W. *J. Catal.* **27**, 374 (1972).
78. Hilliard, J. E. *In* "Phase Transformations", American Society for Metals, Metals Park, Ohio (1970), p. 497.
79. Baetzold, R. C. *J. Catal.* **29**, 129 (1973).
80. Wynblatt, P. and Gjostein, N. A. *Prog. Solid State Chem.* In press.
81. Wynblatt, P. and Gjostein, N. A. *Scripta Met* **7**, 969 (1973).
82. Ruckenstein, E. and Pulvermacher, B. *J. Catal.* **29**, 224 (1973).
83. Gruber, E. E. *J. Appl. Phys.* **38**, 243 (1967).
84. Speight, M. V. *J. Nucl. Mat.* **12**, 216 (1964).
85. Nichols, F. A. and Mullins, W. W. *Trans. AIME*, **233**, 1840 (1965).
86. Nichols, F. A. *J. Appl. Phys.* **37**, 2805 (1966).
87. Ruchenstein, E. and Pulvermacher, B. *A.I.Ch.E. J.* **19**, 356 (1973).
88. Anderson, J. R. and Mainwaring, D. E. *J. Catal.* **35**, 162 (1974).
89. Ross, P. N., Kinoshita, K. and Stonehart, P. *J. Catal.* **32**, 163 (1974).
90. Moore, C. "Atomic Energy Levels", Vol. 3, p. 101, Nat. Bur. Stand. Circ. 467, Washington, D.C. (1958).
91. Collman, J. P., Hegedus, L. S., Cooke, M.P., Norton, J. R., Dolcetti, G. and Marquardt, D. N. *J. Amer. Chem. Soc.* **94**, 1789 (1972).

CHAPTER 6

Measurement Techniques: Surface Area, Particle Size and Pore Structure

	page
1. SURFACE AREA MEASUREMENT	290
Physical Adsorption of Gases—Total Surface Area.	290
Chemisorption of Gases—Component Surface Area	295
Metal Surface Area.	295
Non-Metallic Surface Area	326
Gas Adsorption Methods	328
Volumetric Gas Measurements	328
Gas Volumetric Apparatus—Relatively High Subatmospheric Pressures	332
Gas Volumetric Apparatus—Ultrahigh Vacuum	338
Adsorption in Gas Flow Systems.	344
Gravimetric Measurements.	347
Radiochemical Measurements	352
Adsorption from Solution	353
Calorimetric Methods.	354
Electrochemical Methods	354
2. PARTICLE SIZE	358
Representation of Particle Size	358
Size Distribution and Mean Size	358
Metallic Dispersion.	360
Estimation of Particle Size	363
Electron Microscopy.	363
Via Surface Area	364
X-ray Methods	364
Magnetic Measurements	370
Miscellaneous Methods	376
3. PORE STRUCTURE	376
Physical Adsorption of Gases	376
Mercury Porosimetry.	384
Problems of Pore Structure Estimation.	385
Porosity and Density.	386

1. Surface Area Measurement

In this section we shall assume that the pore structure of the adsorbent is, if present at all, of negligible influence on the adsorption process, that is, any pores are so large that the adsorbent surface behaves as a free surface. Cases where pores do have an influence are dealt with in section 3.

Techniques for the measurement of surface area by adsorption fall into one of two categories: those which rely upon estimating the uptake required to form some specified sort of adsorbed layer so that by multiplying the number of adsorbed molecules at this point by the effective area per molecule, the total surface area is obtained; and those methods which do not involve an explicit use of the effective area per molecule. Adsorption to a monolayer is a particular case of the first of these categories.

The estimation of the monolayer uptake, n_m^s, is always done by fitting the data to a model for the adsorption process. A problem which is frequently met is that of estimating the separate contributing surface areas which may be exposed in a multiphase sample. This is always attempted by making use of the differences in adsorption specificity which the various surfaces may possess. However, the surface of a single phase may well be energetically heterogeneous to adsorption because of its crystallographic and topographic heterogeneity, and big differences in adsorption specificity must therefore be sought when a multiphase sample is examined. This means making measurements with different gases and often relying upon differences in chemical specificity in the adsorption process.

Physical Adsorption of Gases—Total Surface Area

The measurement of the total surface area of a sample requires non-specific physical adsorption. In fact, there is no adsorbate for which the heat of adsorption is not, at least to some extent, dependent on the nature of the adsorbent. However, the desired situation is most closely approached with simple non-polar molecules such as the rare gases or nitrogen, and these have been widely used. However, if a metallic phase is present in the sample, nitrogen can be an unwise choice because it is chemisorbed by some metals. Of the rare gases, argon and krypton are the most convenient in terms of the required temperature and pressure ranges (cf. Table 6.1). In some cases where a low surface area is to be measured there is an advantage with xenon because, other things equal, the gas pressures are lower and so are the dead-space corrections.

The isotherms observed in physical adsorption can vary widely. The basic types are shown in Fig. 6.1. In most cases physical adsorption merges continuously from the sub-monolayer region to the supra-monolayer region. Because of this, the type I isotherm which is fitted by the Langmuir equation is not of extensive practical significance for surface area measurement by physical adsorption. This transition region tends to make attempts to apply the Langmuir adsorption equation to the low pressure part of a physical adsorption isotherm of limited accuracy for surface area measurement except in some rather special cases where the heat of adsorption in the first layer is of considerably greater magnitude than the heat of adsorption in succeeding layers: xenon and krypton adsorption on clean transition metal surfaces are examples. Isotherms of types II and III apply to physical adsorption at a free surface. Types IV and V refer to adsorbents containing pores, and these will be referred to again in a subsequent section.

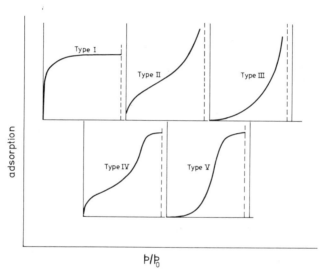

FIG. 6.1 The five general isotherm shapes in physical adsorption. The vertical broken lines indicate $p/p_0 = 1$.

For adsorption at a free surface the BET equation reads

$$n^s = \frac{n^s_m C p}{(p_0-p)[1 + (C-1)p/p_0]} \qquad (6.1)$$

where n^s is the quantity of gas adsorbed at an equilibrium pressure p, p_0 is the vapour pressure of the adsorbate in the condensed state at the adsorption temperature, and C is a constant related to the heat of adsorption into the first layer (ΔH_1) and the heat of condensation of the adsorbate (ΔH_c);

$$C = \mathscr{R} \exp [\Delta H_c - \Delta H_1]/RT] \qquad (6.2)$$

where $\mathscr{R}$ is a constant being approximately j_s/j_c; j_s is the partition function for the internal degrees of freedom for a molecule in the first adsorbed layer, and j_c the corresponding quantity for a molecule in the condensed phase.

The BET equation may be written

$$\frac{p}{n^s(p_0-p)} = \frac{(C-1)}{n_m^s C}\frac{p}{p_0} + \frac{1}{n_m^s C} \tag{6.3}$$

so a straight line results when $p/(p_0-p)n^s$ is plotted against the relative pressure, p/p_0. The value of n_m^s can be obtained from the slope which equals $(C-1)/C\, n_m^s$ and the intercept which equals $1/n_m^s C$.

The surface area, A, of the adsorbent is given by

$$A = a_m\, n_m^s \tag{6.4}$$

where a_m is the effective area per molecule in the monolayer.

The constant C determines the shape of the isotherm. The higher the value of C the more the isotherm tends towards type II (Fig. 6.1), the lower the value of C, the more it tends towards type III. However, only type II isotherms yield reliable values for n_m^s, but fortunately rare gas or nitrogen adsorption usually gives isotherms of this sort. The reason why type III isotherms are less satisfactory is that in this case there is a strong tendency for the second and succeeding layers to begin to form well before the first layer is complete: in other words, monolayer formation is much more severely obscured than is the case with type II. For many practical purposes, the BET equation is generally fitted to data over a range $p/p_0 = 0\cdot05 - 0\cdot3$. However, there are certain cases, notably krypton (77 K) and xenon (90 K) adsorption on clean transition metal surfaces where the relevant portions of the adsorption curves occur at $p/p_0 < 0\cdot05$.

If it happens that $C \gg 1$, equation 6.1 reduces to

$$\frac{p}{n^s(p_0-p)} = \frac{1}{n_m^s}\frac{p}{p_0} \tag{6.5}$$

and then a plot of $p/(p_0-p)n^s$ against p/p_0 passes through the origin, and the slope of the line which yields n_m^s can be evaluated from a single point, whereas a minimum of two points is needed with equation 6.1. Of course, in practice the precision is improved by using more than the minimum number of points.

Some of the parameters for gases which have been used in BET measurements are given in Table 6.1. Although this table lists "recommended general value" for a_m, it should be recognized that a_m is often somewhat dependent on the nature of the adsorbent. The problem has been summarized by Young and Crowell,[3] and has been further commented upon by

TABLE 6.1 Some gases used in BET measurements

Gas	Typical adsorption temperature (K)	p_0, saturated vapour pressure at adsorption temperature (Pa)	a_m, area per adsorbed molecule in monolayer (nm^2 $molec^{-1}$)	
			Range	Best common value*
nitrogen	77	1.013×10^5 at 77.4 K (liquid)	0.13–0.20	0.16_0†
argon	77	2.78×10^4 at 77.4 K (solid)	0.13–0.17	0.15_0
krypton	77 / 90	26.7 at 77.4 K (solid) / 2.70×10^3 at 90.2 K (solid)	0.17–0.22	0.20_0
xenon	90	8.25 at 90.2 K (solid)	0.18–0.27	0.23_0
methane	90	1.08×10^4 at 90.2 K (solid)	0.15–0.17	0.16_0
n-butane	273	1.013×10^5 at 272.7 K (liquid)	0.32–0.57	0.44_0
carbon dioxide	195	1.013×10^5 at 194.7 K (solid)	0.14–0.20	0.20_0

* Recommended general values have been assigned using data provided by references 2 and 158. The values have been rounded off to the number of significant figures consistent with their absolute reliability; the relative precision of area estimation is better than this and requires the retention of a third figure.
† 0.162 nm^2 $molec^{-1}$ has often been used.

Anderson and Baker[4] and by Gregg and Sing.[158] If the potential energy of the adsorbed molecule was independent of its position on the surface, the molecular packing at monolayer coverage would be dictated by the van der Waals size of the adsorbed molecules, and one could reasonably attempt to estimate a_m from the density of the adsorbate in the bulk liquid or solid state. However, there is no certainty that monolayer packing will be the same as in the bulk, and this is particularly so for molecules of non-spherical symmetry. Furthermore, if the interaction energy between the adsorbed molecule and the surface is periodic across the surface and is large compared to RT, the monolayer packing may be dictated by the relative positions of the adsorption sites (lattice packing[4]) and a_m can be larger than the value expected from the van der Waals size of the adsorbed molecule. This is particularly evident in the adsorption of krypton and xenon on *clean* transition metal surfaces at 77 or 90 K, and and in this situation the original literature (mainly refs 4, 216, but also 123, 159) should be consulted to select the most appropriate a_m. The general conclusion is, nevertheless, that it is wise to avoid the use of adsorbate molecules which are markedly non-spherical (e.g. carbon dioxide, n-butane) and to try (if possible) to adjust the temperature to avoid lattice packing. For most purposes nitrogen (77·4 K), argon (77·4 K) or krytpon (90·2 K) are probably the best choices.

The method due to Harkins and Jura[41] makes the use of the adsorption equation

$$\ln(p/p_0) = d - f/(n^s)^2 \qquad (6.6)$$

where d and f are constants. The surface area is proportional to $f^{\frac{1}{2}}$

$$A = k f^{\frac{1}{2}} \qquad (6.7)$$

and in the original method k was evaluated by calibration with a solid of known surface area. In fact, k is mainly dependent on the nature of the adsorbate, but also to some extent on the nature of the adsorbent, in much the same way as is a_m. The relation between the BET and the Harkins–Jura equations has been studied (cf. 3) and

$$\left.\begin{array}{l} n^s_m = F f^{\frac{1}{2}} \\ k = 0\cdot 269 \; F \; a_m \end{array}\right\} \qquad (6.9)$$

where F is about 1, but varies to some extent with the value of the BET parameter C. The Harkins–Jura method is inferior to the BET method[3] because the quantity $1/(n^s)^2$ is sensitive to slight experimental variations in n^s and because the range of p/p_0 in which a linear Harkins-Jura plot is obtained is variable, and the plot may contain more than one linear section.

Other adsorption equations have been used for the evaluation of n^s_m but none has been as extensively used as the BET method. Examples are the Huttig equation[25, 145] the equations due to Lopez-Gonzalez and Deitz,[26] Huttig and Theimer[27] and the procedures due to Crawford and Tompkins[29] and Harkins and Jura.[30] These are all essentially empirical methods, and even at an empirical level have nothing to offer over the BET equation.

Gregg[28] treated the adsorbed layer as a two-dimensional gas in the submonolayer region, but the estimation of surface area is subject to considerable inaccuracy (using the BET surface area as a reference) because of the difficulty of assigning proper values for the effective molecular area.

There are two main approaches in which a value for the effective molecular area is not used explicitly. The first makes use of adsorption in the very low coverage region where uptake is linear with pressure,

$$n^s = k_H \, p \, A \qquad (6.10)$$

where k_H is the Henry law constant. Although attempts have been made at *ab initio* calculations of k_H,[31-40] this is not adequate for practical evaluation of A. The second method due to Barker and Everett[32] is based upon the idea that the volume accessible to a non-adsorbable gas (e.g. helium at a high temperature) is less than the apparent value by an amount equal to AD, where D is the distance of closest approach of the centres of the absorbate and adsorbent atoms. The method, which needs measurement of small deviations from Henry's law, involves very high precision gas volumetric measurement and requires a theoretical evaluation of D, the accuracy of which is not totally satisfactory. On the whole, this method does not give good agreement with the BET method.

Chemisorption of Gases—Component Surface Area

Metal Surface Area

The adsorption selectivity that is required for the measurement of the metal surface area in a multicomponent (e.g. supported) metallic catalyst is achieved by measuring the uptake of a gas which is chemisorbed on the metal surface but which is adsorbed on the non-metallic components only to a relatively small (ideally zero) extent. Of course, with a catalyst which contains only a metal this problem of component differentiation does not exist; the surface area of the metal is equal to the total surface area, and in principle either physical adsorption or chemisorption can be used. However, in this latter situation both methods have their respective drawbacks. Chemisorption requires that the chemical composition of the surface be well defined so that adsorption stoichiometry is also well defined. On the other hand, with samples of small surface area inaccuracies due to dead space corrections are less severe with chemisorption because of the considerably lower gas pressures. The chemisorption of hydrogen, carbon monoxide and oxygen has been extensively examined, while other substances such as nitrous oxide, ethylene, benzene, carbon disulphide, thiophene and thiophenol have on occasions been reported upon.

The use of chemisorption for surface area measurement requires a knowledge of the chemisorption stoichiometry (X_m), at monolayer coverage, *that is, the average number of surface metal atoms associated with the adsorption of each adsorbate molecule,** and also a knowledge of the number of metal atoms per unit area of surface (n_s). Thus, if at monolayer chemisorption, n_m^s is the total adsorbate uptake, the total surface area, A, is given by

$$A = n_m^s X_m n_s^{-1} \qquad (6.11)$$

if n_m^s is expressed in molecules; or this should be multiplied by the factor $2 \cdot 687 \times 10^{22}$ if the gas adsorption is expressed in dm³ (STP). Note that

* This definition of X_m has been adopted because of its convenience when referring to the chemisorption of complex molecules which may undergo fragmentation into chemically different surface species.

the chemisorption stoichiometry is here defined with reference to the adsorbate *molecule* so that if, as with hydrogen, dissociative adsorption occurs, and if each chemisorbed hydrogen atom were associated with one surface metal atom, the chemisorption stoichiometry would equal two.

If the adsorbent is a single crystal surface, the value for n_s may be obtained from the surface crystallography: the necessary information is contained in Table 3.1 and equations 3.1 and 3.2 (Chapter 3) together with the lattice parameter of the metal (Table A.1). For a polycrystalline surface, either as massive metal or dispersed metal, the value for n_s is not completely specified since it depends on the surface structure. However, as a working basis the usual approximation is to assume that such a surface is formed from equal proportions of the main low index planes, and the values so obtained are collected into Table 6.2 for the metals of most common catalytic interest.

TABLE 6.2 Number of surface atoms per unit area of polycrystalline surface (n_s)

Metal	Concentration of surface atoms (10^{19} m^{-2})
chromium	1·63
cobalt	1·51
copper	1·47
gold	1·15
hafnium	1·16
iridium	1·30
iron	1·63
manganese	1·40
molybdenum	1·37
nickel	1·54
niobium	1·24
osmium	1·59
palladium	1·27
platinum	1·25
rhenium	1·54
rhodium	1·33
ruthenium	1·63
silver	1·15
tantalum	1·25
thorium	0·74
titanium	1·35
tungsten	1·35
vanadium	1·47
zirconium	1·14

6. MEASUREMENT TECHNIQUES I

There are a number of factors which in general influence the nature of gas chemisorption behaviour on metals, and many practical situations are thus quite complex. The important factors are: (a) adsorbed surface impurity: (b) crystallographic surface heterogeneity; (c) surface reconstruction during adsorption; (d) incorporation or solution of the adsorbate; (e) intrinsic effect of metal particle size; (f) the presence of a support. It is not our present purpose to give an account of gas chemisorption in extensive detail, we are rather concerned merely to draw attention to the main features of the behaviour which are observed when chemisorption is used to estimate metal surface area.

Hydrogen Absorption by Metals

A number of metals either take up hydrogen in true solution as dissolved hydrogen atoms, or form hydrides. Extensive summaries of the data are available.[44, 162, 163] We are only concerned to summarize the main features which are relevant to the use of hydrogen chemisorption for surface area estimation.

We first consider the metals in which hydrogen forms a true solution, and these certainly include aluminium, chromium, molybdenum, tungsten, iron, cobalt, nickel, manganese, copper, silver and platinum. The

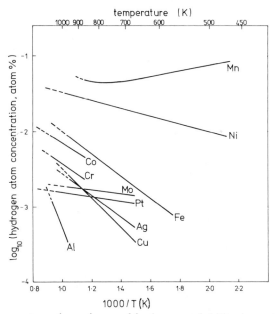

FIG. 6.2 Temperature dependence of hydrogen solubility in metals. Hydrogen pressure 101 kPa (1 atm). Data from reference 44.

solubility in these metals is relatively modest, varies with $p_{H_2}^{\frac{1}{2}}$, and (except in the case of manganese where a complex series of phase changes results in a minimum in the solubility at about 820 K) increases with increasing temperature. Solubility data are summarized in Fig. 6.2. Except for manganese, which is exceptional, the hydrogen solubility in all these metals extrapolated to 273 K is less than about 10^{-3} atom % at a hydrogen pressure of 101 kPa (1 atm).

A number of metals form hydrides. These include alkali and alkaline earth metals which form salt-like hydrides such as NaH and CaH_2. However, there is a substantial number of transition metals which form hydrides, often of variable stoichiometry, and these include titanium, zirconium, hafnium, thorium, vanadium, niobium, tantalum, cerium, lanthanum, rare earth metals, and palladium. With these metals the amount of hydrogen absorbed can often be substantial, and it decreases with increasing temperature. The absorption decreases with decreasing hydrogen pressure, but a dependence on $p_{H_2}^{\frac{1}{2}}$ often is limited to a restricted range of composition and temperature—generally at higher temperatures

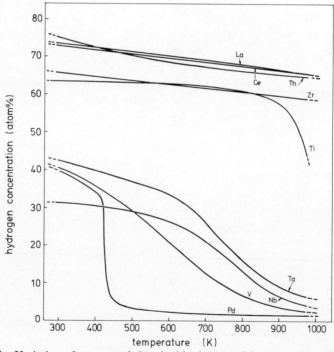

FIG. 6.3 Variation of amount of absorbed hydrogen with temperature. Hydrogen pressure 101 kPa (1 atm). Data from reference 44.

and lower concentrations. Data are summarized in Fig. 6.3. In the vicinity of room temperature and at a hydrogen pressure of 101 kPa the uptake can reach the region of 30–75 atom %.

There appear to be no reliable data for hydrogen absorption by noble metals other than platinum and palladium. Data from the early literature quoted by Dushman[44] are certainly grossly high due to adsorption on the very finely divided metal and to reaction with impurity. We believe it likely that with these metals—rhodium, iridium, ruthenium and osmium—hydrogen absorption will be of the same order as, or perhaps rather less than with platinum.

Main Chemisorption Features: Hydrogen

By way of illustration of the complexities which need to be considered, we shall first discuss adsorption on nickel and platinum.

Hydrogen shows some important crystallographic specificity in chemisorption on platinum, but except in very special situations this is not important from the point of view of surface area estimation. It has been shown[63, 64] that hydrogen does not chemisorb rapidly on platinum (111) or (100) crystal faces. This is due to a low sticking probability, and the result is that at low hydrogen pressures (10^{-2}–10^{-4} Pa; about 10^{-4}–10^{-6} Torr) adsorption is so slow that it is very difficult to observe. On the other hand, it has been shown[65] that if monatomic steps are present in an otherwise flat (111) surface, these provide sites for the ready chemisorption of hydrogen, and since adsorbed hydrogen can be transported from the step sites onto the interstep terraces by surface diffusion, the effect is a large increase in the sticking probability averaged over the whole surface. It appears that (111) steps are more active than (100) steps (cf. 3.17, Chapter 3). However, inasmuch as large (111) or (100) crystal faces with a low concentration of steps or other imperfections will only be found on specially prepared specimens, it can be expected that with all other specimen types where this special situation does not exist, hydrogen chemisorption will occur readily over the entire surface. This is a conclusion that is clearly in agreement with the behaviour on clean polycrystalline specimens of platinum where, as with nickel, adsorption to give a monolayer of hydrogen adatoms is fast, with a time scale of the order of a few seconds to a minute. The term "monolayer" is used here in the sense of referring to the hydrogen uptake in the saturation region of the adsorption isotherm, and where $X_m \approx 2$. There has been protracted discussion on the occurrence and significance of further weak hydrogen adsorption beyond a monolayer at higher hydrogen pressures, the significance of slow uptake under these conditions, and the influence of adsorbent structure when evaporated metal films are used. Although these factors may not be entirely negligible (e.g. on platinum at 77 K[62]) we consider that for the present purpose the above operational definition of monolayer coverage is adequate.

By comparison, on supported dispersed specimens there is always a substantial slow uptake component potentially present which has a time scale of the order of many minutes to several hours: this slow uptake is an activated process and it is observed unless the temperature is so low (for instance 77 K) that it becomes kinetically insignificant. Examples of slow uptake are contained in Figs 6.4 and 6.5. However, it should be empha-

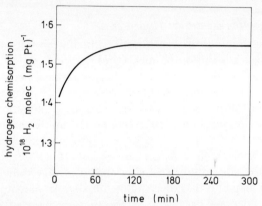

FIG. 6.4 Slow hydrogen chemisorption at 523 K and 26·7 kPa (200 Torr) on 1·1 wt.% platinum/η-alumina catalyst. The catalyst had previously been hydrogen reduced at 770 K followed by evacuation at that temperature. After Gruber, H. L. *J. Phys. Chem.* **66**, 48 (1962).

sized that the actual time scale of this slow adsorption tends to be rather variable depending on the type of sample and its history and, for instance, cases are not uncommon with highly dispersed platinum on γ-alumina where slow hydrogen uptake at 520 K is substantially complete in about 10 min.

In addition to the presence of a slow uptake component, the adsorption isotherm often has a different character on dispersed specimens compared with massive specimens, in that the isotherm tends towards saturation at much higher equilibrium pressures over the dispersed specimens. This sort of behaviour may be inferred from Fig. 6.6, for example, bearing in mind that the pressure range required for saturated adsorption on a nickel film at (say) 300 K is of the same order as that shown for the isotherm at 90 K, and is very much lower in pressure than is required for saturation with a nickel/silica specimen at 291 K.

In most cases, gas diffusion through the pores of the catalyst is not the

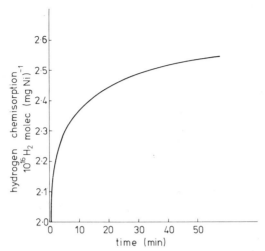

FIG. 6.5 Slow hydrogen chemisorption at 303 K and 26 kPa (196 Torr) on 30 wt.% nickel/alumina catalyst. The catalyst had been previously hydrogen reduced at 660 K, followed by evacuation at that temperature. Data from Narayanan, S. and Yeddanapalli, L. M. *J. Catal.* **21**, 356 (1971).

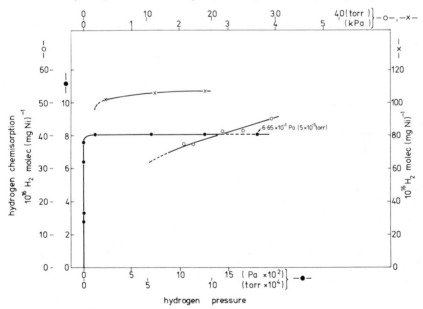

FIG. 6.6 Hydrogen adsorption isotherms on nickel.
●, 90 K on evaporated nickel film (ref. 50)
○, 195 K on 10 wt.% nickel/silica–alumina (ref. 58)
×, 291 K on 10 wt.% nickel/silica (ref. 59).
The nickel/silica–alumina catalyst had been hydrogen reduced at 720 K followed by evacuation at that temperature; the nickel/silica catalyst had been hydrogen reduced at 640 K followed by evacuation at that temperature.

main factor causing slow chemisorption. This can be seen from the following features which are all inconsistent with gas transport being of dominant importance in controlling the rate: the nature of the uptake *vs* time plot (failure of proportionality to time$^{1/2}$), and particularly the existence of a rapid uptake component; the strong dependence of the rate on the chemical identity of the gas; the extended time scale over which slow uptake is often observed; the magnitude of the observed activation energy for slow chemisorption which is often much greater than the value expected for gaseous diffusion (10–15 kJ mol^{-1}).

This characteristic hydrogen adsorption behaviour of supported dispersed metal catalysts could arise in various ways. It could be due to the presence of impurity on the metal surface. There could be a co-operative effect between metal particles and the support, and from the point of view of metal surface area estimation, the main question is then whether the hydrogen so adsorbed is located on the metal or on the support. A slow absorption process may be operating, while there is also the possibility that the behaviour reflects intrinsic properties of very small metal particles. We shall examine these possibilities in turn.

A number of important possible surface contaminants can influence gas adsorption on metals, such as platinum and nickel. Oxygen contamination is an obvious possibility, particularly when specimens have been heated in air or oxygen during processing. With platinum there is good evidence based on LEED and on Auger electron spectroscopy that, provided the specimen temperature can be increased enough, as with a massive platinum specimen, an atomically clean surface can be generated by using gaseous hydrogen to remove surface oxygen.[66] However, not all the oxygen on a platinum surface reacts with equal ease. Thus, when oxygen is added to a clean polycrystalline platinum surface at 195 K, adsorption is fast for most (about 95 %) of the total uptake, but the last 5 % or so is slow with a time scale of a few tens of minutes.[67] This slow component is due to a slow oxygen incorporation process, probably by a place-exchange mechanism. This incorporated oxygen is of diminished chemical reactivity, and for instance does not react at an appreciable rate with gaseous hydrogen at 195 K, although the oxygen of the rapid uptake component does so react: the incorporated oxygen reacts slowly with gaseous hydrogen at 393 K. As the temperature at which the platinum is contracted with oxygen increases, there is an increase in the extent of incorporation, and the temperature required for its removal by hydrogen also increases. Nevertheless, it has been shown that oxygen removal goes to completion at $\geqslant 800$ K by reaction with hydrogen, and this is also sufficient to desorb chemisorbed hydrogen into a vacuum. From this one would conclude that with supported dispersed platinum catalysts, complete removal of surface oxygen by reaction with hydrogen is possible at the highest reduction temperatures which can be used without affecting the pore structure of the support or too severe sintering of the metal particles, namely maximum temperatures in the region of 770–820 K. It is also clear that with *unsupported* dispersed platinum this required reduction temperature would lead to severe sintering. In fact, the temperatures used for hydrogen reduction of dispersed platinum catalysts are often in the region of 470–670 K, and in these circumstances it seems reasonable to infer that while much of the surface oxygen will be removed, some will remain. Sulphur is more difficult to remove by hydrogen treatment from a platinum surface than is oxygen. With massive metal, sulphur may be present as a solute impurity, while with dispersed platinum sulphur may be introduced as an impurity during preparation: some may be generated during hydrogen reduction from any sulphate impurity. Chlorine is likely to be removed by hydrogen reduction with a facility comparable to oxygen. Carbon contamination is of frequent occurrence with massive platinum specimens, and occurs from the presence of carbon solute in the platinum, as well as from the decomposition of carbon-containing residual gases at the metal surface. The problems of carbon removal from a refractory metal surface are relatively severe, and hydrogen treatment is ineffectual. The methods of carbon removal are discussed in Chapter 3 (p. 116). With a metal such as platinum, high temperature treatment with oxygen followed by hydrogen treatment is satisfactory. It thus seems probable that carbon contamination is not a serious problem with dispersed platinum catalysts on oxide supports when these have been subjected to heating in air or oxygen during the course of preparation. However, when carbon is used as a support the likelihood of carbon contamination on the metal surface is quite high:[79, 80] again, much of it can be removed by cautious treatment with oxygen (e.g. about 20 kPa, 150 Torr; 620 K; 15 min) followed by hydrogen reduction.[79] The effect of carbon contamination on gas adsorption is complicated. At least at low levels of contamination, it appears that the sticking probability of hydrogen on *massive* platinum is not much affected:[66, 68] however, oxygen adsorption is highly sensitive and the sticking probability falls dramatically with increasing carbon concentration. On the other hand with a dispersed platinum/carbon catalyst it appears[81, 79] that carbon contamination increases the gross hydrogen uptake. This is because most of the gross uptake in this case is due to spillover hydrogen which is initially adsorbed on the metal but which is transferred to the surface of the carbon support: the carbon contaminant seems to form bridges between particle and support which aid this transfer.

Now while it is true that a slow hydrogen uptake component has frequently been observed on supported dispersed platinum specimens which have been prepared under conditions which would indicate the possible presence of surface impurity, it is also true that a slow uptake has been observed with specimens which were prepared under conditions such that the surface impurity level would be expected to be quite low: examples would be the data due to Gruber[71] (cf. Fig. 6.2) and to Spenadel and Boudart[73] where the specimens had been calcined in air and hydrogen reduced at 770 K. Moreover, the relative amount of slow adsorption decreases as the average platinum particle size increases, and it has been reported to be relatively small over a platinum black specimen,[71] despite the expectation that the level of surface impurity with this type of specimen should be considerable.

In the case of dispersed nickel catalysts the contamination problem is more severe than with platinum. A comparison between hydrogen chemisorption data for evaporated films,[48-51] for some typical supported catalysts[52-54] and for nickel powder[55, 56] has been made by Roberts.[55, 56] On *both* supported nickel and on nickel powder there is a substantial slow uptake component. On both types of specimen the importance of the slow component decreases with increased vigour of surface purification, but is never eliminated.

As for the likely importance of surface contaminant with other dispersed metal catalysts, the best that can be done is to make a qualitative (and rather tentative) judgement based on such thermodynamic data as are available. Some such data are contained in Table 4.1 (Chapter 4). From this we would make the following assessment: the metals palladium, iridium, rhodium, ruthenium, (osmium?) and silver and probably roughly comparable with platinum; nickel, iron and cobalt are much more difficult to obtain in a contamination-free state, while copper (?) and rhenium occupy intermediate positions. Although these conclusions are based on data that refer to the bulk oxides, similar conclusions with respect to relative reactivities are obtained by considering the reaction between gaseous hydrogen and chemisorbed oxygen.[67, 74]

However, in addition to impurity of the sort just considered, there is also the possibility that even with oxide-type supports, some support material migrates onto the metal particle surface where it remains impervious to hydrogen reduction. This is a proposal which has been espoused particularly for supported nickel catalysts (cf. Chapter 4) but it is hard to see why it should not be of general occurrence, and therefore be a generally important factor in controlling the nature of hydrogen adsorption.

We next consider a co-operative effect between the metal and the support. It has been suggested[71] that on a highly dispersed platinum specimen, whereas most of the adsorbing hydrogen molecules will find two vacant adsorption sites on the same metal particle, at high coverage a stage will be reached when an adsorbing molecule has only a single site available on a given particle, so that the second atom must migrate either to the support or across the support to an adjacent platinum particle. Either of these would be slow activated processes, and either all or half of the total slow uptake would be retained by the metal. One could envisage such isolated adsorption sites arising either because there was only one on a given particle, or perhaps because heterogeneity of hydrogen adatom binding energy allows several isolated sites to exist on a given particle. It would certainly seem that imposing a restriction of only one isolated site per particle would result in the model predicting too small an uptake by the slow component. For instance, for a particle of diameter 2 nm, one vacant site would represent only about $0 \cdot 3\%$ of the total number of surface atoms, while for a very small cluster particle of (say) six atoms, one vacant site would (nominally) represent some 17% of the total number of sites, and these figures need to be compared with the fact that a slow uptake component of about 10% of the total is well known for supported platinum catalysts with average particle sizes appreciably in excess of six atoms.

An alternative point of view is to assign the whole of the slow uptake to hydrogen spillover on the support, on which assumption it contributes nothing towards adsorption on the metal. It is well at this juncture to point out that hydrogen spillover is a well established phenomenon. It occurs to a dramatic extent in hydrogen adsorption on platinum/carbon catalysts where, for instance, it has been reported[81] that adsorption at 620 K and 40–80 kPa (about 300–600 Torr) yielded a net number of adsorbed hydrogen atoms (i.e. after applying a correction for the intrinsic adsorption measured on the support alone) which exceeded the *total* number of platinum atoms present in the specimen by a factor of 3–10. Or again, platinum dispersed on tungsten (VI) oxide greatly augments the rate of reduction of the oxide by hydrogen due to spillover of adsorbed hydrogen.[75] Spillover can also occur with oxygen, and the increased rate of carbon oxidation due to the presence of dispersed platinum (e.g. ref. 83) is a typical example. However the surface mobility of adsorbed oxygen is much lower than that of adsorbed hydrogen, and in general spillover of adsorbed oxygen is of negligible importance under conditions where oxygen chemisorption may be used for the estimation of metal surface area.

Although hydrogen spillover can be severe when carbon is used as a support, there is evidence that it is generally of no more than minor significance with oxide supports such as silica, alumina or zeolite. Thus Hall and Lutinski[86] used the temperature dependence of exchange of surface hydrogen with gaseous deuterium to estimate separately the hydrogen on the platinum and on the alumina surfaces of platinum/alumina catalysts. The alumina was a mixture of η- and γ-alumina with a specific surface area about 140 m² g^{-1}. This technique gave much the same value for the amount of hydrogen chemisorbed on the platinum after equilibration and evacuation at 77 K or room temperature as was measured from the uptake at 520 K and 32 kPa (about 240 Torr): if anything the latter measurement gave a somewhat lower uptake figure.

We turn now to hydrogen absorption. It is instructive to examine the data from Figs 6.4 and 6.5 in the light of the hydrogen solubilities given in Fig. 6.2. The slow uptake of hydrogen on the platinum/η-alumina catalyst at 523 K and 26·7 kPa (200 Torr) (Fig. 6.4) amounts to about 2×10^{17} H$_2$ molec (mg Pt)$^{-1}$: from Fig. 6.2 we estimate that the maximum amount of hydrogen dissolved in platinum under these conditions is about 4×10^{-4} atom % that is, about $6·17 \times 10^{12}$ H$_2$ molec (mg Pt)$^{-1}$, which is less than 10^{-2} % of the total slow uptake. Similarly, dissolved hydrogen cannot account for more than about 10^{-2}% of the slow uptake on the nickel/alumina catalyst shown in Fig. 6.5. The solubility data in both these cases thus show that hydrogen absorption by the metal could make no more than a negligible contribution to the slow uptake process, and this conclusion can be extended to all the other metals referred to in Fig. 6.2 (with the possible exception of manganese) and to the other noble metals with the exception of palladium. In the case of palladium there exist sufficient kinetic data (e.g. ref. 44) on the absorption process to make a rough estimate of the time required to reach absorption equilibrium. For a particle of 20 nm diameter in hydrogen at 133 Pa (1 Torr) and at room temperature, the time required is approximately 10 minutes, which is comparable with the time scale of a slow uptake on a dispersed catalyst. The rate of absorption is reduced by surface impurity.

In the light of this discussion we may make the following comments. Provided the catalyst is thoroughly reduced, with a supported dispersed metal like platinum which has a low hydrogen solubility, the characteristic features of hydrogen adsorption must be mainly ascribed to the presence of support material at the surface of the metal particles, and/or to the intrinsic properties of very small metal particles. An assessment of the relative importance of these factors cannot be made with complete certainty, but hydrogen adsorption behaviour on ultrathin metal films is relevant. It is known[164,165] that with this sort of dispersed metal specimen which consisted of platinum particles of average diameter $\geqslant 2·0$ nm on a glass or mica support, hydrogen adsorption at room temperature has a very similar character to that of clean massive metal. The method of preparation of ultrathin films makes it highly improbable that there will be contaminant on the metal surface, either adventitiously adsorbed, or by the migration of support material. For this particle size range then, one may conclude that intrinsic particle size effects are of negligible importance, and the behaviour of conventional supported platinum is more likely to be due to the presence of some support material on the platinum surface. Ultrathin nickel films behaved in the same way, and the same conclusion follows for this metal, except that the impurity present on the surface of conventional supported nickel may well include some other residual impurity in addition to support material.

The situation with respect to smaller metal particles is, however, even more uncertain, and it is possible that an intrinsic particle size effect may operate with particles df (say) <1·0 nm diameter. The evidence which is discussed in Chapter 5 suggests that very small metal particles which consist of no more than a few tens of atoms behave as though they have a greater electron affinity than do the corresponding bulk metals. Thus on this basis alone, if bonding an adatom involved partial or complete electron transfer from the metal, the bond energy would be expected to be lower on such a small particle, so the corresponding dissociative adsorption process would also have a numerically smaller heat of adsorption. Dissociative chemisorption of hydrogen on platinum (as on nickel) causes a work-function increase (cf. ref. 62) and this is consistent with a net movement of electron charge towards the adatoms. At the same time, it is generally found in chemisorption that a decrease in the magnitude of the heat of adsorption is accompanied by an increase in the activation energy for adsorption, and this may be qualitatively accounted for by the change in the crossover position of the potential energy curves for chemisorption and a more weakly adsorbed precursor (cf. ref. 87). The matter requires further illumination at the experimental level.

From the conclusions we have drawn it follows that the entire hydrogen uptake (corrected for intrinsic adsorption on the support and for absorption, if any) that is, both fast and slow components, is needed to give an estimate of the monolayer uptake on clean supported dispersed metals. Nevertheless, the correct chemisorption stoichiometry for the slow component remains somewhat uncertain: in practice it is assumed to take the same value as the fast component.

Main Chemisorption Features: Oxygen and Carbon Monoxide

Chemisorption of oxygen or carbon monoxide is generally less satisfactory than that of hydrogen for surface area estimation with the transition metals because monolayer formation and the chemisorption stoichiometry are less easily controlled. Again, adsorption on clean massive metals is a useful reference point.

With oxygen chemisorption on platinum, a rather similar situation exists with respect to crystallographic specificity as has been described for hydrogen chemisorption: on both (111) and (100) faces the sticking probability for oxygen is low (about 10^{-6} on (111) at 575–775 K.[64, 66] On a (111) face containing (111) or (100) monatomic steps, oxygen adsorption occurs more readily,[65] although temperatures in the region 900–1100 K are needed for substantial oxygen transport away from the step. Thus, although there is considerable evidence that on a polycrystalline platinum surface, oxygen adsorption occurs with an initial sticking probability of $0 \cdot 1$–$0 \cdot 2$ (for a summary of the literature see reference 66), a doubt must remain as to the extent to which the (111) and (100) facets are covered at temperatures <900 K.

Studies with evaporated metal films have shown[106, 160] that there are some transition metals (e.g. rhodium, tungsten, molybdenum, cobalt, nickel) for which rapid oxygen adsorption at 77–90 K and about 10^{-2} Pa (about 10^{-4} Torr) is limited to a monolayer with $X_m \approx 2$. It is safe to assume that the other noble transition metals behave in a similar way. On the other hand, iron gives an oxygen uptake much in excess of a monolayer under these conditions, and another example of similar behaviour is titanium. If the temperature of oxygen adsorption is raised to room temperature, the list of metals to give oxygen adsorption to well in excess of a monolayer is extensive,[161] and includes titanium, chromium manganese, tantalum, iron, cobalt, nickel and niobium, although on the noble metals the rapid oxygen uptake is still limited to about a monolayer.

When one turns to dispersed metals the behaviour of oxygen is more uncertain, and a few cases will serve as illustrations. Poltorak and Boronin[91] found good agreement between the oxygen uptake at 670 K and about 133 Pa (1 Torr) and hydrogen uptake at 77 K on a platinum/silicia catalyst. However, Gruber[71] found that for oxygen uptake at 620 K and about 20 kPa (about 150 Torr) on a platinum/alumina, where the metal particles were so small that almost every atom was a surface atom (platinum dispersion $D_{Pt} \geqslant 0 \cdot 8$), the oxygen uptake was about equal to the hydrogen uptake, while for larger particles ($D_{Pt} \leqslant 0 \cdot 5$) the oxygen uptake exceeded the hydrogen uptake by a factor of about two. This was ascribed to the absence of an incorporation process when the metal particles are sufficiently small. On the other hand, it has been found[88, 89] in oxygen adsorption on both platinum/alumina and platinum/silica catalysts at about room temperature, that the oxygen uptake is *less* than the hydrogen uptake by a factor that approaches about two as the platinum dispersion approaches unity, while for larger particles with $D_{Pt} <$ about $0 \cdot 5$ the oxygen and hydrogen uptakes are comparable, although the oxygen uptake still tends to be variable and relatively irreproducible. The same trend also appears to be valid for platinum/zeolite catalysts.[84] It does appear that for very small platinum particles the extent of oxygen uptake is relatively suppressed and this again may be a reflection of the enhanced electron affinity of such small particles relative to the need for electron transfer *to* an adsorbed oxygen.

A similar unreliability exists in the use of oxygen chemisorption with dispersed rhodium[104] and iridium[108] for surface area measurement. Nevertheless, it is also of interest to note the evidence[90] which suggests that, unlike hydrogen chemisorption, the extent of oxygen uptake on platinum at room temperature may depend on the degree of surface imperfection at a given particle size.

The reaction between a monolayer of chemisorbed oxygen and gaseous hydrogen has been suggested as an alternative method for the estimation of surface area in dispersed platinum[94] and palladium[99, 243] while the converse of this, the reaction between chemisorbed hydrogen and gaseous oxygen, has also been used (e.g. refs 90, 138). These are often referred to as titration processes: that is, titration of chemisorbed species by reaction with the gas.

On platinum at room temperature these reactions are represented by:

$$O_{(s)} + 3/2H_{2(g)} \rightarrow H_{(s)} + H_2O \qquad (6.12)$$

and

$$2H_{(s)} + O_{2(g)} \rightarrow O_{(s)} + H_2O \qquad (6.13)$$

When used with platinum dispersed on a support, the water produced in these reactions is removed by adsorption onto the support. When reaction 6.12 was applied to unsupported palladium[99, 102] this water was collected and measured (in the presence of excess hydrogen the water is thought not to be adsorbed on the palladium) and this served as a check on the reaction stoichiometry.

In using these reactions for surface area estimation, there are two main sources of error: the extent to which the reaction proceeds quantitatively from left to right, and the accuracy with which one knows the number of metal atoms associated with each $O_{(s)}$ and $H_{(s)}$ at monolayer coverage. On platinum, there certainly must be some doubt if reaction 6.12 proceeds to completion at room temperature, and some comments relevant to this have been made in the previous discussion; however, the error is probably not large compared to the likely uncertainty in the chemisorption stoichiometry of $O_{(s)}$. Reactions on palladium are rather more facile than on platinum; other noble metals probably resemble platinum in their reactivity. There is a degree of uncertainty with regard to the chemisorption stoichiometry of $H_{(s)}$ when the metal particles are extremely small, a matter alluded to further in a subsequent paragraph: but other than this, one can say that one surface platinum is associated with each

$H_{(s)}$ (that is, the hydrogen chemisorption stoichiometry is two, referred to the hydrogen molecule). However, we have seen that the oxygen chemisorption stoichiometry has a degree of uncertainty, and for this reason this titration technique cannot be recommended if hydrogen chemisorption is available for surface area estimation, despite the modest increase in sensitivity potentially available (a factor of three). But situations do exist when the titration method should be considered. One such example is due to Sermon[99,102] who examined an unsupported palladium black of high area which would have undergone extensive sintering, had a conventional reduction and hydrogen desorption routine been used prior to hydrogen chemisorption. Indeed, in this case rough agreement was reported for surface areas obtained in this way with values from average particle sizes measured by electron microscopy and X-ray diffraction line broadening. This, and the agreement previously reported with supported platinum,[94] suggest that this titration method is of least uncertainty if the platinum and palladium particles are relatively large, for instance D_{Pt} or $D_{Pd} < 0\cdot 5$, and this is in agreement with more recent studies with supported palladium in which $D_{Pd} \approx 0\cdot 2$.[243]

The reaction stoichiometries between chemisorbed oxygen and gaseous hydrogen, or the converse reaction between chemisorbed hydrogen and gaseous oxygen, depend on the nature of the metal and on the temperature. In the case of oxygen chemisorbed on nickel, reaction with gaseous hydrogen at room temperature is nearer to

$$O_{(s)} + \tfrac{1}{2}H_{2(g)} \rightarrow OH_{(s)} \qquad (6.14)$$

than to reaction 6.12, although further reduction with the formation of water occurs at higher temperatures. The same can be said for iron and cobalt, although these metals have not been studied in detail. Nevertheless, in none of these cases is the reaction as simple as that represented by 6.14; the stoichiometry is fairly uncertain[118] and a quantitative method is not available.

We turn now to carbon monoxide adsorption for surface area estimation. In comparison with the behaviour of oxygen (and hydrogen), carbon monoxide adsorption occurs readily on both low index single crystal and on polycrystalline surfaces: this difference is probably associated with the non-dissociative nature of carbon monoxide adsorption. Although carbon monoxide does not share the propensity of oxygen for a low temperature incorporation process to give an uptake in excess of a monolayer, it does have the difficulty that the chemisorption stoichiometry is variable because the proportion of chemisorbed species in the linear and bridged forms can vary, the former offering a chemisorption stoichiometry of one, the latter of two. Because the linear and bridged forms are bound to the surface with different energies and because their chemisorption stoichiometries are different, the relative proportions of the two forms are temperature and pressure dependent. Furthermore, the proportions appear also to depend on the metal particle size. For instance, in adsorption on supported platinum (e.g. refs 90, 96, 97) at 273–300 K and at pressures up to about 600 Pa (about 5 Torr), the chemisorption stoichiometry at monolayer coverage, X_m falls from about 2 towards 1 as the particle size falls. Moreover, a study with a range of dispersed rhodium[104] and iridium[108,109] resulted in the conclusion that it was generally not possible to rely on agreement between metal surface areas obtained by hydrogen and carbon monoxide adsorption (measurements were made at room temperature and 0·1–2 Pa, about 1–20 Torr; and it was assumed $X_m = 2$). In those cases where agreement has been found (rhodium,[105] ruthenium,[109] osmium,[110]) the metal was present in a high and restricted range of metal loading (5–10 wt. %) and there is no guarantee that this is other than fortuitous.

It should be remembered that some metals form volatile carbonyls easily by direct reaction between the metal and carbon monoxide, and this is facilitated if the metal is finely divided. Such conditions must obviously be avoided in making carbon monoxide adsorption measurements. As an example, nickel carbonyl is readily formed by passing carbon monoxide at a pressure of 101 kPa (1 atm) over finely divided nickel at 350–370 K. However, the situation can be more serious than this. It has been reported[267] that carbon monoxide adsorption on ruthenium/silica (about 1 wt. %) and on ruthenium/carbon (about 0·2 wt. %) catalysts leads to the formation of adsorbed ruthenium carbonyls: this occurs to a substantial extent at 333–423 K, and is still evident at 300 K, as judged both from the magnitude of the gas uptake and from e.s.r. data. It would be wise, therefore, to take the point of view that the use of carbon monoxide adsorption for surface area estimation, with any metal which readily forms molecular carbonyl compounds, should be approached with considerable circumspection.

On the whole, we consider that when dealing with transition metals which readily chemisorb hydrogen dissociatively, this should be the first choice for surface area measurement provided complications due to hydrogen absorption are absent or can be eliminated. This is not to say that oxygen or carbon monoxide chemisorption is necessarily futile, since this supplementary information is often useful, and may be able to act as an indicator to changes in surface structure. Furthermore, situations do exist where the hydrogen chemisorption method is inapplicable. One such situation is with high area unsupported dispersed metal which may be too thermally unstable with respect to sintering to permit adequate cleaning of the surface to be possible prior to hydrogen adsorption: a case in point was a high area palladium black described by Sermon[99,102] and to which a hydrogen titration of chemisorbed oxygen was employed (*vide supra*). Other examples occur with metals which do not chemisorb hydrogen at all or only with difficulty, and metals with which there is a large absorption component present. In these cases gases other than hydrogen must be resorted to, and some examples will be considered in subsequent sections.

Adsorption on the Support

We refer in this section to intrinsic adsorption on the support; that is, we shall not be referring to adsorption by a spillover process, the possible significance of which has been discussed previously. Support materials vary so much in their adsorption behaviour that the intrinsic adsorption always needs to be determined experimentally for each practical situation so that a correction can be applied to measurements made on the supported catalyst itself. Here it is only possible to give some typical examples.

On the whole, the extent of adsorption of hydrogen, oxygen and carbon monoxide on high area silica and alumina stands in increasing order

hydrogen < oxygen < carbon monoxide

and some typical data are contained in Figs 6.7 and 6.8.

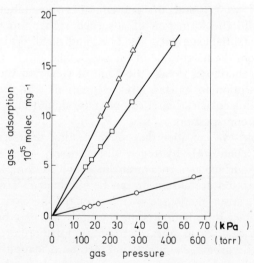

FIG. 6.7 Gas adsorption at 298 K on silica gel (Davison grade 926). ○, hydrogen: □, oxygen: △, carbon monoxide. After Wanke, S. E. and Dougharty, N. A. *J. Catal.* **24**, 367 (1972).

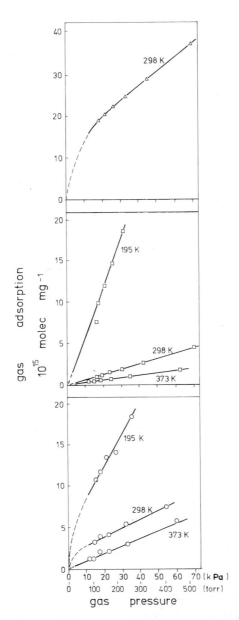

FIG. 6.8 Gas adsorption on η-alumina (Davison grade 992). ○, hydrogen: □, oxgyen: △, carbon monoxide. After Wanke, S. E. and Dougharty, N. A. *J. Catal.* **24**, 367 (1972).

It is often possible to find an optimum temperature at which the adsorption is a minimum. This is illustrated in Fig. 6.9 for hydrogen on a higher area alumina sample reported by Gruber.[71] In this case, there is a reduction by a factor of about three in the intrinsic adsorption at 520 K compared with 370 K. However, it should be noted that quantitatively this effect depends a good deal on the nature of the support. Thus, some data quoted by Wanke and Dougharty[104] for hydrogen adsorption on a η-alumina show a minimum uptake at about 370 K.

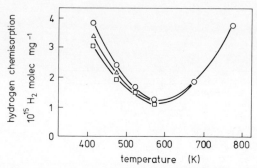

FIG. 6.9 Hydrogen adsorption isobars on η-alumina. The samples had been outgassed at 770 K in vacuum. □, 6·65 kPa (50 Torr); △, 13·3 kPa (100 Torr); 26·7 kPa (200 Torr).

The adsorption of hydrogen has been studied by Benson and Boudart[115] on sodium and calcium-exchanged 13Y-zeolites. Adsorption isotherms were linear over the pressure range 13–90 kPa (about 100–700 Torr) and at temperatures 222–296 K, and adsorption was rapid and reversible. At 273 K, the slopes of the uptake vs pressure relations (i.e. the Henry's law constants) were $1 \cdot 61 \times 10^{11}$ and $2 \cdot 54 \times 10^{11} H_2$ molec mg^{-1}(Pa)$^{-1}$ for sodium and calcium-exchanged specimens respectively ($2 \cdot 14 \times 10^{13}$ and $3 \cdot 38 \times 10^{13} H_2$ molec mg^{-1} Torr^{-1} respectively). Increasing the temperature to 373 K causes a reduction in the uptakes by a factor of around three. No activated chemisorption is observed at $\leqslant 520$ K. Barrer and Sutherland[116] have measured oxygen adsorption on sodium-exchanged 13X-zeolite at 273 K. The isotherm was linear up to about 80 kPa (about 600 Torr) and Henry's law constant was $0 \cdot 95 \times 10^{13} O_2$ molec mg^{-1}(Pa)$^{-1}$ ($1 \cdot 26 \times 10^{15} O_2$ molec mg^{-1} Torr^{-1}). Angell and Schaffer examined carbon monoxide adsorption on a series of ion-exchanged X and Y zeolites.[117] All the zeolites weakly chemisorbed carbon monoxide at room temperature and 26 kPa (about 200 Torr). However, the adsorption was reversible and could be removed by pumping at room temperature.

Gas adsorption on carbon is highly variable depending on the type and history of the specimen. One of the main problems is that severe conditions are required to generate a clean carbon surface: for instance prolonged outgassing at 1300 K is insufficient to obtain more than a partially clean surface. Cleaning conditions are thus inconsistent with the stability of a highly dispersed metal on a carbon support; that is, a carbon processed in the manner required to prepare a dispersed metal catalyst will, even after reduction, be seriously covered with strongly adsorbed oxygen-containing residues which will influence gas chemisorption. As an illustration, that part of a charcoal surface which is clean readily chemisorbs oxygen at room temperature. The heat of adsorption is high (about 330 kJ mol^{-1}) and the adsorption is essentially irreversible. However, adsorption becomes activated before a monolayer of adsorbed oxygen is reached, and slow uptake processes are observed. With hydrogen there is not much adsorption on charcoal between room temperature and 500 K,[114] but at higher temperatures a slow uptake occurs. A typical example is cited by Boudart et al.[81] which shows a slow hydrogen adsorption at 623 K on Spheron 6 carbon (100 m^2 g^{-1}) which reached an uptake of about $8 \cdot 6 \times 10^{15}$ H$_2$ molec mg^{-1} in about 60 min. The hydrogen uptake is influenced by the amount of oxygen-containing residues already present. Carbon monoxide adsorption appears to be small at all temperatures of interest for catalyst characterization.

The importance of a correction for intrinsic adsorption or the advantage to be gained by optimizing the adsorption temperature needs to be viewed in relation to the total hydrogen uptake which is dependent on the metal loading of the catalyst and on its dispersion. As an illustration, a $1 \cdot 0$ wt. % supported platinum catalyst with a platinum disperson (say) $D_{Pt} = 0.8$ will result in a hydrogen uptake of about $1 \cdot 23 \times 10^{16}$ H$_2$ molec per mg of catalyst for the formation of a monolayer on the platinum. If the uptake on the support were $1-2 \times 10^{15}$ H$_2$ molec per mg of support (a value typical for much of the data in Figs 6.5–6.7), the correction for the intrinsic uptake on the support would only amount to about 10 % of the total, which is appreciable but does not represent a serious threat to the accuracy of the metal surface area measurement. On the other hand if the catalyst were such that the dispersion $D_{Pt} = 0 \cdot 2$ as could result from extensive sintering, the correction for intrinsic adsorption would rise to about 40 % of the total.

The extent of gas adsorption on high area supports at 77–90 K is generally so large that accurate chemisorption measurements on a supported metal component become at best very difficult.

Evaluation of the Chemisorbed Monolayer Uptake

The methods for the evaluation of n_m^s are essentially empirical, but all depend on the assumption that the data have been taken into a pressure range where the uptake has become only a slowly varying function of pressure, and that this saturation region corresponds to the close approach of monolayer formation. If in the saturation region the uptake becomes almost independent of pressure, this is often taken as equal to n_m^s without further adjustment (cf. Figs 6.6 and 6.11). Alternatively, if the data encompass a usefully wide range, n_m^s is sometimes evaluated by empirically fitting to a Langmuir adsorption isotherm: assuming that all of the uptake refers to dissociative chemisorption, the appropriate isotherm form is

$$n^s = \frac{n_m^s\, b\, p^{\frac{1}{2}}}{1 + b\, p^{\frac{1}{2}}} \qquad (6.15)$$

Examples of this procedure are given by Spenadel and Boudart[73] and by Hansen and Gruber.[85] One may either plot $(n^s)^{-1}$ against $p^{-\frac{1}{2}}$ and extrapolate to $p^{-\frac{1}{2}} = 0$, when the intercept on the $(n^s)^{-1}$ axis equals $(n_m^s)^{-1}$, or one may plot $p^{\frac{1}{2}}(n^s)^{-1}$ against $p^{\frac{1}{2}}$, when the intercept on the $p^{\frac{1}{2}}(n^s)^{-1}$ axis on extrapolation to $p^{\frac{1}{2}} = 0$ equals $(n_m^s)^{-1}$.

It is not infrequently found that the experimentally measured data show no region where the uptake, after correction for intrinsic adsorption on the support, becomes virtually independent of pressure, but the uptake continues to increase slowly but significantly with increasing pressure in

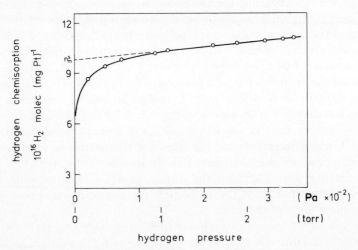

FIG. 6.10 Back-extrapolation of linear part of "saturation" region of hydrogen adsorption isotherm to obtain an estimate of the monolayer uptake, n_m^s. Measured at 298 K on 11·7 wt.% platinum/silica gel catalyst. Corrected for intrinsic adsorption on the support. The catalyst had been hydrogen reduced at 570 K followed by evacuation at 770 K.

the saturation region. In general this problem can be approached by fitting the data to an adsorption isotherm. However, when this region of the graph is approximately linear, a common practice is to back extrapolate to evaluate the intercept on the uptake axis. An example is shown in Fig. 6.10. Although this is a convenient technique, the fact remains that the identification of a linear plot in the saturation region of the isotherm is often dependent on the accuracy with which the data are taken, and the size of the pressure range studied. When the nature of the data permits its use, there is an application of convenience for this technique of linear back-extrapolation, and this is for application to total uptake data, that is, to data uncorrected for intrinsic adsorption on the support. The uptake intercept at zero pressure generally gives as good an estimate of n_m^s as may be had by a two step process involving first correcting the entire isotherm of intrinsic adsorption on the support, followed by back-extrapolation of the corrected isotherm. Provided the correction is not too large, back-extrapolation to zero pressure may also be used to correct for absorption: an example is considered when dealing with palladium.

In a well-reduced metallic catalyst of high dispersion, the metal-support contact area is probably accessible to hydrogen chemisorption by diffusion.

Platinum

Figure 6.11 contains some examples of hydrogen adsorption isotherms measured under varying conditions for clean massive polycrystalline platinum, and for supported dispersed platinum.

On clean specimens of massive polycrystalline platinum, the uptake of chemisorbed hydrogen measured at about 273 K and with an equilibrium hydrogen pressure of about 1 Pa (about 10^{-2} Torr) yields a good estimate of the surface area, assuming $X_m = 2$.

With dispersed platinum various adsorption conditions have been used.

Poltorak and Boronin[70] used hydrogen adsorption at 77 K and at about 1–100 Pa (about 10^{-2}–1 Torr) with platinum/silica catalysts. However, in general at this very low temperature intrinsic hydrogen adsorption on the support is very substantial, so the correction which must be applied to the catalyst adsorption data becomes unacceptably large. Furthermore, under these conditions there is the possibility of further weak molecular chemisorption on the platinum. Although at 77 K slow hydrogen uptake is largely suppressed, these conditions cannot be generally recommended.

Two different sets of conditions have been widely used for platinum dispersed on oxide-type supports: 273–300 K with hydrogen pressures in the range 10–300 Pa (about 0·1–2 Torr); alternatively about 520 K and 1–30 kPa (about 10–300 Torr). With platinum on carbon supports the high

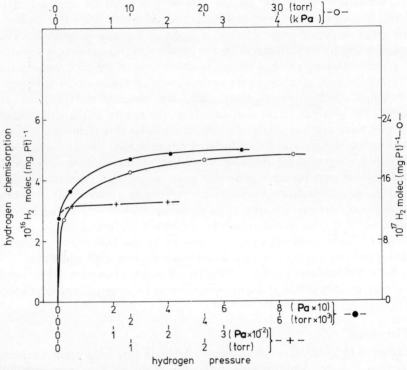

FIG. 6.11 Hydrogen adsorption isotherms on platinum.
●, 273 K on evaporated platinum film (ref. 50)
○, 473 K on 2 wt.% platinum/γ-alumina (ref. 60)
+, 293 K on platinum/silica (ref. 61)

The platinum/γ-alumina catalyst had been hydrogen reduced at 720 K followed by evacuation at that temperature: the platinum/silica catalyst had been hydrogen reduced at 760 K followed by evacuation at that temperature.

temperature conditions are certainly completely inapplicable because of extensive hydrogen spillover.[81] High temperature adsorption conditions (e.g. refs 71, 73) have usually been justified by the need to minimize intrinsic hydrogen adsorption on the support, and by the convenience of an accelerated slow adsorption component. However, there are ample instances of the platinum surface area being successfully measured in dispersed catalysts in which the hydrogen was adsorbed at 273–300K (e.g. refs 88–90, 84) to make it quite clear that adsorption at a high temperature is not necessarily essential.

In general, it may be said that optimum conditions for various systems need to be established experimentally. When it can be done without an unreasonably large contribution from intrinsic adsorption on the support, the measurement of the surface area of dispersed platinum by hydrogen chemisorption is preferably done at 273–300 K (and at pressures up to about 200 Pa; about 1–2 Torr) rather than at 520 K and higher pressures, because the chemisorption stoichiometry at monolayer coverage, X_m, can be set equal to two with greater confidence and furthermore hydrogen spillover will certainly be less important at the lower temperature.

In a monolayer of chemisorbed hydrogen on a dispersed platinum catalyst, it is now clear that a chemisorption stoichiometry, X_m, of two is valid for all particle sizes in the range where an independent size measurement can be made by techniques such as X-ray diffraction line broadening and electron microscopy, that is down to about 1·0 nm by the latter technique.[73, 88-90, 92] This value for X_m is, of course, the one indicated by chemisorption data on massive platinum. A 1·0 nm diameter platinum particle contains of the order of 100 atoms, and the question is whether $X_m = 2$ is also valid for smaller particles and atom clusters. A comparison of hydrogen chemisorption at room temperature with the results of deuterium exchange of chemisorbed hydrogen[84] suggests that $X_m = 2$ is valid for platinum particles containing only about six atoms. However, this part of the problem needs further confirmation before it can be regarded as settled with reasonable certainty, since there are enough cases in the literature reporting $X_m < 2$ for very small particles to raise reasonable doubts (cf. refs 82, 93). It is always easy to suppose that a value of less than two may arise from reaction of hydrogen with some source of oxygen in the system, or from some spillover onto the support, and these effects would be the greater at the lowest platinum loadings and at higher adsorption temperatures, which is where they have usually been reported. Nevertheless, it is worth recalling that weak molecular chemisorption of hydrogen can occur on platinum, and since this is accompanied by a work-function decrease there is some charge transfer to the metal, and so on a very small platinum particle with an enhanced electron affinity compared to the massive metal, this form of hydrogen adsorption may be increased, and this would serve to lower the overall value of X_m.

On balance we believe it is currently best to work on the basis of $X_m = 2$ for the entire platinum size range, with this assumption being regarded as tentative for sizes less than about 1·0 nm diameter.

Palladium

The use of hydrogen chemisorption for the estimation of palladium surface area is influenced by the need to avoid conditions where hydrogen absorption into the metal occurs to an undesirably large extent. This problem has been considered by Aben[98] and Sermon.[99]

From the data in Fig. 6.3 it is obvious that to minimize absorption one needs to work at elevated temperatures and relatively low pressures, but the chosen conditions must remain consistent with the formation of a chemisorbed monolayer, and with the avoidance of thermal damage to the specimen. At 343 K and a hydrogen pressure of 133 Pa (1 Torr) the equilibrium concentration of absorbed hydrogen does not exceed about 0·2 atom %,[101] and these are the conditions used by Aben[98] for surface area measurements on a series of palladium/alumina, palladium/silica and palladium black specimens. The seriousness of this absorption in relation to the monolayer uptake is dependent on the palladium dispersion of the metal. For the absorption component not to exceed (say) 10% of the monolayer uptake, we need $D_{Pd} \geqslant 0·02$.

Aben's samples were mostly reduced in hydrogen at 670 K followed by evacuation (16 h) at this temperature. It was shown that 670 K was the minimum temperature at which removal of the hydrogen was sufficiently complete, about 3% of the surface remaining covered. This residual hydrogen could be readily removed by evacuation at 850 K, but only at the expense of significant sintering. These conclusions are in agreement with those of Scholten and van Montfoort[113] who showed that heating at 570 K in vacuum left some 80% of a palladium surface covered by hydrogen. Hydrogen chemisorption was thus determined after making small corrections for solution in the palladium and for intrinsic adsorption on the support. Measurements made on palladium black samples of known BET surface area confirmed a hydrogen chemisorption stoichiometry $X_m = 2$.

Instead of making a hydrogen uptake measurement at a single pressure which was Aben's procedure, a more accurate method is to obtain data over a range of pressures (say) 60–500 Pa (about 0·5–4 Torr) and extrapolate to zero pressure. So far as hydrogen absorption is concerned, the extrapolation to zero pressure may be done against pressure or (pressure)$^{\frac{1}{2}}$. In principle the (pressure)$^{\frac{1}{2}}$ dependence should be applicable at low hydrogen concentrations for instance $< 0·3$ atom % at 343 K, but in practice it is adequate to use whichever gives the greater linearity. Sermon[99] has shown that this method can be used satisfactorily at adsorption temperatures in the range 303–363 K, although the absorption correction becomes greater the lower the temperature.

The surface area of massive palladium cannot be satisfactorily measured by hydrogen adsorption because of interference from absorption. The best method is a measurement of the rapid oxygen uptake at 77–90 K, with oxygen pressures in the region of about 10^{-2}–10^{-1} Pa (about 10^{-4}–10^{-3} Torr): $X_m = 2$.[106]

Noble Metals other than Platinum and Palladium

Gas adsorption behaviour has been examined in some detail by Wanke and Dougharty[104] on a range of dispersed rhodium catalysts, mainly rhodium/η-alumina. Hydrogen reduction was complete at 770 K in $\geqslant 2$ h, and evacuation under these conditions also appeared to be adequate for quantitative hydrogen desorption. Hydrogen adsorption isotherms were measured at pressures of 6–60 kPa (about 50–500 Torr) and over the temperature range 77–570 K. From 195–470 K the uptake was not strongly dependent on hydrogen pressure, and the net monolayer uptake on the rhodium could be conveniently estimated by back extrapolation to zero pressure.

Provided the rhodium particle size was not too small, there was evidence from agreement between the hydrogen monolayer uptake and BET area data, and from the character of hydrogen adsorption isobars, that a monolayer of chemisorbed hydrogen with $X_m = 2$ was established by measurements made in this way at 298 K, and this agrees with the assumptions made previously by Sinfelt,[105] except that the latter work was confined to the pressure range $0 \cdot 1$–2 kPa (about 1–20 Torr). On the whole it appears that the most useful conditions are at about 298 K and about $0 \cdot 6$–10 kPa (about 5–100 Torr). At 77 K there is evidence for further hydrogen adsorption beyond a monolayer (possibly weakly chemisorbed molecular hydrogen), while at > 298 K the uptake appears to be submonolayer. However, there is also evidence[104, 85] that for very highly dispersed rhodium (probably particle diameters $< 1 \cdot 0$ nm) the chemisorption stoichiometry falls below two, and this is an inadequacy in the area estimation technique that so far remains unresolved.

On clean massive rhodium specimens, it has been concluded[106, 107] that monolayer hydrogen uptake with $X_m = 2$ is achieved at 90 K and about $0 \cdot 2$ Pa (about 1–2×10^{-3} Torr). This is not necessarily inconsistent with the monolayer criteria suggested for dispersed catalysts, but a quantitative correspondence has, nevertheless, not yet been established.

Hydrogen chemisorption has been studied on dispersed iridium catalysts by Brooks (iridium/γ-alumina[108]) and Sinfelt (iridium/silica[109]) with much the same sort of results as have been described for rhodium, and the conditions recommended for the latter also apply to iridium.

Surface areas of dispersed ruthenium[109] and osmium[110] catalysts (silica supports) have been estimated from hydrogen adsorption isotherms, obtained under essentially the same conditions as recommended with rhodium and iridium.

Surface areas of massive specimens of iridium, ruthenium and osmium can probably be best estimated from the hydrogen uptake in the region of

1 Pa (about 10^{-2} Torr) with $X_m = 2$. There are insufficient data to offer a reliable estimate of the optimum adsorption temperature, that is 77–90 K or 273–300 K; we incline to the latter.

The surface area of gold cannot be measured by gas chemisorption methods.

Non-noble Transition Metals

We first consider the metals in this group which do not absorb large amounts of hydrogen, the surface areas of clean, massive specimens of which are best determined from the fast monolayer hydrogen uptake at about 1 Pa (about 10^{-2} Torr) and 273–300 K. The use of temperatures 77–90 K is also satisfactory. The chemisorption stoichiometry is $X_m = 2$.

With massive specimens of metals which absorb large amounts of hydrogen, chemisorption of this gas is not a satisfactory method. These metals also react strongly with oxygen, but the chemisorption of oxygen has only been examined on a few of them. Nevertheless, it is known with titanium, for instance, that even at 77–90 K oxygen uptake proceeds to well in excess of a monolayer; therefore, unless there is explicit evidence to the contrary, it is probably wise to assume that oxygen adsorption is inapplicable for surface area estimation. In this situation the use of carbon monoxide adsorption may be considered, since at least with titanium no incorporation or solution processes occur, and the monolayer uptake can be conveniently measured at about room temperature and about 10^{-2} Pa (about 10^{-4} Torr). However, the chemisorption stoichiometry is not known with any accuracy and appears to be in the region of 1·5 with titanium.[123]

Of the metals in this group which can be generated in a dispersed form (usually supported) by reduction of a metallic derivative, that is, principally nickel, cobalt, iron and rhenium, the behaviour of nickel has been the most thoroughly studied and serves as a useful illustration.

Compared to the behaviour of clean massive nickel, dispersed nickel specimens require considerably higher hydrogen pressures to reach the saturation region of the adsorption isotherm, as indicated in Fig. 6.6. We consider the most appropriate temperature for the estimation of nickel surface area in dispersed catalysts by hydrogen chemisorption is 273–300 K and at pressures up to about 10–20 kPa (about 100–200 Torr) with $X_m = 2$. These conditions, or something fairly close to them have been frequently used (e.g. refs 119, 59) although adsorption at 195 K,[58,120] and at 520 K[121] has also been used.

For cobalt, iron and rhenium the situation is probably rather similar to that described for nickel.[59,122] As with nickel, the difficulty of freeing the surface of these metals of adsorbed oxygen by hydrogen reduction at

⩽770–800 K should not be overlooked. Nevertheless, compared to nickel the reduction of surface oxide is considerably more difficult on iron and probably more difficult on cobalt, and this probably accounts for the abnormally slow and small hydrogen uptakes reported by Adrian and Smith[266] for a cobalt/kieselguhr catalyst (pre-reduced at 690 K for 15 h).

Finally, it is worth noting that the adsorption of carbon monoxide at 77 K has had a long history for the estimation of the iron surface area in iron synthetic ammonia catalysts,[124–127] in which the assumption was made $X_m = 1$. The technique was to increase the carbon monoxide pressure to about 60 kPa, but a correction was made for physical adsorption. Bearing in mind the variability of the chemisorption stoichiometry with carbon monoxide pressure, this value of $X_m = 1$ is not necessarily inconsistent with the value of $X_m = 1 \cdot 6$ found[106] with evaporated iron films at 90 K and with a carbon monoxide pressure of about 10^{-2} Pa (about 10^{-4} Torr).

Non-transition Metals

The metals we are mainly concerned with here are copper and silver.

Hydrogen is not chemisorbed on silver, while on copper chemisorption is an activated process and the resulting surface coverages tend to be low. Thus hydrogen is not suitable for surface area measurement with these metals. On the other hand, although oxygen is readily chemisorbed on copper, even at 77 K it is difficult to ensure that $X_m = 2$ because of a tendency for adsorption to proceed beyond a monolayer. Scholten et al.[129] devised a technique for copper area estimation based on the decomposition of nitrous oxide at the copper surface. The reaction at 360–370 K and about 27 kPa (about 200 Torr) gave a surface coverage of one oxygen atom for each surface copper atom, the nitrogen of course being returned to the gas phase. This method gave tolerable agreement with values expected from X-ray diffraction line broadening measurements of copper particle average diameters.

Emmett and Skau[130] used the amount of weakly chemisorbed carbon monoxide, measured at 77 K to estimate the copper surface area. However, this is difficult to separate from physically adsorbed carbon monoxide, and if a support is present the physical adsorption component may well be so large as to make the method unworkable.

With silver, oxygen chemisorption has been used for surface area measurement, but with conflicting claims for success. Thus, although Smeltzer et al.[128] found that the oxygen uptake at 470 K in the pressure range 27–92 kPa (about 200–700 Torr) gave a monolayer with $X_m = 2$, Kholyavenko[131] claimed to produce monolayer coverage at 470 K and at 270–400 Pa (about 2–3 Torr) while Sandler and Hickam[132] claimed that these conditions led to uptakes considerably in excess of a monolayer, a conclusion recently confirmed.[133]

A reliable technique, due to Scholten et al.,[133] for silver surface estimation has appeared which, like copper, depends on the decomposition of nitrous oxide to generate a monolayer of chemisorbed oxygen with an overall composition of one oxygen atom per surface silver atom. Optimum conditions are about 420–430 K and a nitrous oxide pressure of about 20 kPa (about 150 Torr).

Bimetallic Systems

A catalyst containing two metallic components may have much more complex gas adsorption behaviour than when only one component is present, because the magnitude of gas uptake may well be a function both of the composition of the metallic surface with respect to the two components, and of the total metallic surface area. Clearly, if an estimate of the total exposed metal surface area is required, the adsorbing gas should be chosen (if possible) so that its behaviour is independent of the surface composition. As a rough guide, if the two metals are such that each in the pure state can have its surface area estimated by hydrogen chemisorption, a catalyst containing two such components can also have its total surface area estimated from hydrogen chemisorption. On the other hand, if one of the components in the pure state does not chemisorb hydrogen, the estimation of gross surface area requires a search for an alternative adsorbate gas which does not suffer from this specificity. A transition metal with a group IB metal is the sort of combination most commonly encountered. Oxygen is the gas which first comes to mind in this situation, but it brings with it the problems of variable chemisorption stoichiometry which have been previously mentioned. However, it must be said that in a situation where information is scarce, one should be prepared to accept a technique of diminished accuracy, rather than have no information at all.

If a particular alloy specimen is monophasic and at equilibrium, the question of a subclassification of the gross surface area does not occur: one is only left with the question of the surface composition relative to that of the bulk and this is a question dealt with in a subsequent section. However, if more than one phase is present there is the problem of the extent to which each phase contributes to the total surface area, and this problem exists for massive metal as well as dispersed specimens. Chemisorption data alone can only give information relevant to this problem to the extent that chemisorption specificity can be achieved with reference to the various phases. With regard to chemisorption on single component phases which may be present, the behaviour can be assessed by reference to the corresponding single component specimens. For instance, a specimen thought to expose only nickel and copper surfaces may have the extent of the nickel surface estimated from rapid hydrogen chemisorption at room temperature or below,[134] since there is no uptake on copper under these conditions.* However, the chemisorption behaviour on a two component phase may well be dependent on the phase composition: this is well known for example, with nickel–copper[135, 136] and platinum–copper.[137] More often than not, the situation is such that gas chemisorption *alone* does not offer sufficient specificity with regard to surface composition to allow anything other than partial characterization of a bimetallic catalyst. In principle, the specificity of chemisorption may be increased by measuring a property of the adsorbed species which may depend on the nature of the surface: for instance, the infrared absorption spectrum, the binding energy to the surface (e.g. temperature programmed desorption) or the chemical reactivity (e.g. temperature programmed isotope exchange; reaction of chemisorbed oxygen with gaseous hydrogen, or of chemisorbed hydrogen with gaseous oxygen). One of the few examples of this sort of approach yet to be recorded is due to Menon *et al.*[138] who noted that, in contrast to platinum, oxygen chemisorbed on rhenium could not be reduced at all by gaseous hydrogen at room temperature, so that *if* one were to assume a platinum–rhenium/η-alumina catalyst contained only separate particles of pure platinum and pure rhenium (i.e. no alloy) the total oxygen uptake could be used to estimate the total metal area, while the extent to which adsorbed oxygen reacted with hydrogen gas at room temperature according to reaction 6.12 gives an estimate of the platinum area, and the rhenium area would then be available by difference. However, if an alloy component surface were present, this simple interpretation would not be available because of the need to know and allow for any reaction occurring on this part of the surface too. The temperature at which a reaction occurs is not the only parameter which may change on going from one type of metal to another, since the reaction stoichiometry may change. As an example, we may compare the differing stoichiometries for the reaction of chemisorbed oxygen with gaseous hydrogen at about room temperature on platinum and nickel which are represented by reactions 6.12 and 6.14 respectively.

With bimetallic catalysts, a combination of techniques is mandatory, and particularly in the case of supported dispersed catalysts, the whole question remains one which requires much more work for a proper understanding.

---

* Assuming there is no impurity present which may modify the chemisorptive behaviour of copper, cf. Cadenhead and Wagner.[166]

Non-Metallic Surface Area

While with ordinary supported metal catalysts an estimate of the non-metallic surface area may be had from the difference between the total surface area such as may be measured by the BET method and the metal surface area measured by selective chemisorption methods, situations exist where a direct estimate of the surface area of a non-metallic component is desirable. This need occurs most often when studying catalysts containing several distinct phases. However, the number of cases where simple volumetric uptake measurements have been successfully applied to this end are relatively rare, since chemisorption on a metal is usually easier than on a non-metallic component, and it is difficult to find a sufficient difference in chemisorption specificity to establish a qualitative distinction between different non-metallic components. Greater success may be had by applying various diagnostic techniques to the adsorbed species.

The best known application of selective chemisorption for the estimation of the surface area of a non-metallic component is the use of carbon dioxide chemisorption for the potassium oxide surface area in stabilized iron synthetic ammonia catalysts.[42] In these catalysts the oxide component contained (for example) up to about 1·6 wt.% potassium oxide as a chemical promotor, and up to about 10 wt.% aluminium oxide as a stabilizer. Adsorption was carried out at 195 K up to a carbon dioxide pressure of about 80 kPa (about 600 Torr), but the equilibrium uptake under these conditions includes both physically adsorbed as well as chemisorbed gas, and the latter was evaluated as that portion which could not be desorbed by pumping at 273 K. However, in the light of later knowledge[140, 141] the original assumption that carbon dioxide was only chemisorbed on the potassium oxide surface is open to serious doubt, because carbon dioxide is known to be strongly and rapidly chemisorbed on clean iron at 195 K: even allowing for the fact that at monolayer coverage of carbon dioxide on iron the chemisorption stoichiometry $X_m \approx 10$, it is difficult to escape the conclusion that at least some of the chemisorbed carbon dioxide previously attributed to potassium oxide must have occurred on the iron. It may be possible to minimize chemisorption of carbon dioxide on the iron surface by preadsorption of oxygen.

Metal oxides show some degree of selectivity in chemisorption, and a convenient rough generalization is that when the adsorption process involves electron transfer from the adsorbent towards the adsorbate (e.g. oxygen), the degree of chemisorption is relatively small on n-type oxides but considerably greater on p-type oxides in which metal ions may be converted to a higher oxidation number in the process of supplying electrons to the oxygen; on the other hand, when adsorption involves electron transfer from the adsorbate towards the adsorbent (e.g. hydrogen or carbon monoxide) the reverse specificity is found with respect to n- or p-typeness.

Table 6.3 lists some of the common oxides according to their semiconducting behaviour.

TABLE 6.3 Classification of metal oxides*

metal deficit, p-type conductivity	Ag_2O, CoO, Cr_2O_3(<1520 K), Cu_2O, FeO, MnO, Mn_3O_4, Mn_2O_3, NiO, PdO, UO_2, $CoCr_2O_4$, $FeCr_2O_4$, $MgCr_2O_4$, $ZnCr_2O_4$, $CoAl_2O_4$, $NiAl_2O_4$
metal excess, n-type conductivity	(Al_2O_3), BaO, BeO, CaO, CdO, CeO_2, Fe_2O_3, MnO_2, MoO_3, Nb_2O_5, PbO_2, (SiO_2), SnO_2, SrO, Ta_2O_5, (TiO_2), Tl_2O_3, (ThO_2), U_3O_8, UO_3, V_2O_5, WO_3, ZnO, (ZrO_2), $MgFe_2O_4$, $NiFe_2O_4$, $ZnFe_2O_4$, $ZnCo_2O_4$, ($MgAl_2O_4$), $ZnAl_2O_4$.
amphoteric conductors	Cr_2O_3(>1520 K), MoO_2, PbO, RuO_2

* Parentheses indicate substances with a particularly low conductivity.

Nevertheless the magnitude of the gas uptake even in a favourable situation is highly variable, depending on the chemical nature of the oxide and the temperature, and an activated adsorption component is usually present. Thus, while cuprous oxide[156] and cobaltous oxide[157] readily chemisorb in excess of a monolayer of oxygen at about room temperature, nickel oxide only chemisorbs typically to 10–20% of a monolayer under similar conditions, and this behaviour seems to be associated with the difficulty of freeing the nickel oxide surface of preadsorbed oxygen. The surface coverage of hydrogen or carbon monoxide at the surface of n-type oxides also varies widely. In summary we may say that while chemisorption of gases such as oxygen, hydrogen or carbon monoxide offers some scope for selective surface area estimation on oxides, individual behaviour is so variable that detailed data from individual components is essential before quantitative measurements can be contemplated.

One of the few examples of selective surface area measurement in two component oxide systems is provided by Bridges et al.[142] who used oxygen chemisorption for the estimation of the chromia surface area in a chromia/alumina catalyst.

Although the chemisorption of molecules such as ammonia, pyridine, hydrogen sulphide or boron trifluoride is specific for certain types of adsorbents, this behaviour is generally related to the occurrence of specific surface sites (e.g. acidic sites: ammonia, pyridine; basic sites: hydrogen sulphide, boron trifluoride) which are present in a surface concentration that is not known in advance. Data of this kind are therefore useful for the identification and estimation of specific sorts of adsorption sites, but of little value for surface area estimation as such.

Gas Adsorption Methods

Volumetric Gas Measurements

We have seen that a number of measurements made for the purpose of catalyst characterization involve volumetric and manometric gas techniques. Typical examples are the measurement of dead space or of real catalyst density and these amount to the measurement of an unknown volume, or gas adsorption measurement in which the volume is known and the amount adsorbed is obtained as the difference between the total amount added and the amount left in the dead space at adsorption equilibrium. In any case, what is involved is expansion of a quantity of gas which is initially contained in a known volume into a larger volume, the initial and final gas pressures being measured. Schematically the situation is illustrated in Fig. 6.12.

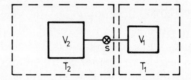

FIG. 6.12 Schematic illustration of basic apparatus for gas volumetry.

If ideal gas behaviour is assumed and if the entire system is at a constant temperature T ($=T_1=T_2$), the pressure–volume–temperature relationships are easily written down in terms of the ideal gas equation

$$pV = nRT \qquad (6.16)$$

If the apparatus is not at constant temperature so that $T_1 \neq T_2$, a more complex situation arises. Let us assume, as would be the case in practice, that the junction between the two regions of differing temperature is established in the connecting tubing as shown in Fig. 6.12 so that the volume which is at an indefinite temperature is negligible in relation to V_1 or V_2. Suppose a quantity of gas has been expanded throughout the apparatus by opening the tap S. We wish to know the relevant pressure, volume, temperature relationships. Provided the pressure is high enough so that the mean free molecular path is very small compared to the diameter of the connecting tubing, d_t, the measured pressure will be uniform throughout the apparatus and one simply has

$$\left. \begin{array}{l} pV_1 = n_1RT_1 \\ PV_2 = n_2RT_2 \\ \text{and } n_1 + n_2 = n \end{array} \right\} \qquad (6.17)$$

where n is known, and again ideal gas behaviour is assumed. As a rough but conservative criterion, this uniform pressure regime will hold provided the pressure (Pa) is more than about $500/d_t$, or the pressure (Torr) is more than about $4/d_t$, in either case d_t being measured in mm.

If the pressure is such that the mean free molecular path is much greater than d_t, thermomolecular flow between the two regions of different temperature will establish a steady-state pressure difference, and the pressures p_1 and p_2 corresponding to T_1 and T_2 are related by

$$p_1/p_2 = (T_1/T_2)^{\frac{1}{2}} \qquad (6.18)$$

Again as a rough but conservative guide, equation 6.18 holds if the pressure (Pa) is less than about $10^{-2}/d_t$, or the pressure (Torr) is less than about $10^{-4}/d_t$, in either case d_t being measured in mm.

If the pressure lies between the extreme limits corresponding to uniform pressure on the one hand or the applicability of equation 6.18 on the other, a more complex procedure is needed to evaluate p_1/p_2. Following Miller[19] who developed the treatment due originally to Liang,[23]

$$p_1/p_2 = \frac{[(T_1/T_2)^{\frac{1}{2}} - 1]}{\left[\dfrac{\pi y^2}{128} + \dfrac{\pi y}{12} + \left(\dfrac{1 + 5y/2}{1 + 2y} \right) \right]} + 1$$

$$y = [(p_2 d_t) \, 2 \cdot 58 \times 10^3 r^2]/(T_1 + T_2) \qquad (6.19)$$
$$(T_2 > T_1)$$

when r, the elastic sphere molecular radius is expressed in nm, and p_2 is in Pa and d_t in mm. If p_2 is in Torr the numerical factor $2 \cdot 58 \times 10^3$ is replaced by $3 \cdot 43 \times 10^5$. Values for the elastic sphere molecular radius are available from gas viscosity data, and values have been tabulated.[20,21] Except for the very light gases (hydrogen, neon and probably helium) the predictions of equation 6.19 are in good agreement with the careful experimental data of Podgurski and Davis.[22] A selection of data giving p_1/p_2 in terms of the characteristic parameter $p_2 d_t$ is shown in Fig. 6.13.

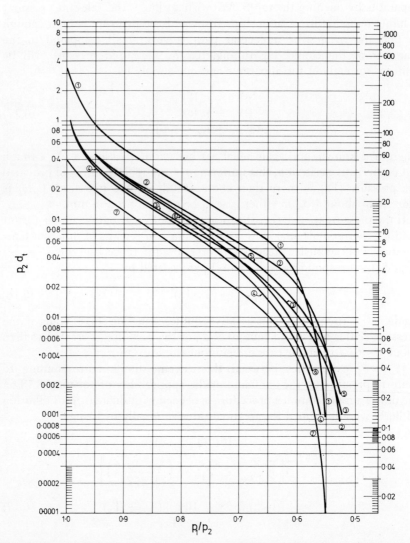

FIG. 6.13 Thermomolecular flow corrections. Each curve gives p_1/p_2 as a function of $p_2 d_t$, where p_1 and p_2 are the pressures in the two vessels at temperatures T_1 and T_2 (K) respectively, separated by a connecting tube of diameter d_t (mm).

Curve 1, hydrogen, $T_1 = 77$ K, $T_2 = 299$ K
Curve 2, argon, $T_1 = 77$ K, $T_2 = 299$ K
Curve 3, krypton, $T_1 = 77$ K, $T_2 = 299$ K
Curve 4, krypton, $T_1 = 90$ K, $T_2 = 299$ K
Curve 5, nitrogen, $T_1 = 77$ K, $T_2 = 299$ K
Curve 6, nitrogen, $T_1 = 90$ K, $T_2 = 299$ K
Curve 7, xenon, $T_1 = 90$ K, $T_2 = 299$ K.

Curves 1, 2, 7 are smoothed experimental results due to Podgurski, H. H. and Davis, F. N. *J. Phys. Chem.* **65**, 1343 (1961). Curves 3–6 were calculated using equations 2.4. The left vertical scale is for p_2 in Torr; the right vertical scale for p_2 in Pa.

Accurate thermomolecular flow corrections are important as will be obvious from the extent to which p_1/p_2 may depart from unity.

There have been numerous experimental studies of thermomolecular flow, and a concise summary of references to the literature is given by Young and Crowell.[3] However, a significant proportion of the recorded experimental work is inaccurate and in general the data need to be approached with caution.

Volumetric and manometric gas measurements of the sort with which we are mostly concerned are usually carried out at gas pressures less than one atmosphere. Deviations from ideal gas behaviour are usually not severe but may be appreciable if the temperature is low enough. In practice, correction for non-ideal gas behaviour beyond the use of the second virial coefficient is never required, so that where necessary the corrected gas equation may be written

$$pV = nRT + np\ B(T) = nRT + n^2RT\ B(T)/V \quad (6.20)$$

where $B(T)$ is the second virial coefficient. Values for $B(T)$ for some gases of practical interest are listed in Table 6.4. To the accuracy with which measurements are needed for catalyst characterization, corrections for nonideal gas behaviour are not worth making unless the magnitude of the term $np\ B(T)$ (or its equivalent) amounts to more than 0.5% of pV and

TABLE 6.4 Second virial coefficients for gases*

	$B(T)$, (10^{-3} dm^3 mol^{-1}) at indicated temperature, (K)						
	77	90	195	273	293	298	573
helium	+10·76	+11·66		+11·77		+11·68	+9·91
argon	−278†	−213†		−22·2		−16·1	
krypton				−61·4		−50·5	
xenon				−155·8		−131·2	
hydrogen	−9·68	−5·47		+13·7		+14·3	
nitrogen	−255†	−194†		−10·1		−4·46	
methane		−398‡		−53·5		−43·0	
carbon dioxide			−360†	−149·3		−124·4	
n-butane				−1030†	−867·0	−834·7	

* Landolt-Bornstein, II. Band, I. Teil, Springer-Verlag (1970).
† Extrapolated from listed values using procedure given in Landolt-Bornstein.
‡ From the data quoted by Loebenstein and Deitz.[17]

this means that the correction is only worth making for the gas which is maintained at the temperature of physical adsorption. The correction factor α introduced by Emmett and Brunauer[42] and listed (inaccurately) by Young and Crowell[3] is used in a term of the sort $V(1 + \alpha p)$ in which V is the amount of ideal gas needed to fill a given volume to a pressure p, and the corrected term is the amount of non-ideal gas needed. The factor α equals $B(T)/RT$.

Gas Volumetric Apparatus—Relatively High Subatmospheric Pressures

The basic apparatus for static gas measurement consists of a gas dosing device, a pressure measuring device, a specimen chamber and associated gas handing and vacuum plumbing. The amount of gas adsorbed is determined by taking the difference between the amount of gas dosed into the volume containing the sample, and the amount retained in the dead-space. The latter is available from the measured pressure if the dead-space volume is known, and the dead-space volume is usually measured using helium. The main features of this sort of apparatus are illustrated schematically in Fig. 6.14, while Fig. 6.15 shows an apparatus which is

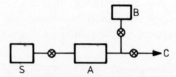

FIG. 6.14 Schematic illustration of apparatus for gas adsorption measurement. S is sample; A a known volume in which the pressure can be measured; B supply of gas to be adsorbed; C to pump.

quite adequate for most applications over a wide range of gas pressures down to a region of $0 \cdot 02$ Pa: it is a modest extension of the classical design (e.g. ref. 6).*

In studying chemisorption, particularly on metals, it is essential that access of mercury vapour to the specimen be eliminated. This can be done with a liquid nitrogen trap or a gold foil trap as indicated in Fig. 6.15. It should be noted that the film of mercury condensed in a liquid nitrogen trap can itself chemisorb some oxygen. Although in physical adsorption the influence of mercury vapour is not so serious, it is wise to protect the specimen nonetheless. When working at adsorption temperatures <195 K it is customary to rely on the mercury vapour being condensed in the connecting tubing where it first enters the refrigerant bath. For measure-

* Commercial units are available, for instance, Numec Instruments and Control Corpn.

ments above 195 K a gold foil trap should be used. Gold foil traps may be regenerated by heating to about 500 K in vacuum.

The experimental details of volumetric adsorption techniques (e.g. ref. 5) and the BET method (e.g. refs 6, 7, 24) are well known and do not need to be repeated here. However, the following points should be noted. From the point of view of practical surface area measurement, there is little to be gained by going to great experimental elaboration to improve the precision with which the gas uptake can be measured beyond what is available with the type of simple apparatus exemplified in Fig. 6.15. The

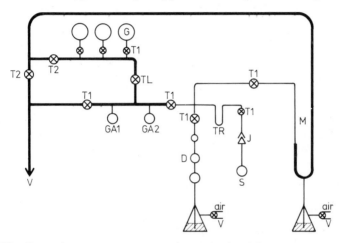

FIG. 6.15 General purpose apparatus for gas adsorption measurements. The apparatus is readily fabricated in glass. Heavy lines indicate wide-bore tubulation, light lines indicate 2 mm bore tubulation. M is a manometer, D a gas doser. Taps T1 are small-bore greaseless, T2 are large-bore greaseless. TL is a venier-controlled leak valve. TR is a trap for mercury vapour. Gases are stored in bulbs G. GA1 and GA2 are a high-pressure ionization gauge and a precision Pirani gauge respectively. The sample S may be removed via the demountable joint J. V leads to vacuum.

accuracy with which the surface area is obtained is dominated by the adequacy of the model from which n_m^s is calculated, and by the uncertainties in a_m. For gas pressures in the range above about 600 Pa (4–5 Torr, say), the pressure is conveniently measured with a simple U-tube mercury manometer, and the gas is dosed into the system with a simple gas burette. The apparatus shown in Fig. 6.15 is quite adequate for most applications. A U-tube manometer with a mirror scale can be read by eye to a precision of about $\pm\,0\cdot 3$ mm, while a cathetometer will yield a precision of about $\leqslant 10^{-2}$ mm. The gas burette may be calibrated by mercury weighing before assembly. Some more detailed information on this type of apparatus is given by Joyner.[24] Although in the classical design greased taps were used, these should now be replaced by modern greaseless taps. In a glass apparatus it is very convenient to use a tap with a glass body and the range made by J. Young Ltd., London (small bore) and Ace Glass Inc., New Jersey (large bore) are highly satisfactory. For systems made in metal it is convenient to use a tap with a metal body, such as is made by Hoke Mfg. Company, New Jersey. There are a number of advantages: there is no grease in which vapours can dissolve, the transfer of grease to other parts of the apparatus (particularly mercury surfaces) is eliminated, the vacuum and sealing properties are superior, and they can be used above room temperature if necessary. With the advent of the greaseless tap the mercury cut-off is superfluous.

For gas pressures below about 600 Pa (about 5 Torr) the U-tube mercury manometer is unsatisfactory. Although a U-tube oil manometer can, in principle, be used down to about 50 Pa (about 0·4 Torr), oil has the disadvantages of appreciable gas solubility and drainage problems due to the oil film on the glass walls. A combination oil-and-mercury manometer has been described[5] in which the gas to be measured only comes into contact with mercury, but this is a relatively complex design and is not simple to use. In practice a McLeod gauge is often used for pressure measurement in the range $> 0\cdot01$ Pa ($>$ about 10^{-4} Torr), and this can also serve as a gas dosing device. If a multirange McLeod gauge is used, the upper practical pressure limit can be made to overlap the lower limit of a U-tube manometer. It is also convenient to have a high quality Pirani, thermistor or thermocouple gauge in the system since this is very convenient for rapid measurement, and is quite accurate enough for much gas adsorption work provided its calibration is checked against the McLeod gauge (note the contribution from the vapour pressure of mercury). The use of such a gauge makes it possible to eliminate the large volume of the McLeod gauge from the system when this becomes necessary. If a thermistor gauge is used, a glass-encapsulated thermistor bead is essential.

In Fig. 6.15 the connecting tubing and the active side of the U-tube manometer are of reduced internal diameter (about 2 mm) in order to keep the dead-space to a minimum, to avoid large volume variations as the mercury level of the manometer changes, and to provide a well defined boundary between the refrigerated volume and the rest of the apparatus. From the point of view of the volumetric measurement, it is necessary to assume that the sample is sufficiently well outgassed and the system sufficiently well evacuated that residual gas is negligible compared to the equilibrium adsorption pressures to be measured. For xenon adsorption at 90 K or krypton adsorption at 77 K this requires achieving a preadsorption background pressure of < 0.01 Pa ($<$ about 10^{-4} Torr); the presence of connecting tubing of only 2 mm internal diameter makes this a slow operation, and if these measurements are to be made often it is convenient to use a separate volumetric apparatus in which the connecting tubing is of larger bore—say 5 mm internal diameter. A compromise has to be struck between pumping speed and dead-space.

The accuracy with which the gas uptake n^s can be measured is dependent on the magnitude of the actual uptake relative to the amount of gas which remains unadsorbed and to the amount which (in physical adsorption) is adsorbed on the walls of the container which are at sample temperature. Thus, other things being equal, the accuracy deteriorates the smaller the sample surface area. In physical adsorption the problem due to gas remaining unadsorbed in the dead-space can be overcome to some extent by using an adsorbate gas which has a lower p_0 and which is more strongly adsorbed, thus reducing the equilibrium gas pressure over the sample. Xenon at 90 K or krypton at 77 K are obvious choices (cf. Table 6.1). Correction can be made for adsorption on the refrigerated walls of the sample container by using data from a blank experiment. The relative size of this correction depends not only on the size of the container wall area relative to the sample surface area, but also on the strength of adsorption on the two surfaces. For instance, the heat of adsorption of xenon or krypton (which would normally be chosen for use with very low area samples) on transition metals is numerically greater than on glass so that at 90–77 K the coverage on glass is only some 10–15 % of that on the clean metal at the same equilibrium pressure. On the other hand, adsorptions on metal oxides and glass are more nearly comparable. The minimum sample surface area which can be conveniently estimated with reasonable accuracy depends on the sample configuration since this influences the size of the dead-space.

If the specimen is massive metal, the total specimen surface area may be limited. Typical values would be 0.5–10^{-2} m^2 for evaporated films, 10^{-2}–10^{-5} m^2 for wires, ribbons, slabs, foils, buttons, etc., and 1000–0.1 m^2 for metal powders. The size of the apparatus dead-space as dictated by the specimen configuration means that most evaporated metal film and metal powder specimens can be measured by the BET method: films and low area powder specimens require the use of xenon at 90 K or krypton at 77 K, high area powders can be measured with argon at 77 K. On the other hand, most specimens in the form of wires, ribbons, slabs, foils, buttons, etc., cannot be measured by physical adsorption and chemisorption is needed where the gas pressure at monolayer coverage is lower so that the dead-space correction is lower.

An alternative elaboration of the basic volumetric adsorption apparatus

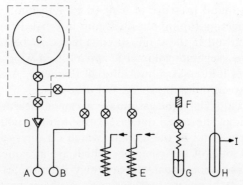

FIG. 6.16 Gas adsorption apparatus: A, sample bulb; B, nitrogen thermometer; C, Wallace and Tiernan gauge; D, 10/30 standard taper; E, copper spiral cold traps connected to nitrogen and helium tanks; F, gold powder trap; G, safety manometer; H, cold trap; I, to vacuum pump. Reproduced with permission from Benson, J. E. and Garten, R. L. *J. Catal.* **20**, 416 (1971).

has been given by Benson and Garten[46] which has the advantages of ruggedness and simplicity of operation associated with the elimination of mercury-containing components. The design is shown in Fig. 6.16. The pressure gauge serves as a dosing volume. The apparatus may be constructed mostly in metal if desired.

A simple micro-system has been described by Hayes[47] which relies entirely on a thermistor gauge and is shown in Fig. 6.17. This is particularly suited to small samples, and the author quotes a BET surface area measurement of good precision using nitrogen adsorption at 77 K on a sample with a total area of $0 \cdot 45$ m^2 (at $0 \cdot 15$ m^2 g^{-1}).

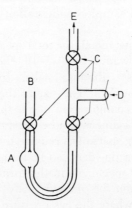

FIG. 6.17 Micro BET apparatus. A, sample bulb; B, vent; C, Hoke valves, 46/6 M4N; D, thermistor gauge; E. vacuum system, gas handling, manometers, etc. After Hayes, K. E. *J. Catal.* **20**, 414 (1971).

A number of improvements have been made to the BET method for better speed and convenience. A computational technique has been described[9] which eliminates the need for dead-space calibration. When carrying out routine measurements, it is worth examining how far pre-adsorption outgassing can be curtailed without affecting the accuracy of area measurement.

With the conventional BET apparatus which relies upon batchwise addition or removal of gas, it is relatively difficult to provide for automation and recording of the results. Nevertheless, automated equipment of this sort has been constructed (e.g. refs 13, 14) but at the penalty of very considerable instrumental complexity.

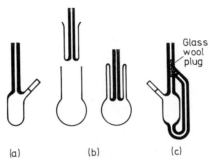

FIG. 6.18 Various sample chambers for gas adsorption measurement. After Joyner, L. G. "Scientific and Industrial Glass Blowing" (W. E. Barr and V. J. Anhorn, eds) Instruments Publishing Co., Pittsburgh (1949).

A number of designs for sample chambers for gas adsorption measurements have been suggested: Fig. 6.18 shows a selection. The point at issue is that of providing a means of introducing the sample into the chamber, remembering that the sample may range from a fine powder to large lumps, while minimizing the dead space and eliminating any demountable joint from the region which is to be heated or cooled. There is no easy way of providing a vacuum-tight demountable joint which is stable on cooling to liquid nitrogen temperature and heating to (say) 700 K. Viton O-ring seals fail on cooling due to shrinkage of the elastomer. The design shown in Fig. 6.19 is by far the simplest and most generally satisfactory. Special types of samples such as evaporated films, wires, ribbons, etc., often dictate their own chamber geometry by virtue of requirements imposed by techniques of preparation, cleaning or use.

The sample chamber design shown in Fig. 6.19 is also suitable for the measurement of real catalyst density since the chamber volume is fixed (that is, not subject to change during the sealing operations required with designs of Fig. 6.18).

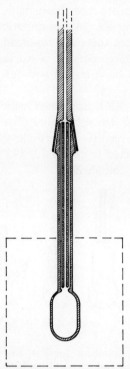

FIG. 6.19 Preferred sample chamber for gas adsorption measurement. The capillary tubing insert reduces the dead space and can be removed for filling.

Gas Volumetric Apparatus—Ultrahigh Vacuum

If metal specimens of limited surface area (e.g. most massive metal specimens) are to be maintained for an appreciable time with an atomically clean surface, it is necessary to use techniques which will reduce the pressure of adsorbable gas to less than about 10^{-7} Pa (about 10^{-9} Torr). Ultrahigh vacuum techniques are frequently employed for this purpose. A detailed account of ultrahigh vacuum techniques has been provided by a number of authors (e.g. refs 43, 44, 45) and for the present purpose we only comment upon the main points.

Ideally the system should contain only glass, high density stable ceramic and UHV-compatible metal: a small exposure of high stability elastomer such as Viton is permissible in components such as gaskets, but this is accompanied by some deterioration in UHV capability, and should be eliminated if possible. A UHV system requires outgassing by baking under vacuum. An elastomer-free system is baked at 600–700 K, while if Viton is present the temperature is limited to about 500 K. Pumping is normally done with a trapped diffusion pump, some type of sputter ion pump, or getter pump. A wide range of bakeable all-metal valves is commercially available,* as well as valves containing a Viton gasket. For small laboratory systems extensive use is made of those with tubulation diameters in the range 25–5 mm bore. Vacuum plumbing is normally in glass or stainless steel, or a combination of these.

For pressures $< 10^{-3}$ Pa ($<$ about 10^{-5} Torr) the Bayart-Alpert ionization gauge is the normal measuring device for total pressure. Pressures in the range 10^2–10^{-4} Pa (about 1–10^{-6} Torr) may be measured with a Schultz-Phelps ionization gauge, while bakeable Pirani, thermocouple or thermistor gauges can be used in the range 10^{-2}–10^{-1} Pa (about 1–10^{-3} Torr).

Static gas adsorption measurements can be made in an apparatus of UHV capability, using essentially the same style of procedure as we have previously discussed, that is, expansion of a known amount of gas into a volume which contains the specimen under study, and this is a not infrequent procedure when studying chemisorption on an atomically clean surface. Mention should be made of the need to keep tubulation as large in diameter as possible to provide for adequate gas flow rates (e.g. pumping speed) in the UHV pressure range: the dead space in a UHV apparatus will thus usually be substantial.

The UHV pressure range lends itself to the use of mass spectrometric partial pressure gas analysis, and a number of small appendage mass spectrometers are commercially available of the magnetic deflection, time of flight or quadrupole analyser design. For most purposes the latter is very convenient because it does not require the use of a bulky magnet, is of high sensitivity and has a rapid mass scan.

The process of thorough cleaning of the surface of a metal specimen (thermal and/or ion bombardment) inevitably involves the removal of some metal which is deposited in the vacuum chamber. Even though the amount of metal deposited in this way may be small, very little is needed to be of considerable consequence in adsorption (or catalytic) studies. For instance, a very thin metal film of density about 10^{-2} g m^{-2} consists of a sparse array of discrete and very small metal crystallites, yet on a given area of support carrying such a film, there may well exist a total metal surface area of a magnitude equal to that of the surface on which the crystallites are deposited. It may be possible to keep the deposited metal at a temperature that is too low for adsorption (or catalysis) to occur. Inasmuch as many adsorption processes with gases such as oxygen, hydrogen

---

* Well-known suppliers of UHV components include Varian, Granville-Phillips, Vacuum Generators, Riber.

or carbon monoxide are fast even at temperatures as low as 77 K on transition metals, the extent to which temperature can be used to suppress unwanted adsorption activity may be rather limited: it is of more general applicability if the specimens are being used for catalytic studies, since there are not many catalytic reactions which are fast at 77 K. If suppression of unwanted activity by temperature differentiation is not possible, it will be essential to isolate the cleaned metal specimen from the metal which was deposited during cleaning. Either the cleaned specimen must be transferred through a closeable port into another part of the vacuum system, or else the deposited metal must be transferred. The selection from these alternatives will depend on the nature of the reaction being studied, and the type of metal specimen used. Thus, if the specimen were in the form of a wire or ribbon which relied on resistive heating, it may prove more convenient to collect the deposited metal on a moveable shield

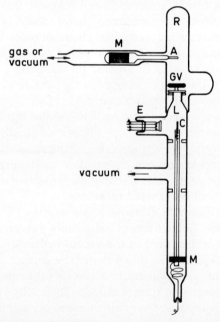

FIG. 6.20 Apparatus incorporating a specimen transfer device for studying adsorption or catalysis over a single crystal catalyst. The reaction chamber R may be separated from the cleaning chamber L by means of the non-lubricated ground glass valve, GV. The specimen C, after cleaning, is transferred to the support A. Movement is effected magnetically via the encapsulated iron slugs M. The filament E is used for electron or ion bombardment of the specimen. In an obvious manner, E could alternatively be used as a metal evaporation source for the deposition of an evaporated film on C if, for instance, the catalyst was to be an epitaxed metal film. After Shooter, D. and Farnsworth, H. E. *J. Phys. Chem. Solids.* **21**, 219 (1961).

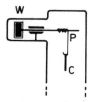

FIG. 6.21 Simple transfer device. The specimen C is supported by the wire P which is wound up and down by the magnetically operated windlass W. P can conveniently be made from 0·05 mm diameter platinum wire.

than to arrange to move the specimen since the latter, although feasible, would require arranging for heavy flexible electrical leads. On the other hand, it might be more convenient to move the specimen if it consisted of a slab or button, which did not rely on resistive heating. A large number of experimental arrangements are clearly possible. Figures 6.20–6.22 show three typical UHV-compatible transfer devices. As shown, Figs 6.20

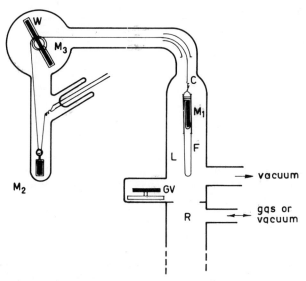

FIG. 6.22 Apparatus incorporating a moving filament transfer device. The filament cleaning chamber L is separated from the working chamber R by a non-lubricated ground glass valve GV. The lead C to the filament F consists of two multistrand copper conductors, each in a thin woven-silica insulating sleeve. W is a rotary windlass by means of which F may be raised and lowered. M are encapsulated iron slugs to obtain mechanical movement magnetically: M_1 allows for lateral positioning of F; M_2 is a counterweight, and M_3 operates the windlass. Three rather than two conductors may be used if it is desired to operate twin filaments. After Anderson, J. R. and Macdonald, R. J. *J. Catal.* **19**, 227 (1970).

and 6.21 are alternatives for moving non-resistively heated specimens. Figure 6.22 shows an arrangement for moving a resistively heated filament. The all-glass isolation valves shown in Figs 6.20–6.22 will not support a large pressure differential. If this is deemed necessary, the glass valves could readily be replaced by straight-through UHV all-metal valves, although the activity of the metal in the valve would need to be investigated. If necessary, the exposed metal surface of the valve could be coated with a layer of gold.

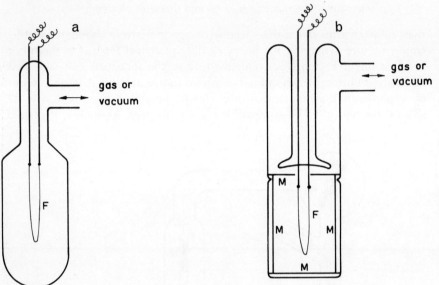

FIG. 6.23 Vessels for carrying out adsorption or catalysis with evaporated metal films, (a) Conventional reaction vessel in which the glass walls are the film substrate; (b) Vessel in which the film is deposited on the mica substrate M, and in which a fringe film is mostly eliminated. In each case F is the evaporation filament.

There is now available, in most commercial equipment designed for LEED and Auger spectroscopy, the option of an all-metal specimen isolation chamber, which allows the establishment of a relatively high gas pressure around the specimen without filling the entire apparatus with high pressure gas: this allows high pressure adsorption studies to be more readily made (and is virtually essential if catalytic reaction rates are to be measured). Nevertheless, the exposed metal surface of the isolation chamber (and any specimen metal deposited thereon during specimen cleaning) needs to be examined for possible activity.

Much of the adsorption (and catalytic) work with thick continuous evaporated metal films has used a cylindrical reaction vessel (Fig. 6.23a).

This cylindrical geometry permits a mica liner to be inserted as a film support and Fig. 6.23b shows an arrangement in which virtually all the evaporated metal is confined to the mica. A very thin discontinuous film will always occur at the upper edge of a thick film deposited in a reaction vessel of the sort shown in Fig. 6.23a. This may be a problem if the very thin and the thick films have different properties. This fringe film is minimized by the design of the sort in Fig. 6.23b and in any case its presence can be made innocuous if its activity can be suppressed by keeping it at a suitably low temperature.

If an epitaxed film catalyst is prepared by metal deposition onto a support consisting of a flat plate or slab of single crystal material, it will be essential to transfer the film after deposition into a separate reaction chamber, because during deposition it will be impossible to confine the evaporated metal to the single crystal support alone. This may be done by the use of transfer devices of the sort shown in Figs 6.20–6.22 in a more or less obvious manner. Cleavable crystalline supports may be cleaved in vacuum by application of a bending torque (e.g. ref. 245), or by the use of a knife-edge under pressure: the latter may be driven magnetically or

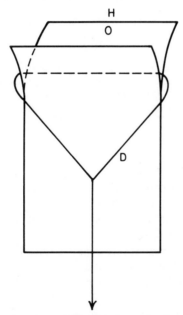

FIG. 6.24 Schematic arrangement for cleaving mica in vacuum. The wire draw-bar D may be actuated either magnetically or via bellows. A hole H allows a cleaved sheet to be hooked to a transfer device. The end of the original sheet needs to be air-cleaved for a short distance as shown before mounting.

through bellows (e.g. ref. 246). Mica may be cleaved in vacuum using a draw-wire and one arrangement is shown in Fig. 6.24.

In this sort of work, much depends on the ingenuity of the investigator.

Adsorption in Gas Flow Systems

Measurements by the BET method have been made using a procedure which involves the continuous slow addition of adsorbate gas to the adsorbent (e.g. refs. 11, 12) and this is a very convenient technique if continuous recording of the adsorption isotherm is required.* A schematic description of the apparatus due to Lange[12] is shown in Fig. 6.25. The adsorbate gas (typically nitrogen) emerges from the capillary leak (flattened copper tubing) at a rate of 5×10^{-5}–10^{-4} dm^3 s^{-1} STP, the actual value

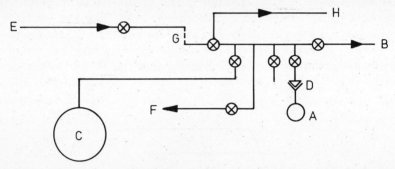

FIG. 6.25 Schematic representation of flow apparatus for physical adsorption of nitrogen. A, sample bulb; B, to vacuum pumps; C, pressure gauge; D, standard taper; E, nitrogen supply, 200–400 kPa (about 2–4 atm); F, to McLeod gauge; G, capillary; H, to flow meter. Reproduced with permission from Lange, K. R. *J. Coll. Sci* **18**, 65 (1963).

being governed by the fore-pressure which is controlled in the range 200–400 kPa (about 2–4 atm) with a precision gas regulator valve. A constant flow rate is essential because time is used to indicate the total amount of gas added to the sample volume. The flow rate is calibrated by measuring the rate of pressure rise in a known volume. Flow methods are essentially non-equilibrium in nature, but it has been shown[11,12] that at these flow rates the gas adsorption *vs* pressure relation does not depart too seriously from the equilibrium relation at least so far as the evaluation of the BET surface area is concerned: the surface areas obtained by the two techniques agree to better than 2%.

Two techniques have been described in which nitrogen adsorption is

---

* Manufacturers of commercial units include American Instrument Co., Fisher Scientific, Numec Instruments and Control Corpn., Perkin-Elmer.

studied using a helium–nitrogen gas mixture. One uses a technique resembling some of the features of gas chromatography.[16] The schematic apparatus is shown in Fig. 6.26. The gas mixture is passed through the sample and the composition of the effluent is monitored by a thermal conductivity detector connected to a recorder. When the sample is cooled in liquid nitrogen, the adsorption of nitrogen is indicated by a peak on the recorder chart. After adsorption equilibrium is established the recorder pen returns to the baseline. The sample tube is allowed to warm by removing the refrigerant, and the desorbed nitrogen produces another peak on the recorder. The area under each peak is proportional to the amount of nitrogen adsorbed or desorbed. These peak areas can be calibrated by adding a known amount of nitrogen to the gas stream at the point normally occupied by the sample, or alternatively a calibration may be effected by using an adsorbent sample of known surface area. To obtain more than one point on the BET isotherm requires measurements with gases of varying helium/nitrogen ratios. Typical operating conditions are: total gas flow rate about 10^{-3} dm^3 s^{-1}, proportion of nitrogen 5–30% (vol/vol), sample

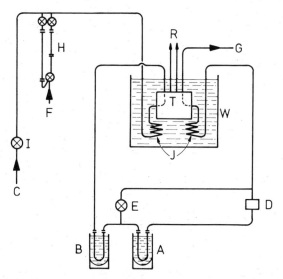

Fig. 6.26 Schematic representation of flow apparatus for adsorption of nitrogen pulse from helium flow. A, sample tube; B, dummy tube (both A and B are immersed in liquid nitrogen); C, helium supply; I, helium flow control; D, rotameter; E, by-pass valve for rotameter; F, nitrogen supply; H, U-volume to contain nitrogen sample (this U-volume may be varied); G, gas exit and precision flow measurement; J, gas temperature equilibration coils; T, thermal conductivity sensors; R, recorder coupled to T; W, water bath. Reproduced with permission from Nelson, F. M. and Eggertson, F. T. *Anal. Chem.* **30**, 1387 (1958). Copyright: the American Chemical Society.

surface area 5–25 m². Values for the surface area were obtained within about 2% of those from conventional BET methods.

A second technique using a helium–nitrogen gas mixture has been described by Loebenstein and Deitz[17] in which a known quantity of the gas mixture is cycled repeatedly through the adsorbent until no further nitrogen adsorption occurs: the uptake is measured volumetrically and manometrically. The need for a vacuum system is eliminated.

Chemisorption measurements can also be made in a flow system. This is a standard technique when using UHV equipment with pressures of the adsorbing gas in the range $< 10^{-6}$ Pa ($<$ about 10^{-8} Torr). The rate of adsorption is measured in terms of the difference between gas flow rates upstream and downstream from the specimen. This has the advantage of providing better control over gas purity because it minimizes the influence of residual gas in the adsorption chamber. For details the reader is referred to Ehrlich.[45] Chemisorption measurements on dispersed metal catalysts can also be made using gas flow methods, and a technique derived from gas chromatography is very convenient. Freel[8] has described an apparatus which is a simple modification of a commercial gas chromatograph. The normal column is replaced by a short length (ca. 200 mm) of tubing (ca. 6 mm internal diameter) which contains the catalyst under

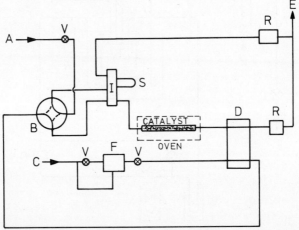

FIG. 6.27 Schematic representation of flow system for chemisorption measurement, based on a commerical gas chromatograph. A, source of gas to be adsorbed, e.g. hydrogen or oxygen; B, standard four-port glc "back-flush" valve, used for routing gas flows; C, carrier gas, e.g. nitrogen; D, thermal conductivity detector; F, flow controller; V, needle valve; I, standard six-port glc sample injector valve; S, gas sample loop; R, rotameter; E, vent. The alternative position of valve B provides a hydrogen stream over the catalyst for prereduction. Adsorbed hydrogen is removed from the catalyst after reduction using a stream of nitrogen at about 770 K. Reproduced with permission from Freel, J. J. Catal. **25**, 139 (1972).

study, and the arrangement is shown schematically in Fig. 6.27. In this method, the gas to be adsorbed (e.g. hydrogen or oxygen) is inserted as a pulse into the stream of carrier gas. In Freel's description with a supported platinum catalyst nitrogen was used as the carrier, and can be reasonably assumed to be inert. However, some metals chemisorb nitrogen, and should this be so another carrier gas (e.g. argon) would have to be considered. Gas flow rates in the region of 5×10^{-4}–10^{-3} dm^3 s^{-1} were used, with the catalyst particle size 40–60 mesh. In the simplest method, a single pulse is made somewhat larger than the anticipated uptake; the quantity of gas in the pulse is measured before and after contact with the catalyst and the amount adsorbed is given by difference. The quantity of gas in the pulse is obtained by integrating the area under the peak recorded from the detector output. By a minor modification several pulses may be used, each individually smaller than the expected uptake, but a sensitivity calibration is still required by using a fully eluted pulse. This method of measuring the chemisorption uptake defines conditions that may differ from those used in conventional experiments, since the pulse method as outlined measures only chemisorption which occurs rapidly and irreversibly. The static method, in contrast, allows time for equilibration, and the final adsorbate pressure may well be quite high, whereas in this dynamic method the final adsorbate pressure is extremely low.

Nevertheless, complete adsorption isotherms can be obtained from pulse adsorption methods, and the method of analysis is outlined by Hansen and Gruber[85] following Gregg and Stock[113] and others[50,139] and is based on the shape of the elution profile.

It is an obvious extension to use this type of gas pulse technique for carrying out titration reactions with chemisorbed gas in addition to the chemisorption process itself, and an example of this is described by Menon et al.[138]

Gravimetric Measurement

By this method the mass of the adsorbed material is measured directly. Assuming that a buoyancy correction is made, this has the advantage over volumetric methods that the sensitivity is not diminished at high gas pressures. However, microbalances are relatively delicate pieces of apparatus and need to be treated accordingly.

The applicability of microbalance adsorption measurement needs to be judged in relation to the expected mass increment and the available sensitivity. For instance, a monolayer of hydrogen chemisorbed on a platinum surface for which $n_s = 1 \cdot 25 \times 10^{19}$ m^{-2} and with $X_m = 2$, corresponds to a mass of $20 \cdot 8$ μg. An $0 \cdot 5$ g sample of a 1 wt. % supported platinum catalyst with $D_{Pt} = 0 \cdot 5$ would expose $0 \cdot 77 \times 10^{19}$ Pt atoms in the surface, and the expected mass of a chemisorbed hydrogen monolayer would be $12 \cdot 8$ μg. To make useful quantitative adsorption measurements at coverages down to $n_m^s/10$ would require a balance sensitivity of $0 \cdot 1$ μg. On the other hand, with this same $0 \cdot 5$ g of catalyst, if we suppose the support had a specific surface area of 100 m^2 g^{-1}, the mass increment accompanying the formation a monolayer of physically adsorbed nitrogen would be about $14 \cdot 5$ mg, and a balance sensitivity of (say) 10 μg would be adequate.

There are a substantial number of different microbalance designs in the literature (for a comprehensive review cf. reference 167), but there are only two which have extensive use in adsorption studies—coiled spring and beam-torsion suspension—and which are also commercially available, and to these should perhaps be added the quartz crystal resonator which can also be used in adsorption work in certain circumstances. However, the quartz crystal resonator can only be used for adsorption on an adsorbent which can be deposited in a thin layer onto a face of the crystal (e.g. an evaporated metal film), and although its inherent sensitivity for mass change is high, about 10^{-1} to 10^{-2} μg, the technique is fairly restricted in its applicability.

The most satisfactory and useful beam microbalance for adsorption work depends on the suspension of a light rigid beam, centrally supported by a fibre of circular or ribbon cross-section. A hang-down from one end of the beam carries the sample (in a bucket if a catalyst in a divided form is being studied). The balance may be symmetrical in the sense that the other end of the beam carries a second hangdown to a counterweight and this is the usual design for balances fabricated in the laboratory, and it is also used in two of the most common commercial designs (Cahn Instrument Company and Sartorius-Werke AG.). However, there are commercial designs in which the second hangdown is eliminated (e.g. Mettler Instrument A.G.). For fabrication in the laboratory (not to be undertaken lightly) silica remains the preferred material for beam construction, with the suspension usually made of fine tungsten wire for the sake of robustness. A silica fibre suspension may alternatively be used, and the case is made of glass (e.g. refs 173, 174). The case contains some form of demountable joint in the region of the hangdown to provide access to the load.

Balances may be used either in a deflection or null mode: the latter, which is much to be preferred, usually involves magnetic coupling and the magnitude of the applied magnetic field (via the current in a coil) provides a means for automatic recording. Sensing the position of the beam may be done by photoelectric (e.g. refs 175–181), capacitance (e.g. refs 182, 183) or inductive (e.g. ref. 184) methods. The photoelectric method is preferable if the vacuum properties of the system are to be optimized, since it need not involve the introduction of extra hardware into the balance case. In the Cahn, Sartorius and Mettler instruments, all of the balance components including the sensing and null adjustment mechanisms are contained within the balance case. These balances are suitable for use in unbaked vacuum systems only. Commercial equipment is also available for use in UHV or quasi-UHV conditions (e.g. Worden Quartz Products Inc. and Rodder Instrument) in which all of the sensing and null adjustment equipment is exterior to the case, and the beam and suspension are made of silica.

These beam microbalances are available in models with varying load capacities and sensitivities, but as a rough guide the range varies from a maximum load of about 100 g with a sensitivity of about 2 μg to a maximum load of about 2 g with a sensitivity of about 0·1 μg.

In the coiled spring type, the load is attached to the free end of a helically wound spring which is usually made of silica fibre, the top end of the spring being attached to the glass balance case. The mass change is indicated from the vertical position of the load, and with care and a good cathetometer this may be done to $\pm 10^{-3}$ mm. For springs of practical dimensions the sensitivity is approximately inversely proportional to the load, and the most usual values for the load-sensitivity product lie in the range 40–500 mm, with loads in the region of 0·1 to 0·5 g. In this situation, sensitivities lie in the range 10^2 to 10^3 mm g^{-1}. In practice it is extremely difficult to achieve a sensitivity better than about 1 μg, and the best figure quoted for a commercial balance is about 5 μg.

Coiled spring microbalances may be made recording using capacitance,[168] photoelectric[169, 170] or transducer[171] coupling, but it appears that this is not available on commercial equipment.

On the whole, a beam balance is much to be preferred to the coiled spring balance. It is much less susceptible to vibration, the sensitivity is less dependent on the load capacity, at least partial compensation for gas buoyancy is possible, and in a null instrument the position of the sample is invariant. The latter can be quite an important feature if some sort of simultaneous examination of the specimen is being made (e.g. infrared absorption spectrometry).

Gas buoyancy corrections are readily made provided the real volume of the displaced gas is known. In a beam balance, compensation for gas buoyance can be attempted by making the real volumes of the sample and counterweight equal. Complete compensation is usually difficult to achieve in practice, but partial compensation is relatively easy. If the sample and counterweight are at different temperatures the mass imbalance (Δm) due to a differential buoyance effect is given by

$$\Delta m = \frac{pM}{R}\left[\frac{V_s}{T_s} - \frac{V_c}{T_c}\right] \qquad (6.21)$$

where V_s and V_c are the real volumes of sample and counterweight at temperatures T_s and T_c respectively, and p is the pressure of the gas of molecular weight M. Clearly, $V_s = V_c$ only results in $\Delta m = 0$ when $T_s = T_c$.

With a coiled spring type balance, no compensation can be provided for gas buoyance. Although gas buoyance corrections can be sizeable, they can be accurately calculated and generally present no problem.

Despite the attractive sensitivities which the better of these microbalances provide (particularly of the beam type), there are some potent factors which in practice act to make this sensitivity quite difficult to realize.

The most serious problems arise from temperature inequalities and are due to effects of thermal transpiration and to convection of the gas. These become important in different pressure ranges. A detailed analysis has been provided by Massen and Poulis.[185] As pointed out previously (cf. p. 330) the thermomolecular flow region exists where in a region of thermal gradient, the gas molecular mean free path λ is much greater than d_t, the smallest dimension of the vessel; if $\lambda \ll d_t$ convection results. In either case a body placed in the gas at the temperature gradient will experience a net force due to momentum transfer from gas molecules striking it, and the magnitude of this force depends on the size and shape of the body. In the thermomolecular flow region this force arises because the kinetic energy of the gas molecules depends on whether they come from the hot or cold end of the gradient, while in the convective region the body is situated in a macroscopic gas flow. In practical situations the forces which result from these effects cannot be computed with sufficient accuracy to allow quantitative corrections to be made. The best that can be done is to understand the behaviour so that the effects can be minimized, and to allow an assessment of the ultimate accuracy with which microbalance mass measurements may be made.

The force F on a sample of diameter d_s is given by

$$F = d_s d_t p(\Delta T)/4T \qquad (\lambda \gg d_t) \qquad (6.22)$$

and

$$F = \pi \Lambda^2 [2p\, \ln(d_t/d_s - 1)]^{-1} T(\Delta T) \qquad (\lambda \ll d_t) \qquad (6.23)$$

where d_t is the diameter of the vessel, ΔT equals $T_{hot} - T_{cold}$, T is an average temperature, and Λ is a parameter equal to $p\lambda/T$: in the ordinary ideal gas approximation for the mean free path, λ, we have

$$\Lambda = k/2^{1/2} \pi \sigma^2 \qquad (6.24)$$

where k is Boltzman's constant, and σ is the hard sphere molecular diameter. The force F is defined to be positive in the direction in which it acts, that is towards the cold end in each case.

A model calculation by equations 6.22 and 6.23 gives an indication of the magnitude of the expected effect. The results are contained in Fig. 6.28 for two cases the details of which are given in the figure. The effect is at its maximum at a pressure in the region of 10 Pa (about 0·1 Torr), and its magnitude is substantial, Δm being in the region of 100–200 μg under these conditions. This discussion and the data in Fig. 6.28 also show the

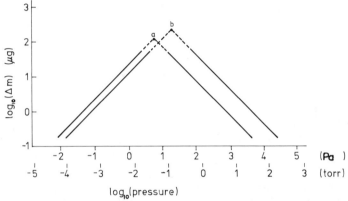

FIG. 6.28 Apparent mass change (Δm) caused by thermomolecular and convective flow. Region 1 is for thermomolecular flow ($\lambda \gg d_t$), region 2 for convective flow ($\lambda \ll d_t$). Curve a is for a sample of diameter $d_s = 2$ mm surrounded by a tube of diameter $d_t = 10$ mm, with $\Delta T = 5$ K, $T = 500$ K. Curve b is for a hangdown wire of diameter $d_s = 20$ μm, with $d_t = 10$ mm, $\Delta T = 200$ K, $T = 400$ K. The gas is assumed to be oxygen in each case. After Massen, C. H. and Poulis, J. A. *In* "Ultramicro Weight Determination in Controlled Environments" (S. P. Wolsky and E. J. Zdanuk, eds) Interscience, New York (1969), p. 107.

difficulty of utilizing the inherent ultimate sensitivity of a microbalance due to small thermal inhomogeneities. For instance, if there were an adventitious thermal gradient of $\Delta T = 0 \cdot 1$ K across a specimen, with $p = 13 \cdot 3$ Pa (0·1 Torr) and with the other parameters as in Fig. 6.28, this would give $\Delta m \approx 2 \cdot 4$ μg. Of course the situation improves as the pressure moves away from this most critical region.

When working with the sample at a significantly different temperature from that of the balance itself, some degree of compensation for spurious thermal effects can be achieved by making the counterweight of a beam balance at the same temperature as the sample. However, in practice complete compensation is impossible to achieve, and there is the added difficulty that these thermal effects tend to fluctuate with time, particularly the convective component. The outcome of this is that when working in the pressure range where thermal effects are serious, say from 10^{-1} to 10^3 Pa (about 10^{-3}–10 Torr), an accuracy better than about 1 μg will not be possible except when operating the entire balance under isothermal conditions, and only then if considerable attention is paid to thermostating the apparatus: if the sample volume is maintained at a substantially different temperature from the rest of the balance in this pressure range, it is doubtful if this accuracy can ever be achieved.

Effects of electrostatic charging can introduce large spurious mass changes of the order of several hundred μg, but this effect can be eliminated by the use of conductive coatings and suitable grounding arrangements.

All microbalances are sensitive to vibration and the extent to which special antivibration mountings are required can only be assessed in the light of local conditions.

Radiochemical Measurements

In gas adsorption studies, radiochemical measurements may be made as a method of monitoring the gas pressure and as a method of monitoring the quantity of gas adsorbed.

For physical adsorption, krypton labelled with ^{85}Kr ($t_{1/2} = 10.6$ years, 0.67 MeVβ, 0.51 MeVγ) has been used. Aylmore and Jepson[192] used a thin end-window GM counter to monitor the β-activity of the gas phase as a measure of the krypton pressure. Considering the instrumental and safety problems associated with the use of a radioactive gas, this method has nothing to recommend it in comparison with a direct pressure measurement by conventional means, since the precision at high coverage is still dependent on the dead space correction. On the other hand, the amount of labelled krypton adsorbed by the sample has been monitored directly[193,194] by counting the γ-radiation: since the sample was refrigerated, the activity was measured by a scintillation counter placed as near to the sample as possible, but exterior to the Dewar containing the refrigerant. The equilibrium krypton pressure may be measured by conventional means[193] but radiochemical monitoring is also possible.[194] Of course, all these activity measurements give only a relative measure of concentration. Calibration is necessary and one needs a krypton-containing sample of known specific activity. In the case of gas phase activity, calibration is done against a direct measurement of gas pressure. For the activity of the adsorbed krypton, calibration may be done using another adsorbent sample of known specific surface area, or indeed using the unknown sample itself by measuring the absolute krypton uptake by pressure measurement at some convenient point on the isotherm. The latter assumes that the unknown area is large enough for an accurate uptake measurement to be possible by manometric means. Assuming one is measuring the activity of the adsorbed krypton, the specific activity required to obtain a satisfactory accuracy in BET surface area measurement depends on the specific surface area of the sample and on the geometry of the detector relative to the sample. However, as a guide it has been reported[194] that a sample of $1 m^2 g^{-1}$ needs a specific activity of about 42 mCi cm^{-3}, and this falls to about 0·42 mCi cm^{-3} for 100 $m^2 g^{-1}$. We believe that the use of radioactive krypton is only worth considering if one is forced to deal with samples of such low specific surface area that manometric uptake measurements become impossibly inaccurate. The adsorption of xenon labelled with ^{133}Xe ($t_{1/2}$ 5·27 days, 0·347 MeVβ, 0·081 MeVγ, the last two being the main modes) has also been studied.[195]

Chemisorption of carbon monoxide labelled with ^{14}CO ($t_{1/2}$ 5730 years, 0·156 MeVβ) has been studied on platinum, rhodium, nickel and chromium oxide catalysts.[196,197] In the technique used by Hughes *et al.*,[196] the labelled carbon monoxide was used in a gas flow, being present in helium carrier gas. The method thus closely resembled that described in a previous section (p. 345) except that the concentration of adsorbable gas was here monitored radiochemically. For the present purpose, that is quantitative adsorption measurements, it is doubtful if the use of labelled carbon monoxide has sufficient advantage over conventional methods to make it worth extensive use.

Adsorption from Solution

The use of a solution implies that there will always be two components present, the solute and the solvent. Surface area estimation by adsorption from solution requires that the solute be preferentially adsorbed and, as in the case of adsorption from the gas phase, that the effective area per adsorbed species be known. Both physical adsorption or chemisorption from solution may be studied.

In the case of physical adsorption from solution, the interaction energies of the solute and solvent molecules with the adsorbent surface are often not sufficiently different to obtain complete preferential solute adsorption. As a consequence, for a given solute the nature of the isotherm, the value of the monolayer uptake, and the area per molecule are often dependent on the nature of the solvent: an example of this behaviour occurs in the adsorption of stearic acid on Spheron carbon black.[149] As would be expected, the extent to which this occurs is also dependent on the nature of the adsorbent surface. Physical adsorption of dyestuff has been used, the advantage being the ease with which a spectrophotometric measurement of the solution concentration may be made. Examples are nitrophenol from either aqueous or nonaqueous solvent[150,151] and methylene blue from aqueous solvent,[152,153] the former having been selected as the most generally satisfactory molecule out of an extensive list of examples. However, dyestuff solutes suffer from the same problems as mentioned above.

Although solute chemisorption will usually eliminate competitive (physical) adsorption by the solvent, the use of this method for surface area estimation is still subject to substantial uncertainty which arises from a lack of knowledge about the nature of the chemisorption reaction, and the effective area occupied by each chemisorbed molecule. Chemisorption of fatty acid on oxide surfaces has been examined, but the extent of the reaction (that is, the effective area per chemisorbed molecule) is known to be dependent on the nature of the adsorbent, the chain length of the fatty acid, as well as the presence of impurity such as moisture which may facilitate progress beyond a monolayer and the formation of bulk corrosion product (cf. refs 154, 155).

The upshot of these uncertainties is that adsorption from solution cannot be used as a primary method for surface area measurement. It can only be considered where a given solution/adsorbent system has been calibrated by previous surface area measurement on the adsorbent by (say) the BET method, thus establishing the effective area per adsorbed solute molecule. Obviously the technique must be restricted to the region within which the calibration is valid, and this means in particular that the nature of the adsorbent surface must not vary. The extent of adsorption from solution may be most readily followed by measuring the solution concentration. Radiochemical methods may also be employed.

Calorimetric Methods

The heat of wetting can be used to estimate surface area provided the value of the heat per unit area is known in advance. The results are much dependent on the prior state of the surface, particularly prior adsorbed water, and the method thus needs careful standardization of procedure. On the whole the technique is inferior to gas adsorption methods, and often requires fairly complex apparatus.

Microcalorimetry has also been used[242] to monitor oxygen–hydrogen titration reactions on dispersed platinum catalysts, but the technique remains essentially a research tool rather than a method for routine surface area measurement.

Electrochemical Methods

The surface area of a massive metal specimen may be estimated electrochemically by measuring the quantity of charge corresponding to the formation (or removal) of a monolayer of adsorbed hydrogen atoms, $H_{(s)}$. The electrochemical processes are

$$2H_{(s)} \rightleftharpoons 2H^+_{(aq)} + 2e \qquad (6.25a)$$

$$H_{2(g)} \rightleftharpoons 2H^+_{(aq)} \qquad (6.25b)$$

In principle, electrochemically adsorbed oxygen may also be used, but the electrochemical processes in this case are imperfectly understood, and hydrogen adsorption is the more generally used technique. Nevertheless, the method is limited in practice to noble metals to avoid corrosion of the metal under the conditions such that reaction 6.25a goes from right to left.

Basically the measurement consists of recording the current–voltage characteristic for an electrochemical cell formed from the metal specimen and a reversible electrode in a suitable electrolyte (e.g. 1 mol dm^{-3} H_2SO_4 at 298 K), measurements being made in the range 0–0·5 V.* A typical voltammogram is shown in Fig. 6.29a for platinum. The cathodic charge is obtained by integrating the current–voltage curve, and the result is shown in Fig. 6.29b. In carrying out this integration, the baseline of zero current is inserted by back extrapolation from the adjacent current minimum (cf. dotted line in Fig. 6.29a), thus making an approximate correction for double layer charging and for the presence of some adsorbed oxygen which is residual from more anodic regions of the voltammogram.

Since the measurement depends on confining the measured charge to that occurring in reaction 6.25a, the problem remains of terminating the current integration at a voltage such that reaction 6.24b makes a negligible contribution. There is, in fact, considerable overlap normally, and full coverage by $H_{(s)}$ is not reached until about $-0\cdot01$V by which voltage considerable discharge of H_2 occurs.[143] The suggestion[144] that the integration be terminated at the potential of the minimum A in Fig. 6.29a apparently leads to a serious underestimation of the charge. A separation of these two processes is available from a detailed analysis of the voltage dependence of the electrode pseudocapacitance[144] but this is an inconveniently protracted procedure for easy use, and for practical purposes it is adequate to integrate to a fixed potential, and to multiply the value of the integrated charge so obtained by a correction factor which has been evaluated from the detailed pseudocapacitance analysis. Both the integration limit and the correction factor depend on the chemical identity of the electrode metal, but they are taken as independent of the morphology of the electrode surface. Data are available to allow the method to be used

* Potential given *vs* reversible hydrogen electrode.

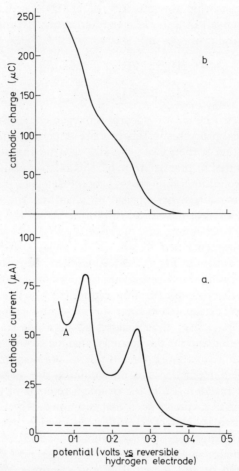

FIG. 6.29 Cathodic voltammogram (a) and its current integral (b). Platinum in 1 M aqueous H_2SO_4 at 298 K. Linear potential sweep at 40 mV s^{-1}. The dotted line in (a) is the baseline inserted for current integration.

with platinum, rhodium and iridium electrodes, and values for the integration limits and the correction factors are given in Table 6.5.[146]
In principle this method is available for other noble metals and their alloys, but sufficient data are not yet available to define the monolayer charge with sufficient accuracy.

The method has been used for the estimation of the platinum surface area in a catalyst in which the metal was dispersed on highly graphitized carbon:[265] the support in this case is of course an electrical conductor.

TABLE 6.5 Parameters for the integration of hydrogen adsorption voltammograms for electrode surface area estimation

	Integration limit (volt *vs* reversible hydrogen electrode	Correction factor
platinum	0·08	1·28
rhodium	0·07	1·70
iridium	0·06	1·54

Hydrogen is not adsorbed electrochemically on gold, but a method for surface area estimation based on the electrochemical adsorption of oxygen has been described.[146-148] It is proposed that a monolayer is formed at 1·8 V in 100 s, and the amount adsorbed can be obtained by desorbing the layer on a cathodic sweep and integrating the charge passed.

In all cases, the surface area is obtained by converting the total charge for a monolayer to the number of adatoms, and then by assuming one adatom is associated with each exposed surface metal atom. In the case of hydrogen, where $H_{(s)}$ and $H^+_{(aq)}$ are related by a single electron transfer, 1 μC of charge is equivalent to $6·24 \times 10^{12}$ $H_{(s)}$, while in the case of $O_{(s)}$ which requires a two electron transfer, the corresponding number of adatoms is $3·12 \times 10^{12}$ $O_{(s)}$. If the specimen is polycrystalline, the number of surface metal atoms per unit area (n_s) is given in Table 6.2. For single crystal surfaces, the necessary information is contained in Table 3.1 and equations 3.1 and 3.2 (Chapter 3) together with the lattice parameter of the metal (Appendix 1, Table A.1). On the basis of the n_s values given in Table 6.2, a monolayer of $H_{(s)}$ is associated with 201, 214 and 209 μC for platinum, rhodium and iridium respectively, while for gold a monolayer of $O_{(s)}$ is associated with 369 μC.

In practice considerable care needs to be taken with regard to the purity of the electrochemical cell. In summary, the presence of ions (adventitious or otherwise) in the electrolyte which strongly complex with electrode metal cations and so result in electrode corrosion, must be avoided: halide ions appear to be particularly potent in this respect. Other dissolved electrochemically active impurity (e.g. dissolved oxygen and metallic ions) need to be eliminated. There are three common methods which have been used to purify the electrolyte solution: passage over active carbon, pre-electrolysis and the introduction of a getter electrode. Pre-electrolysis suffers from the disadvantage that the electrode which is to be examined may have its surface modified. The other two methods are to be preferred.

2. Particle Size

Representation of Particle Size
Size Distribution and Mean Size

In practical situations, a finely divided solid usually consists of particles the size of which spans a range of values: that is, there is a particle size distribution. Inasmuch as the particle size is bounded by zero at the lower limit and is unbounded at the upper limit, this distribution will generally be skewed. A schematic illustration is given in Fig. 6.30 which also indi-

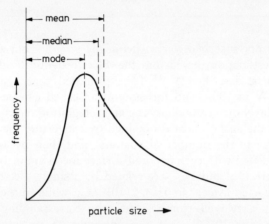

FIG. 6.30 Schematic particle size distribution indicating three common measures of central tendency.

cates the three common measures of central tendency. The mode line passes through the peak of the distibution curve, the median line divides the area under the curve into equal parts, and the mean, $\bar{d}$, is defined by

$$\bar{d}_{LN} = \frac{\sum_i f_i d_i}{\sum_i f_i} \qquad (6.26)$$

and is the point on the diameter scale the deviations from which sum to zero, i.e. $\sum_i f_i(d_i - \bar{d}) = 0$, and for which the sum of the squares of the deviations is minimized. The subscript LN in $\bar{d}_{LN}$ indicates that the mean is defined as a number average of a length.

We shall use the mean exclusively as a measure of central tendency.

The parameter d_i itself is not the only one which may be used to define a mean diameter. For instance, the quantity d_i^2 may be used. Since d_i^2 is related to the particle surface area, A_i we have

$$A_i = \alpha f_i d_i^2 \tag{6.27}$$

where α is a dimensionless constant depending on the particle geometry: if all the particles are spheres and d_i is the sphere diameter, $\alpha = \pi$; if all are cubes and d_i is the cube edge, $\alpha = 6$. Thus

$$\sum_i A_i = \alpha_{SN} \sum_i f_i d_i^2 \tag{6.28}$$

where $\sum_i A_i$ is the total surface area, and

$$\bar{d}_{AN} = \sqrt{\frac{\sum_i f_i d_i^2}{\sum_i f_i}} = \alpha_{AN}^{-\frac{1}{2}} \sqrt{\frac{\sum_i A_i}{\sum_i f_i}} \tag{6.29}$$

This defines an area-number average diameter as indicated by the subscript to $\bar{d}_{AN}$.

There are thus a number of different ways of defining a mean particle diameter, and their utility is determined by what parameters it is possible to measure in practice. The more important means are given in Table 6.6. If the particles in a sample are of varied irregular shapes, α is not defined geometrically. It is usual to assign a value to α based on an assumed spherical particle equivalent, and these values are included in Table 6.6.

Our discussion assumes of course that the particles are non-porous so that the surface area refers only to the external surface.

TABLE 6.6 Some definitions of mean particle size

Type	Definition of mean	Value of α for spherical particle equivalent
length-number	$\bar{d}_{LN} = \dfrac{\sum_i f_i d_i}{\sum_i f_i}$	
area-number	$\bar{d}_{AN} = \sqrt{\dfrac{\sum_i f_i d_i^2}{\sum_i f_i}} = \alpha_{AN}^{-\frac{1}{2}} \sqrt{\dfrac{\sum_i A_i}{\sum_i f_i}}$	$\alpha_{AN} = \pi$
volume-number	$\bar{d}_{VN} = \sqrt[3]{\dfrac{\sum_i f_i d_i^3}{\sum_i f_i}} = \alpha_{VN}^{-\frac{1}{3}} \sqrt[3]{\dfrac{\sum_i V_i}{\sum_i f_i}}$	$\alpha_{VN} = \dfrac{\pi}{6}$
volume-area	$\bar{d}_{VA} = \dfrac{\sum_i f_i d_i^3}{\sum_i f_i d_i^2} = \left(\dfrac{\alpha_{AN}}{\alpha_{VN}}\right) \left(\dfrac{\sum_i V_i}{\sum_i A_i}\right)$	$\dfrac{\alpha_{AN}}{\alpha_{VN}} = 6$

Metallic Dispersion

Because dispersed metallic catalysts often have the extent of their surface characterized by gas chemisorption methods which count the number of surface metal atoms, it is convenient to define the state of subdivision of the metal in terms of the ratio of the total number of surface atoms to the total number of metal atoms present. Thus the metallic dispersion of the metal M, D_M, is defined by

$$D_M = N_{(S)M}/N_{(T)M} \qquad (6.30)$$

This definition is obviously related to the volume-area mean diameter, $\bar{d}_{VA}$, since in the spherical particle equivalent approximation

$$\bar{d}_{VA} = 6\left(\frac{\sum_i V_i}{\sum_i A_i}\right) = 6\frac{v_M}{a_M}\frac{N_{(T)M}}{N_{(S)M}} = 6\left(\frac{v_M}{a_M}\right)\bigg/D_M \qquad (6.31)$$

where a_M and v_M are respectively the effective average area occupied by a metal atom in the surface, and the volume per metal atom in the bulk. Values for a_M can be obtained from the data listed in Table 6.2, while v_M is given by $M_w/\rho N_0$ where M_w is the atomic weight, ρ the density and N_0 is Avogadro's number.

For a particle of specified geometric shape the ratio of the number of surface atoms to the total number of atoms is well defined and calculable. Values for this ratio (called F_S) are given in Figs 5.4–5.6 for f.c.c., b.c.c. and h.c.p. metals for a variety of crystal shapes. Thus if an actual specimen were to contain metal particles all of identical size and of geometrically regular shape, D_M would be identical with the appropriate F_S. In practice this situation never occurs, and F_S acts as an ideal model with which to compare reality.

For practical purposes it is convenient to construct an average representation of the metallic dispersion in terms of particle size, and this relation is shown in Fig. 6.31: it has been constructed by averaging the F_S data from Figs 5.4–5.6. Although somewhat different dispersion–size curves may be constructed by averaging over other sets of crystallite shapes, and somewhat different averages will be obtained for the three crystal structures (f.c.c., b.c.c., h.c.p.) the practical accuracy for which this curve is needed makes further elaboration pointless, since in practice the actual crystal shapes are seldom known at all.

If the catalyst is bimetallic the definition of metallic dispersion becomes more complicated. Suppose the two metals present are L and M. If there are $N_{(S)L}$ and $N_{(S)M}$ surface atoms of L and M respectively, the total metallic dispersion ($D_{L,M}$) may be defined by (cf. ref. 268)

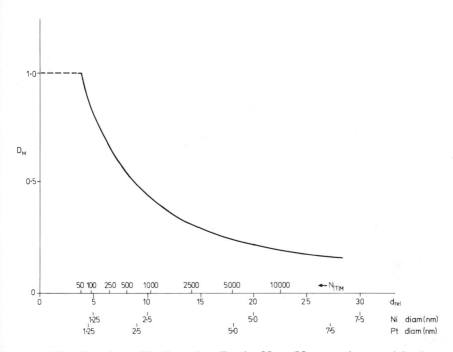

FIG. 6.31 Plot of metallic dispersion, D_M ($= N_{(S)M}/N_{(T)M}$ against particle size, the latter being expressed as (i) $N_{(T)M}$, the total number of atoms per particle; (ii) d_{rel}, the ratio of the diameter of a sphere with a volume equal to $N_{(T)M}$ times the volume occupied by an atom in the unit cell, to the atom diameter; (iii) diameter for nickel particles; (iv) diameter for platinum particles. The curve assumes spherical particles. To convert d_{rel} to actual values for the apparent spherical particle diameter, multiply d_{rel} by the nearest neighbour distance in the metal.

$$D_{L,M} = \frac{N_{(S)L} + N_{(S)M}}{N_{(T)L,M}} \quad (6.32)$$

where
$$N_{(T)L,M} = N_{(T)L} + N_{(T)M} \quad (6.33)$$

However, a dispersion parameter may also be defined for L and M individually

$$D_L = N_{(S)L}/N_{(T)L} \quad (6.34a)$$

$$D_M = N_{(S)M}/N_{(T)M} \quad (6.34b)$$

$N_{(T)L}$ and $N_{(T)M}$ are related by the known composition of the system. Thus if $x_{(T)}$ is the mole fraction referring to the total composition

$$x_{(T)L} = \frac{N_{(T)L}}{N_{(T)L} + N_{(T)M}} \qquad (6.35a)$$

$$x_{(T)M} = \frac{N_{(T)M}}{N_{(T)L} + N_{(T)M}} \qquad (6.35b)$$

D_L, D_M and $D_{L,M}$ are thus related by

$$D_{L,M} = x_{(T)L} D_L + x_{(T)M} D_M \qquad (6.36)$$

It is also useful to define the overall surface composition in terms of a surface mole fraction $x_{(S)}$

$$x_{(S)L} = \frac{N_{(S)L}}{N_{(S)L} + N_{(S)M}} \qquad (6.37a)$$

$$x_{(S)M} = \frac{N_{(S)M}}{N_{(S)L} + N_{(S)M}} \qquad (6.37b)$$

and it follows that

$$x_{(S)L} = \frac{x_{(T)L} D_L}{x_{(T)L} D_L + x_{(T)M} D_M} \qquad (6.38a)$$

$$x_{(S)M} = \frac{x_{(T)M} D_M}{x_{(T)L} D_L + x_{(T)M} D_M} \qquad (6.38b)$$

The relationship between the metallic dispersion and average particle size now depends on knowing something more about the way the two metals are distributed. In any case the total dispersion, $D_{L,M}$ is valid if the difference between the metallic components is ignored, and an average particle size may be obtained from $D_{L,M}$ via the relation in Fig. 6.31. However, in general a more detailed description of the behaviour of the individual components requires extra information beyond a knowledge of $N_{(S)L}$ and $N_{(S)M}$ (assuming these are available from, for instance, specific gas chemisorption). The two limiting cases are illustrative. If it is known that L and M remain completely unmixed in the metal particles so that only particles of pure L and pure M are present, D_L and D_M are independently defined by equations 6.34a and b, subject only to the constraint of the relative proportions of L and M: the average particle sizes of the L and M components may be obtained from D_L and D_M. On the other hand, if each particle is a mixture of L and M, each of the same overall composition, $D_{L,M}$ gives the average particle size directly. However, there are more complex situations between these extremes. For instance, if the array of metallic particles consisted of some pure L, some of pure M and some of mixtures of L and M with the composition variable from particle to particle, all that can be obtained from a measurement of $N_{(S)L}$ and $N_{(S)M}$ is $D_{L,M}$, and a more detailed description of the distribution of particles of various types needs extra experimental data.

Estimation of Particle Size
Electron Microscopy

The metal particles in a supported catalyst can be observed and measured directly by using an electron microscope in transmission (e.g. ref. 200). The techniques of sample preparation are summarized in a subsequent section (pp. 404–407) which also gives some comments on available resolution.

Crystal size in thick evaporated metal films and in polycrystalline massive metal specimens may be estimated from transmission electron microscopy, and this type of observation is now routine (e.g. refs 201, 202) and has been thoroughly described.

In most cases, 100 keV electrons should be used, and there is some advantage in using 200 keV for thick specimens.

A good modern electron microscope offers an effective resolution of about 0·3 nm: this means that there is little point in trying to measure particles of size less than 1·0–1·5 nm, although the presence of particles below this size can often be observed.

The scanning electron microscope has its usefulness restricted by its limited resolution which lies in the region of 15 nm for normal commercial instruments. In other words, the scanning microscope is only useful for particles bigger than about 100 nm (but cf. p. 404).

The main problem in the use of electron microscopy for a quantitative estimation of particle size lies in ensuring that the results are representative of the catalyst. It is imperative that several samples be examined, and a large number of particles be measured. The number of samples needed can only be judged by a statistical examination of their variability. However, the total number of particles measured always runs to at least several hundred in order to obtain an accuracy of $\pm\,5\,\%$ on the mean diameter.

If the image of a particle as seen in an electron microscope is of irregular shape, there remains the problem of selecting a measurement to represent its size. The image is, of course, two dimensional and it should be remembered that a false estimate of the mean particle size can be obtained if the particles are not randomly oriented in the specimen being examined. There are a number of common representations of particle size. The most satisfactory is the projected area diameter which is the diameter of a circle having the same area as the two-dimensional image of the particle. Without instrumentation this is very tedious to measure, but the Zeiss-Endter particle size analyser has been developed to allow a direct comparison between the projected particle area and the area of a reference circle, the latter being adjustable.[218, 219] The second and perhaps the most common method is the measurement of the intercept which a line, drawn across the array of particles, makes on each particle.[217] Thirdly, there is the Feret diameter which is the distance between two tangents on opposite sides of the particle, parallel to some fixed direction which is the same for all particles in the array. Fourthly, there is the mean of the maximum and minimum width of each particle. A mean diameter for the particle array may be defined for each of these measured parameters in the usual way.

On the whole, the projected area diameter, d_P, lies between the intercept diameter, d_I, and the Feret d_F, and the maximum/minimum mean, d_M, diameters: that is $d_I < d_P < d_F$, d_M. The difference between these measures obviously depends on particle geometry, but it has been shown in a number of cases with crushed material that the difference between d_I and d_F, d_M is 20–30%.

The electron beam itself may not be without influence on the specimen. Undesirable thermal effects are minimized when anticontamination cooling blades are used, and the latter should in any case be used routinely so that deposition of contaminant does not impair the performance.

The direct measurement of particle size generates a length-number mean diameter, $\bar{d}_{LN}$ (cf. Table 6.6). It also has the advantage of producing a diameter distribution curve.

Via Surface Area

If the total surface area, A, and the total volume, V, of a dispersed metal are known, the mean value of the equivalent spherical particle diameter, $\bar{d}_{VA}$ (cf. Table 6.7) is given by

$$\bar{d}_{VA} = 6V/A \tag{6.39}$$

The determination of A has been extensively discussed in previous sections, while V is usually obtained by knowing the mass of metal present and its density.

X-Ray Methods

There are two main methods for the estimation of mean particle size which use X-ray techniques. These are: (i) diffraction line broadening which makes use of information contained in the peak shape of one or more diffraction lines from the substance in question—this method is therefore specific for a particular component; (ii) low angle scattering—in principle all the particles in a divided solid contribute, but in practice the scattering power of a component is dependent on its chemical nature, so some degree of specificity is possible and this can be increased by special procedures. X-ray absorption edge spectroscopy and the determination of the radial electron distribution function (cf. refs 237, 238) are techniques which can give additional information relevant to the dispersion of a metal catalyst, but they do not generate quantitative data for particle size. A detailed account of X-ray techniques is provided by Klug and Alexander.[220]

X-ray Diffraction Line Broadening

An X-ray diffraction line broadens when the crystallite size falls below about 100 nm. The technique is particularly applicable to metal crystallites of size 3·0–50 nm: below 3·0 nm the line is so broad and diffuse as to become lost, while above about 50 nm the change in peak shape is small and the method therefore insensitive.

There are factors other than particle size which can contribute to the observed peak width. In addition to purely instrumental factors there are matters such as strain, and stacking and twin faults. For a supported metal catalyst it is probable that only particle size will contribute significantly to the line width in excess of the instrumental width, and we shall consider this situation first since the treatment is very simple.

If one assumes that the line shapes are Gaussian, then the squares of the contributing width factors are additive. Thus if B_{inst} is the instrumental line width and B_{obs} the observed width, the value of B_d, the line width due to particle size broadening is given by:[203]

$$B_d^2 = B_{obs}^2 - B_{inst}^2 \qquad (6.40)$$

B_{inst} is obtained by a calibration procedure using a material consisting of large crystals of good crystalline perfection. Sodium chloride or quartz powder (100 nm < d < 1000 nm) is often used. The value of B_{inst} is a function of the Bragg angle θ so a value must be interpolated from the calibration data to the θ value of the reflection from the sample being studied.

An alternative method for correcting for instrumental line broadening has been given by Jones (ref. 221 and cf. ref. 220).

Having evaluated B_d, the mean crystallite diameter, $\bar{d}_B$ is given by the Scherrer equation

$$\bar{d}_B = \frac{K\lambda}{B_d \cos \theta} \qquad (4.61)$$

where λ is the X-ray wavelength, K is the Scherrer constant, and B_d is the angular width expressed in terms of $\Delta(2\theta)$ (radian). The mean diameter, $\bar{d}_B$, thus obtained, is defined by $\sum_i f_i d_i^4 / \sum_i f_i d_i^3$.

The value of K depends on how the peak width is measured, and for this there are two methods. If the width is measured at half the maximum peak height, K takes values between $0\cdot84$ and $0\cdot89$ depending on the assumed particle shape. In the absence of detailed knowledge on this point, and this is the usual situation, a spherical particle is assumed, and more often than not K has been taken as $0\cdot90$. A value of $0\cdot89$ is preferable, but the difference is not of much significance in relation to the absolute accuracy of the method.

An alternative method for the measurement of peak width is to divide the integrated peak area by the peak height: on this basis K takes values between $1\cdot00$ and $1\cdot16$ depending on particle shape, and a value of $1\cdot00$ is most commonly used.

In the simple case of particle size broadening only, the simple method of using the width at half peak height is the obvious choice. However, there are some practical precautions to be observed.

In the first place, there may be sensitivity limitations with supported catalysts if the metal loading is low. Difficulties may well be encountered, for instance, at $< 0\cdot5$ wt. % platinum. Since the intensity of diffracted radiation is proportional to the square of the atomic number, this problem becomes more severe for lower atomic number elements.

We shall assume that a recording diffractometer is being used: with modern instrumentation no other method should be contemplated. To measure the peak height requires the peak baseline to be established. This is done using the level parts of the diffractometer trace adjacent to the peak in question. However, problems may arise if there is an adjacent peak so that the true baseline is not established between them. If it is not possible to select a peak which is sufficiently isolated for this not to occur, a correction from the overlap can be made by fitting the outer portions of each peak to a Cauchy function (of the form $1/(1 + k^2x^2)$). There is always a tendency to underestimate peak height. It is essential to adjust the rate of scan so that the record is not influenced by the time constant of the diffractometer counting and recording equipment. This will vary with the detailed experimental situation, but scan rates in the range $0\cdot05°–0\cdot5°$ min^{-1} (in 2θ) are commonly used.

When working with a multicomponent system such as a supported catalyst it is essential to select a metal diffraction peak that is well separated from peaks which may originate from the support. This is often a serious limitation on the number of metal peaks which can be used. For supported catalysts of metals such as platinum, palladium and nickel, the 111 reflection has often been used, but others are also possible. However, the platinum 111 reflection is obscured by that from the support in platinum/ γ-alumina and in this situation the platinum 311 line is often used. Nevertheless, the sensitivity is considerably lower when using the 311 line compared with what is available if the platinum 111 line can be used.

Sample preparation deserves care. Supported catalysts should be finely ground to about 300 mesh and the powder spread on the sample holder, care being taken that the surface of the powder sample is flat. Unsupported metal powders are spread in the same way. Although pressing the powder to a wafer is sometimes done, it has no extra merit, and should be avoided with unsupported powder because of the introduction of strain broadening.

The primary X-ray beam should be selected to avoid fluorescence problems, and should be as monochromatic as possible. For instance, nickel-filtered Cu K_α is often used with platinum, palladium and nickel and iron-filtered Co K_α or zirconium-filtered Mo K_α with iron. In fact, nickel-filtered Cu K_α radiation consists of the α_1 plus α_2 doublet, and this distorts the peak shape. This is less important the wider the peak, but should be corrected for using the graphical method described by Rachinger.[205]

The wavelengths of some commonly used X-rays are given in Table 6.7.

TABLE 6.7 Some common X-ray wavelengths

	$K_{\alpha 1}$(nm)	$K_{\alpha 2}$(nm)	Weighted mean for unresolved $K_{\alpha 1}$ and $K_{\alpha 2}$(nm)
molybdenum	0·070926	0·071354	0·07107
copper	0·154050	0·154434	0·15418
nickel	0·165783	0·166168	0·16591
cobalt	0·178890	0·179279	0·17902
iron	0·193597	0·193991	0·19373
chromium	0·228962	0·229352	0·22909

In situations where the peak broadening contains contributions from strain, or from faulting, the analysis is more complex. This can be the case with unsupported dispersed metal specimens, particularly if they have been subjected to mechanical disturbance. The simplest procedure is to note that the strain broadening B_s is given by

$$B_s = K' \tan \theta \qquad (6.42)$$

where K' is a parameter proportional to the strain in the sample. Assuming Cauchy line profiles so that the line widths are directly additive,

$$B_{obs} = \frac{K\lambda}{\bar{d}_B \cos \theta} + K' \tan \theta \qquad (6.43)$$

so $\bar{d}_B$ may be obtained from the intercept of a plot of $B_{obs} \cos \theta$ vs $\sin \theta$.

A more sophisticated analysis based upon a Fourier analysis of the line profile is also available.[206, 204]

The absolute accuracy with which the average particle diameter can be obtained by line broadening methods should not be overestimated, and the influence of particle shape and size distribution factors probably limits this to an accuracy of about 30%.

An advantage of the method is that the area under the diffraction peak curve is proportional to the amount of diffracting material, and it is useful to make use of this feature as a check on the amount of metal which may be present in particles too small for detection: that is, by comparing the amount estimated from the diffraction line area with the total known to be present from the chemical composition of the specimen. This method requires a calibration. In principle this is possible by the use of another catalyst of the same type in which it is known that all of the metal is detectable by X-ray diffraction. However, this assumes a high degree of constancy in the equivalence between the examination of the two specimens, and it is customary to overcome this by the use of an internal standard such as the use of magnesium oxide with palladium/silica.[207]

Small Angle X-ray Scattering

This technique uses information contained in radiation scattered within a few degrees ($< 5°$) of the primary beam. Use of the method for metal particle size estimation with catalysts has been limited (e.g. ref. 208) however, the principles of the method have been well described.[220, 209-215, 222-224] The theory applies to a dilute particle array.

Let I(s) be the intensity of the scattered radiation: this is a function of a variable s defined by $s = 2\theta/\lambda$, where 2θ (radians) is the scattering angle and λ is the X-ray wavelength. If the sample consists of particles all of the same size, I(s) follows the relation

$$\ln I(s) = \ln (N_p N_E) - \frac{4\pi^2}{5} R_G^2 s^2 \qquad (6.47)$$

where N_p is the number of particles in the sample, and N_E is the number of electrons per particle, while R_G is a constant called the Guinier radius. R_G can be evaluated from the slope of the line relating $\ln I(s)$ and s^2, and is a measure of the particle radius. However, if the particles are of non-uniform size, a plot of $\ln I(s)$ vs s^2 is nonlinear. The simplest general procedure then is that due to Jellinek *et al.*[225] who relied on drawing tangents to the curve to obtain the contributions from particle fractions of various sizes, while Shull and Roess[222] adjusted a particle size distribution function to generate a match to the original $\ln I(s)$ vs s^2 curve. A more convenient method of analysis has been suggested by Harkness et al.,[211] *provided* the particle size distribution function is a reasonable approximation to a log-normal function. The latter is

$$P(r) = \frac{1}{(2\pi)^{1/2} r \ln \sigma} \exp\left[-\frac{1}{2}\left(\frac{\ln r - \ln \bar{r}_g}{\ln \sigma}\right)^2\right] \quad (6.48)$$

where r is the particle radius, $\bar{r}_g$ is the geometric mean of the distribution, and σ is the square root of the variance of the distribution. For a particle size distribution which follows a log-normal function, the plot of ln I(s) *vs* s^2 retains a linear portion from which an R_G may still be evaluated. Furthermore a Porod radius, R_P is defined by

$$R_P = \frac{3\int_0^\infty s\,I(s)\,ds}{8\pi(1-f)\lim_{s\to\infty} s^3 I(s)} \quad (6.49)$$

where the numerator is readily evaluated numerically, while the denominator is obtained from the tail of the plot of $s^3 I(s)$ *vs* s at large s. The factor f is the volume fraction of scattering particles in the sample (for a supported catalyst $1 - f \approx 1$). In practice there may be some difficulty in the accurate evaluation of R_P because of diminished sensitivity at large s. Nevertheless, for this situation the distribution parameters $\bar{r}_g$ and σ can be related to the scattering parameters R_G and R_P by

$$\ln \bar{r}_g = \ln R_G - 1\cdot 714 \ln (R_G/R_P) \quad (6.50)$$

$$(\ln \sigma)^2 = 0\cdot 286 \ln (R_G/R_P) \quad (6.51)$$

In practice, a Kratky block-slit camera[215] which is a means of producing a finely collimated slit-shaped X-ray beam is desirable, and models are available commercially. Nevertheless, the use of a slightly modified commercial X-ray diffractometer has been described;[226] the main modifications being the replacement of the normal collimating slit by one giving a horizontal divergence of $0\cdot 1°$, and making provision for rotating the counter while keeping the specimen stationary. Specimens are examined in transmission, and thicknesses of the order of $0\cdot 1$ mm are typical. The literature gives further details for optimizing performance and resolution.[226-228] Monochromatic radiation is required. The K_α component of the characteristic X-ray doublet can be resolved by using single crystal dispersion, a balanced double-filter[244] or the use of a single K_β absorber (e.g. a nickel filter with copper radiation) together with a pulse-height analyser. The last is the easier.

A major advantage of small angle X-ray scattering for particle size estimation is that the lower limit is of the order of 1·0 nm particle diameter, and the effective range up to about 100 nm.

For application to supported catalysts, scattering from the pores of the support has to be eliminated. Heinemann et al.[208] following a suggestion by Gunn,[229] describe a method for achieving this by impregnating the catalyst specimen with a liquid having the same electron density as the support, so the metal particles are left as the only main scattering centres. For platinum/η-alumina catalysts, impregnation with methylene iodide was used, although the liquid concentration has to be adjusted for optimum performance.

X-ray Absorption Edge Spectroscopy

An X-ray absorption edge of an element arises when an incident photon has just enough energy to eject an electron from an atomic level. There is fine structure associated with the absorption edge, and this depends on the fate of the excited electron. For instance, with regard to the L_3 absorption edge of platinum which is one of the three L-absorption edges associated with the excitation of a $2p$ electron, the fine structure on the longer wavelength side (Kossel absorption) is due to excitation into a partly filled $5d$ and $6s$ level (there is a selection rule $\Delta l = \pm 1$) while fine structure on the shorter wavelength side (Kronig absorption) is due to interaction of the ejected electron with neighbouring atoms. Thus the nature of the fine structure becomes dependent on the chemical environment of the atom.

It is known[232, 233] that the appearance of an absorption edge can be affected by the absorber thickness, while Lewis[234, 235] has observed a change in the absorption near the L_3 edge of platinum due to the small particle size in platinum/η-alumina and in platinum/X- and Y-zeolite catalysts. Nevertheless, particle size effects in X-ray absorption edge spectroscopy do not appear to have been systematically explored, although the technique should, in principle, prove of value in identifying the influence of particle size on electronic properties. The effect of gas adsorption on the K-absorption edge of nickel in some supported catalysts has been reported.[230, 231]

Experimental details based upon the use of a modified commercial diffractometer have been described.[230, 231, 234, 235, 236]

Magnetic Measurements

This type of measurement has been mainly developed by Selwood.[237] If one has a particle of a normally ferromagnetic metal which is smaller than the ferromagnetic domain size (10–30 nm) the particle when placed in an external field will behave magnetically like a paramagnetic atom with a very large magnetic moment, and one may treat an assembly of such particles under the approximation of assuming that they behave as an assembly of paramagnetic atoms. Thus if the assembly consisted of n_i particles each of volume v_i, the magnetic moment m_i would be given by

$$m_i = M_{sp} n_i v_i \left[\coth\left(\frac{M_{sp} v_i H}{kT}\right) - \frac{kT}{M_{sp} v_i H} \right] \tag{6.52}$$

where M_{sp} is the spontaneous magnetization per unit volume, and H the field strength. Magnetic quantities are here written in terms of their magnitudes. Both low and high H approximations are available to equation 6.52. This small particle behaviour of normally ferromagnetic metals is referred to as collective paramagnetism.

At low values of H

$$m_i = \frac{M_{sp}^2 n_i v_i^2 H}{3 \, kT} \tag{6.53}$$

If one has a collection of particles of various sizes, the total magnetic moment $m = \sum_i m_i$, so that

$$m = \frac{M_{sp}^2 H}{3kT} \sum_i n_i v_i^2 \tag{6.54}$$

However, the saturation moment m_s is given by

$$m_s = M_{sp} \sum_i n_i v_i \tag{6.55}$$

Thus, by equating m/m_s to σ/σ_s or to M/M_s

$$\frac{M}{M_s} = \frac{\sigma}{\sigma_s} = \frac{M_{sp}}{3k} \left(\frac{H}{T}\right) \left(\frac{\sum_i n_i v_i^2}{\sum_i n_i v_i}\right) \tag{6.56}$$

On the other hand, at high values of H

$$\frac{M}{M_s} = \frac{\sigma}{\sigma_s} = 1 - \frac{k}{M_{sp}} \left(\frac{T}{H}\right) \left(\frac{\sum_i n_i v_i}{\sum_i n_i}\right)^{-1} \tag{6.57}$$

where M is the magnetization per unit volume, σ the magnetization per unit mass, and M_s and σ_s are the corresponding saturation values. The saturation magnetization is assumed to be independent of particle size. A mean particle volume (but not the number mean used in 6.57) is defined by $\sum_i n_i v_i^2 / \sum_i n_i v_i$. Thus the magnetically determined mean particle diameter $\bar{d}_m$ is given by

$$\bar{d}_m = (6\bar{v}_m/\pi)^{1/3} \tag{6.58}$$

where $\bar{v}_m$ is the mean particle volume.

M_{sp} and σ_s both are constants for a given specimen which should be determined at high H and low T ($\leqslant 4.2$ K). However, provided the type and amount of the metallic phase is known, approximations are usually made by assuming that for the sample consisting of small particles, σ_s (or M_s) and M_{sp} take the same values as are known for the metal in the massive form and, for instance, with some dispersed nickel catalysts this has given $\bar{d}_m$ values which are consistent with average particle sizes measured by other methods. Both of the quantities, M_s and M_{sp}, are taken to equal the saturation magnetization at 0 K, M_{so}: σ_{so} is related to M_{so} by $\sigma_{so} = M_{so}/\rho$ where ρ is the density. Values for M_{so} and σ_{so} are listed in Table 6.8[189] for the common ferromagnetic metals. If the amount or composition of the metallic phase is not known, σ_s (or M_s) should be determined experimentally via an equation of the form 6.57 as $H^{-1} \to 0$ at $T \leqslant 4.2$ K. At higher temperatures (e.g. 77 K) a very long extrapolation is needed: empirical relations, for instance $\sigma_s/\sigma = 1 + \alpha H^{-m}$ (m = 0.9 for nickel[264]) have been used to help extrapolate, but the accuracy is still not particularly high.

TABLE 6.8 Values for saturation magnetizations M_{so} and σ_{so} at 0 K

	$M_{so}(A\ m^{-1})$*	$\sigma_{so}(A\ m^2\ kg^{-1})$*
iron	$1\cdot735 \times 10^6$	$2\cdot219 \times 10^2$
cobalt	$1\cdot444 \times 10^6$	$1\cdot625 \times 10^2$
nickel	$0\cdot509 \times 10^6$	$0\cdot575 \times 10^2$

* To convert M_{so} in MKS units ($A\ m^{-1}$) to cgs emu, divide by 10^3: σ_{so} values are numerically the same in both systems.

In the low field approximation, which is the more commonly used, one needs the slope of the plot of (σ/σ_s) vs (H/T) at $H \to 0$ (cf. equation 6.56). Measurements are typically made up to a field strength of $3-4 \times 10^5$ A m^{-1},* at temperatures lying in the range 77 K to room temperature. For the use of the high field approximation, measurements are made to about 1×10^6 A m^{-1}, and preferably at low temperatures, so $\bar{v}_m$ may be evaluated from the slope of a plot of (σ/σ_s) vs (T/H) (cf. equation 6.57). Examples of the low and high field methods are due to Carter and Sinfelt[240] and MacNab and Anderson[238] respectively.

Particulate metallic specimens which exhibit collective paramagnetism do not generally give magnetization curves which can be fitted to a simple Langevin function (cf. equation 6.52) over the entire field range. It is

* To convert magnetic field strength in MKS units ($A\ m^{-1}$) to cgs emu (Oe), divide by $10^3/4\pi$.

usually presumed that this has its origin in the fact that there is a distribution of particle sizes. One would expect that the low field susceptibility is most affected by the largest particles, and the high field susceptibility by the smallest particles. The magnetization curve may be analysed in terms of the particle size distribution. Thus if f(v) is the fraction of particles having a volume between and $v + dv$, the normalized volume distribution function is $\int_0^\infty f(v)dv = 1$, and then

$$\frac{\sigma}{\sigma_s} = \int_0^\infty \left[\coth\left(\frac{I_{sp}vH}{kT}\right) - \frac{kT}{I_{sp}vH} \right] f(v)dv \qquad (6.59)$$

The form of f(v) can be adjusted to reproduce the experimental magnetization curve, or parameters defining the distribution can be obtained via some algebraic manipulation: this approach has been described by a number of workers (e.g. refs 262–264), but it should be recognized that the magnetization curve is not a very sensitive function of the detailed shape of the size distribution.

There are two main methods of magnetic measurement commonly employed with catalysts. In the static Faraday method the sample is entirely immersed in a non-uniform field and the magnetic force on the specimen is measured in some convenient way such as with a microbalance (cf. p. 348). In the most common procedure the desired field is generated with a tapered magnet gap which gives a field of constant gradient: such magnets are commercially available. The sample should be the only metal-containing component in the field, and silica is the standard material for the construction of the sample container, supporting fibre, etc. Nevertheless, correction must be made for the magnetic force due to the diamagnetic susceptibility of the silica, and also for the support material upon which the metal may be dispersed. The sample needs to be small enough to be totally immersed, but amounts up to a gram or two are easily accommodated. The magnetic force, f, on the specimen measured by the microbalance is given by

$$f = m\chi H(\partial H/\partial l) \qquad (6.60)$$

where H and $\partial H/\partial l$ are the field and field gradient at the specimen, χ is the mass susceptibility, m the mass of magnetic material in the specimen, and it is assumed that the specimen length is small enough for the change in H over the length to be small. By the use of a calibration experiment with a known amount of material of known susceptibility it is obviously possible to set up a scale relating measured f and χ, and σ is given by

$$\sigma = \chi H \qquad (6.61)$$

Ferrous ammonium sulphate hydrate $((NH_4)_2Fe(SO_4)_2 \cdot 6H_2O)$ is often used as a calibrating substance for which the kg susceptibility in MKS units is $1.194 \times 10^{-4}/(T + 1)$ m³ kg⁻¹, where T is the temperature in K. Alternatively, in cgs emu, the gram susceptibility of $(NH_4)_2Fe(SO_4)_2 \cdot 6H_2O$ is $9.50 \times 10^{-3}/(T + 1)$ cm³ g⁻¹.

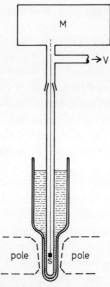

FIG. 6.32 Schematic illustration of Faraday magnetobalance. The sample S is suspended from the vacuum microbalance M in a magnetic field of constant field gradient. V leads to vacuum and gas handling. The Dewar allows the specimen to be cooled to 77 K with liquid nitrogen and may be removed for measurements at room temperature or for thermal treatment of the specimen above room temperature.

The Faraday method is illustrated in Fig. 6.32. For this type of specimen the Faraday method is preferable to the Gouy method, although the latter can be used. With the Gouy method which requires the sample to be part in and part out of a uniform field, if the amount of specimen is small it has to be arranged in a long and narrow configuration, for instance powder packed in a narrow tube, and it then may be difficult to obtain uniformity of packing and it is difficult to arrange for the entire specimen to be at the same temperature if measurements are to be made at other than room temperature. Furthermore the configuration of the calibrating specimen has to be made to match that of the unknown specimen. However, the Faraday method has its own problems. The main one is that ideally the product $H(\partial H/\partial l)$ should be constant over the length of the specimen. In practice this is usually approximated by making the specimen sufficiently short in the direction of l. For precise work specially shaped magnet gaps have been designed for this purpose.

An alternative method of measurement is the use of a vibrating sample magnetometer, and equipment based on the Foner design[239] is commercially available. The sample, which is totally immersed in a uniform field, is vibrated normally to the lines of force, and the oscillating specimen field induces an alternating voltage in a pair of stationary detector coils: this voltage is compared with another alternating voltage generated by the simultaneous movement of a permanently magnetized reference specimen with respect to a second set of detector coils outside the static magnetic field. Calibration against a substance of known susceptibility is still required, although commercial units provide an approximate calibration.

With either of these methods, it is not difficult to arrange for a magnet gap wide enough to take a Dewar which will allow the sample temperature to be adjusted down to 77 K while still maintaining the ability to have the specimen in vacuum or in a controlled atmosphere, and still allowing the field to reach about 1×10^6 A m^{-1}. Figure 6.33 illustrates the basis of the vibrating magnetometer. In all these arrangements, the provision of a sample heater to allow for (say) hydrogen reduction of the specimen can be made in the obvious way.

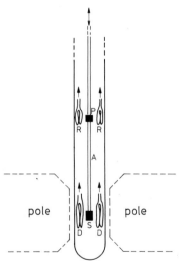

FIG. 6.33 Vibrating magnetometer. The sample S and magnetic reference sample P are vibrated vertically together via the support rod A. The alternating voltages induced by S in the detector coils D and by P in the reference coils R, are compared.

Miscellaneous Methods

There are a number of other general methods for particle size estimation based on a wide range of techniques: these include sieving, optical microscopy, sedimentation, centrifugation, elutriation, light scattering and permeametry. However, these methods are usually useful with particles considerably larger than those of metals in most dispersed metal catalysts, although they may be of use in the characterization of some divided support materials. Nevertheless, these methods have been well described in the literature (e.g. ref. 241), and the description will not be repeated here.

3. Pore Structure

Pore structure* is important to metallic catalysts in two ways. Supports which are commonly used to carry dispersed metallic particles are usually porous, and this pore structure influences the way in which the metallic particles are deposited, as well as reactant access when the catalyst is in use. Alternatively, a dispersed unsupported or stabilized metallic catalyst contains pores by virtue of the particulate nature of the metal itself.

There are a number of experimental methods for studying pore structure, but for our purposes there are two of dominant importance: from gas physical adsorption isotherm data, and from mercury porosimetry.

Physical Adsorption of Gases

We first consider the case of solids having pores of diameters in the range from a few nm to tens of nm, that is, mainly mesopores. The effect of the pore structure on multilayer adsorption is to convert an isotherm which would otherwise be BET type II into type IV, or a type III into type V (cf. Fig. 6.1). This occurs because the finite pore width places a limit on the extent of multilayer formation before the pore becomes filled with liquid-like condensate, and this will occur with the formation of a highly curved miniscus which results in a depression of the liquid vapour pressure below its normal value. Adsorption–desorption hysteresis also results because pores are filled and emptied by different processes: emptying occurs by evaporation from the liquid miniscus which retreats down the pore in the process. In practice, except sometimes with micropores (diameter less than about 1 nm) which are discussed subsequently, physical adsorption with capillary condensation usually gives an isotherm

---

* Where appropriate we shall use the classification of pores: macropore, width greater than about 50 nm; micropore, width not greater than about 2 nm; mesopore, intermediate widths.

with a hysteresis loop, and Fig. 6.34 illustrates this feature schematically for isotherms of basic types IV and V. In a type IV isotherm (Fig. 6.34b), adsorption to the vicinity of a monolayer occurs by point B, and the monolayer uptake and thus the surface area may be obtained by fitting the data in the adsorption branch to the BET equation in the usual region of its validity ($0 \cdot 05 < p/p_0 < 0 \cdot 3$). However, an attempt to evaluate the monolayer uptake from a type V isotherm is just as hazardous as it is with a type III isotherm, and the reasons which were discussed previously are the same for both. For intermediate situations, the accuracy with which the monolayer uptake may be evaluated deteriorates as the isotherm tends from type IV towards type V.

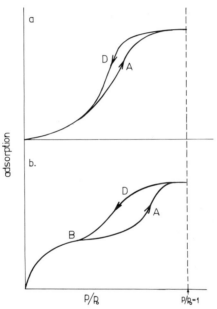

FIG. 6.34 Schematic illustration of type V and type IV physical adsorption isotherms, a and b respectively, with hysteresis loops.

The qualitative shape of the hysteresis loop contains information about pore shape. Loop shapes have been classified by de Boer[247] and the five basic types are shown in Fig. 6.35: these are to be regarded as archetypes and in practice loop shapes are often of mixed character. Nevertheless it is useful to summarize the various pore shapes corresponding to the various loop shapes. Type A: tubular pores open at both ends, with various forms of cross-section and with no or only slightly widened parts; ink-

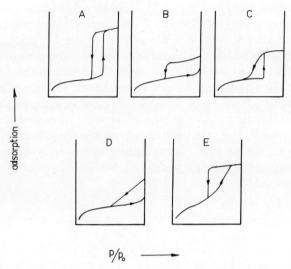

FIG. 6.35 Basic types of hysteresis loops in physical adsorption on porous adsorbents. After de Boer, J. H. *In* "The Structure and Properties of Porous Materials" (D. H. Everett and F. S. Stone, eds) Butterworths, London (1958), p. 68.

bottle pores with wide or very short necks; trough-shaped pores. Type B: slit-shaped pores with parallel walls; very wide pores with narrow short openings. Type C: heterogeneous distribution of pores which would show type A behaviour if present in a restricted range of dimensions; tapered tubular capillaries; certain tapered slit pores. Type D: most tapered slit pores; pores with wide bodies and a distribution of narrow necks. Type E: distribution of pores which would lead to type A but for which the dimensions responsible for the adsorption branch are heterogeneously distributed, while the dimensions responsible for the desorption branch are of equal size—for example spherical cavities of varying radii but constant neck width. In practice, types A, B and E are the most important, and type E has been the most frequently associated with ink-bottle pores, although as we have seen they are not exclusively restricted to this type. The problem of hysteresis loop analysis for the characterization of ink-bottle pores has been discussed[257] and will not be treated further here.

Although there is a considerable earlier literature much of which has been summarized,[158, 188-191] the analysis of pore structure from physical sorption isotherms has more recently been treated in extensive detail by de Boer and his collaborators,[186, 248-261] and the following discussion is based upon this later work. Methods are available which depend on the use of either the descending desorption[186, 258] or ascending adsorption[255-257] parts of the hysteresis loop, and which treat cylindrical[255, 256] and slit-shaped[186, 258] pores. However, we propose to limit the following discussion to slit-shaped pores and the desorption region. There are a number of reasons. The analysis for slit-shaped pores is mathematically rather simpler, particularly if a modified Kelvin equation is used; the structure of real porous materials usually is not sufficiently well described by either a cylindrical or slit pore model to make the difference between the analyses significant, although in fact an assumption of slit-shaped pores is a better general description of reality for inorganic porous materials which are crystalline (including high area alumina) than is an assumption of cylindrical pores; finally, analyses based on desorption data have been widely used in the past.

On the desorption branch the vapour is in equilibrium with liquid in the pores, and the Kelvin equation is the classical expression which relates the vapour pressure of a liquid to the radius of curvature of its surface. For a liquid constricted between walls, the result is that

$$\tau = \frac{2\overline{V}\gamma \cos\theta}{RT \ln(p_0/p)} \quad (6.62)$$

where p is the vapour pressure relative to the vapour pressure of the liquid in a flat surface p_0, $\overline{V}$ is the molar volume, θ is the contact angle and τ is a function of the dimension of the vessel in which the surface is constrained. For a liquid surface in a cylindrical capillary of diameter d, $\tau = d/2$; for a pair of plane parallel walls separated by a distance, d, $\tau = d$.

It is usual to assume $\theta = 0$ and this will be done henceforth.

It should be noted that the Kelvin equation is inaccurate because the adsorption forces modify the shape of the meniscus. We shall return to the question of correcting for this subsequently, but for the present it is useful to consider the simple case where the Kelvin equation is assumed. Starting at the top end of the desorption branch, the removal of a small quantity of adsorbate corresponding to a volume ΔV of liquid occurs at a certain relative pressure, and to this there corresponds a certain pore size which can be calculated from the Kelvin equation. If no other complications were to intervene, the pore size distribution could obviously be characterized by the collection of values of ΔV and the corresponding pore sizes. However, the problem is complicated by the fact that the pore wall from which the liquid meniscus has retreated is not bare but is covered by a multilayer of adsorbate, the thickness of which is a function of the relative pressure. The pore size parameter which is evaluated from the Kelvin equation is thus less than the true pore size. Correction for the thickness, t, of this adsorbed multilayer needs to be made during the course of the calculation.

The dependence of the multilayer thickness, t, on p/p_0 is obtained by making use of the observation that for physical adsorption on a free surface the ratio n^s/n^s_m for a given adsorbate at a given temperature is not much dependent on the chemical nature of the adsorbent at coverages above a monolayer. Thus a master plot of n^s/n^s_m vs p/p_0 may be prepared and the required relation obtained by converting n^s/n^s_m to t. Since nitrogen adsorption at 77 K has been almost universally used, we shall confine the following discussion to this material.

As pointed out by Lippens *et al.*[186] it is most realistic to consider the molecular packing in the nitrogen multilayer as approximating to a close-packed structure with a density equal to that of liquid nitrogen, and these authors provide a tabulation of t vs p/p_0, based on an averaged n^s/n^s_m vs p/p_0 curve compiled from adsorption behaviour on a number of substances. Several empirical relations have been proposed to represent the variation of adsorbed layer thickness with relative pressure over the range of interest ($0 \cdot 3 < p/p_0 < 1$), and two are worth noting here. Remembering that they refer to adsorbed nitrogen at 77 K, we have[187,188]

$$t = 0 \cdot 459 [\log_{10} (p_0/p)]^{-1/3} \qquad (6.63)$$

or alternatively[256]

$$\log_{10} (p_0/p) = 13 \cdot 99/t^2 - 0 \cdot 034$$
$$\text{for } t < 1 \text{ nm} \qquad (6.64a)$$

$$\log_{10} (p_0/p) = 16 \cdot 11/t^2 - 0 \cdot 1682 \exp(-0 \cdot 1137 t)$$
$$\text{for } t > 0 \cdot 55 \text{ nm} \qquad (6.64b)$$

The use of an analytic expression is very convenient if the computational problem is being coded for a computer. The net result in practice is that when applying the Kelvin equation the appropriate value of τ is (d–2t) for a slit pore (and (d/2)–t for a cylindrical pore).

The data in the desorption branch of the isotherm are analysed stepwise, starting at a point where all the pores are full of liquid. Let us assume we have reached the i'th step which starts at a relative pressure $((p/p_0)_i + \Delta)$ and finishes at $((p/p_0)_i - \Delta)$. The value $(p/p_0)_i$ is thus at the midpoint of the step. For the sake of simplicity we shall use the notation i, $(i + \Delta)$ and $(i - \Delta)$ to designate situations when the relative pressures are $(p/p_0)_i$, $((p/p_0)_i + \Delta)$ and $((p/p_0)_i - \Delta)$ respectively.

The pore volume emptied by the process of going from $(i + \Delta)$ to $(i - \Delta)$ can be characterized by the midpoint parameters at i; thus

$$d_i = d_{Ki} + 2t_i \qquad (6.65)$$

where d_i is the distance between the slit walls and d_{Ki} is the distance that would be evaluated from the Kelvin equation (6.62) at i. Furthermore,

$$\Delta V_i = d_i \Delta A_i / 2 \qquad (6.66)$$

where ΔV_i and ΔA_i are the volume and surface area of the i'th pores. During the i'th step the total desorbed volume is given by

$$\Delta X_i = X_{(i+\Delta)} - X_{(i-\Delta)} \qquad (6.67)$$

where $X_{(i+\Delta)}$ and $X_{(i-\Delta)}$ are the total volumes of adsorbed nitrogen (expressed as volume of liquid) at $(i + \Delta)$ and $(i - \Delta)$ respectively. ΔX_i comes from (ia) capillary evaporation from the i'th pores, and (ib) the decrease in the thickness of the adsorbed multilayer in the emptied part of the i'th pores during this process: this volume is equal to

$$(d_i - 2t_{(i-\Delta)}) A_i / 2 \qquad (6.68)$$

and (ii) to the decrease in the thickness of the adsorbed multilayer in those pores which were already empty at $(i + \Delta)$ and this equals

$$(t_{(i+\Delta)} - t_{(i-\Delta)}) A_{(i+\Delta)} \qquad (6.69)$$

In evaluating the contribution from (ib), it was recognized that the area of multilayer exposed in the i'th pores varied and an average was needed: this was set equal to $(t_i - t_{(i-\Delta)}) \Delta A_i$. However, the area of the empty pores at the beginning of the i'th step (i.e. at $(i + \Delta)$) which is $A_{(i+\Delta)}$, obviously equals the area at the end of the (i–1)'th step, $A_{(i-1)-\Delta}$, and the latter equals the sum of all the ΔA's up to and including the (i–1)'th step. We shall call this sum $\sum\limits_{(i-1)} \Delta A$

$$\Delta X_i = (d_i - 2t_{(i-\Delta)}) A_i / 2 + (t_{(i+\Delta)} - t_{(i-\Delta)}) \sum\limits_{(i-1)} \Delta A \qquad (6.70)$$

and combination of equations 6.66 and 6.70 gives

$$\Delta V_i = \frac{d_i}{d_i - 2t_{(i-\Delta)}} \left[\Delta X_i - (t_{(i+\Delta)} - t_{(i+\Delta)}) \sum\limits_{(i-1)} \Delta A \right] \qquad (6.71)$$

Equation 6.71 is applied sequentially, remembering that for the first step $\sum\limits_{(i-1)} \Delta A = 0$: when ΔV_1 has been calculated this enables an area to be calculated via equation 6.66 which then is used for $\sum\limits_{(i-1)} \Delta A$ for the second step, and so on. V_i, that is the total volume from pores having a size greater than and equal to d_i, is obtained by successively subtracting each ΔV_i as it is evaluated from the previous $V_{(i+1)}$. The resulting V_i vs d_i relationship is a cumulative distribution curve which may be differentiated if needed. Of course, the calculation also provides a A_i vs d_i distribution.

If the effect of the adsorption forces on the shape of the liquid miniscus are taken into account, the computation becomes more complex. In the case of slit pores, it has been shown[260] that the Kelvin equation 6.62 is modified by the addition of a second term on the right hand side which, when evaluated with the help of equations 6.64 leads to equations 6.72 and the latter are used in place of the simple Kelvin equation

$$\tau \equiv (d-2t) = \frac{4\cdot 05 + 27\cdot 98[(1/t)-(2/d)] - 0\cdot 068\,[(d/2)-t]}{\log_{10}(p_0/p)} \quad (6.72a)$$

and

$$\tau \equiv (d-2t) = \frac{4\cdot 05 + 32\cdot 22[(1/t)-(2/d)] + 2\cdot 966[\exp(-0\cdot 056d)-\exp(-0\cdot 1137t)]}{\log_{10}(p_0/p)} \quad (6.72b)$$

where 6.72a and 6.72b are valid in the same ranges as 6.62a and 6.62b respectively are valid. The use of the modified Kelvin equation results in calculated pore sizes that are larger than those calculated using the Kelvin equation itself. The difference becomes more significant the smaller the pore size: for instance, in comparative calculations given by Brockhoff and de Boer[260] the use of the Kelvin equation underestimates the size of 100 nm pores by about 8%, while 2.5 nm pores are underestimated by about 20%.

In practice, the analysis of physical adsorption hysteresis loops can be used for pore size determination in the range 1–20 nm.

de Boer and his collaborators have also used a plot of the total gas uptake n^s vs t to give information about the pore structure of the adsorbent.[251, 252] If the adsorbent is such that there is no restriction on multilayer formation and there is no capillary condensation, such a plot gives a simple straight line through the origin, and in this regime the slope of the plot of n^s vs t is a measure of the adsorbent surface area, and the latter, A, is given by

$$A = 5\cdot 721 \times 10^{-20} n^s/t \quad (6.73)$$

where A is in m^2, t in nm and n^s in molecules of nitrogen adsorbed at 77 K. The area values obtained in this way are generally in good agreement with those from the BET method[251] except for certain adsorbents such as highly graphitized carbon[254] where stepwise adsorption occurs.

However, if at a certain relative pressure capillary condensation occurs, the adsorbent takes up more adsorbate than corresponds only to multilayer formation and the slope of the n^s vs t plot increases. On the other hand, if capillary condensation is not possible until a quite high relative pressure is reached, as adsorption increases below the onset of condensation, the free space in the pores becomes smaller owing to the growth of the adsorbed

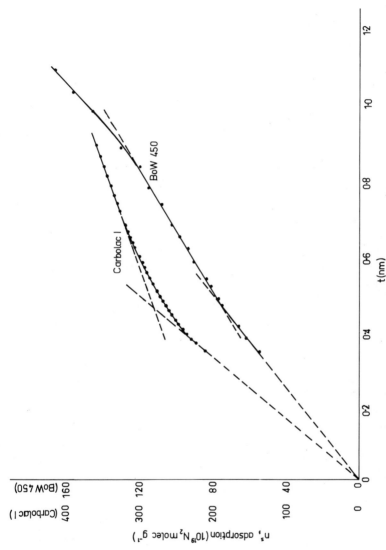

Fig. 6.36 Plots of adsorption (n^s) vs average adsorbed layer thickness (t) for nitrogen physical adsorption at 77 K on a microcrystalline boehmite (BoW 450), and on a Carbolac 1 carbon. After Lippens, B. C. and de Boer, J. H. *J. Catal.* **4**, 319 (1965), and de Boer, J. H., Linsen, B. G., van der Plas, Th. and Zondervan, G. J. *J. Catal.* **4**, 649 (1965).

layer. If there are parallel slit shaped pores present, these may become completely filled by the meeting of the adsorbed layers on both parallel walls: in this case, such pores are no longer accessible for uptake and the n^s vs t plot assumes a smaller slope. The curve for the sample BoW 450 in Fig. 6.36 illustrates both a downward and then an upward break in the n^s vs t plot corresponding to these phenomena. On this basis, if t_1 is the value of t at the first downward break in the n^s vs t plot, $2t_1$ is an estimate of the width of the first group of slit-shaped pores (group 1) which have just become filled. The volume V_1 of these group 1 pores can be calculated from

$$V_1 = A_1 t_1 \tag{6.74}$$

where the group 1 area, A_1, can be calculated from the difference between the slopes of the two straight parts of the n^s vs t plot before and after the break at t_1. The area A_2 of the wider group of pores (group 2) corresponding to the linear part of the plot above the t_1 break can be estimated as the difference between the total area and A_1 or, what amounts to the same thing, the application of equation 6.73 to the second linear part of the plot.

Sometimes the downward break may not be sharp, and this is illustrated with Carbolac 1 in Fig. 6.36. This is due to a distribution of slit widths. In this case the area of the pores in the curved part of the n^s vs t plot can be handled by dividing the plot into a number of sections incremental in t, that is, each of width Δt. If we refer to a particular section j and draw tangents dn^s/dt at the beginning and end of that section, the surface area ΔA_j of the pores in the j section will be given by

$$\Delta A_j = 5 \cdot 721 \times 10^{-20} \Delta (dn^s/dt)_j \tag{6.75}$$

Mercury Porosimetry

The contact angle between mercury and non-wetted solids exceeds 90°, and the result is that mercury can only enter a pore under these conditions by the application of a positive pressure. The relation between the required pressure difference p_c, and the pore diameter (assumed cylindrical), d, is given by

$$p_c = \frac{4\gamma \cos \theta}{d} \tag{6.76}$$

For various non-wetted solids the contact angle, θ, with mercury varies from about 110° to about 140° and 130° is often taken as a typical value frequently encountered. Assuming $\theta = 130°$ and taking the surface tension of mercury to be $0 \cdot 474$ N m^{-1} the pore diameter which will just permit mercury penetration at a pressure of 10^5 kPa (987 atm) is $12 \cdot 20$ nm. The maximum pressure at which a mecury porosimeter will reliably function is about 3×10^5 kPa (about 3000 atm) which corresponds to a minimum pore diameter of about 4 nm. In practice, the pore diameter range for which it is commonly used is about 150 μm–10 nm, so the useful range just overlaps that available from physical adsorption measurements.

The determination of the pore diameter distribution[198,199] requires a measurement of how much mercury has been forced into the pore space of a material as a function of the applied pressure. There are a number of pieces of commercial equipment available all of which record a volume–pressure (or volume–pore diameter) trace more or less automatically. Mercury is slightly compressible, so a blank calibration is required.

Mercury porosimeter data can be used to calculate that part of the specific surface area of a porous specimen accessible to mercury. The work dW required to immerse an area dA of a non-wetting material in mercury is

$$dW = \gamma \cos \theta \, dA \qquad (6.77)$$

where γ is the surface tension and θ is the contact angle. Thus, assuming γ and θ are independent of pressure

$$A = - \frac{\int_0^{V_{max}} p \, dV}{\gamma \cos \theta} \qquad (6.78)$$

and for $\theta = 130°$ and $\gamma = 0 \cdot 474$ N m^{-1}, this reduces to

$$A = 3 \cdot 28 \times 10^{-6} \int_0^{V_{max}} p \, dV \qquad (6.79)$$

where A is in m^2 g^{-1}, p in Pa, and V in dm^3 kg^{-1}.

Problems of Pore Structure Estimation

Micropores of widths less than about 1 nm are difficult to study by conventional methods. Mercury penetration in this size range does not occur at pressures that are readily achievable experimentally, while physical adsorption of nitrogen usually gives an isotherm in which the hysteresis loop is, relative to the accuracy of the measurement, vestigial or not detectable at all. In favourable cases the method based on an analysis of the n^s vs t plot gives some information about very narrow slit-shaped pores of width < 1 nm.

An even more serious problem is the influence that pore geometry may have on the significance of pore structure measurement either by nitrogen adsorption or mercury porosimetry. The methods of data analysis from both of these techniques assume, in effect, that (i) all pores are directly connected to the exterior surface of the specimen and so function independently, and (ii) pores are not interconnected. In practice, both of these assumptions may be in error. The result is that the accuracy of these experimental methods of pore structure estimation should not be overestimated: they give results which often cannot be taken as better than a semiquantitative representation of reality.

Porosity and Density

If a body is porous it is useful to distinguish the parameters real density, apparent unit density and apparent bulk density. The real density is the actual average density of the material of which the catalyst is composed. It is best measured by helium displacement since this gas has an effective atomic diameter of only some $0\cdot2$ nm so even extremely small pores can be readily penetrated. The heat of adsorption of helium is also so low that adsorption is only extensive well below room temperature. Nevertheless, with high area absorbents it is possible for there to be sufficient adsorption at room temperature for this to make an appreciable error in the measurement of real density. An example of this has been cited by Maggs et al.[1] for a range of charcoals. Adsorption was always negligible at 580 K, but at 298 K adsorption could result in an error in the real density of up to about 14% (apparent density too high).

The apparent unit density is the apparent density of an individual catalyst unit (tablet, bead, grain, etc.), taking the volume of the unit that which is defined by the external geometric surface. If the units are reasonably large and of regular geometric shape, this volume can be readily estimated from direct measurement. However, if the units are irregular or small, this will not be possible, and a mercury displacement technique must be used. If v_p is the specific pore volume (per unit mass) and ρ_R and ρ_A are the real and apparent densities respectively,

$$v_p = \frac{1}{\rho_A} - \frac{1}{\rho_R} \tag{6.80}$$

The apparent bulk density refers to the mass of catalyst which packs into unit volume. It thus depends on the apparent unit density and on the efficiency with which these pack in space. The apparent bulk density is somewhat arbitrary since its value depends on the method of packing the catalyst into the volume; nevertheless, it is an important parameter in practice because it determines the mass of catalyst that will pack into a reactor.

References

1. Maggs, F. A. P., Schwabe, P. H. and Williams, J. H. *Nature* **196**, 957 (1960).
2. Livingstone, H. K. *J. Coll. Sci.* **4**, 447 (1949).
3. Young, D. M. and Crowell, A. D. "Physical Adsorption of Gases", Butterworths, London (1962).
4. Anderson, J. R. and Baker, B. G. *J. Phys. Chem.* **66**, 482 (1962).
5. Ross, S. and Olivier, J. P. "On Physical Adsorption", Interscience, New York (1964).
6. Brunauer, S. "The Adsorption of Gases and Vapours", Princeton University Press, Princeton (1943).
7. Emmett, P. H. *Advances in Colloid Sci.* **1**, 3 (1942).
8. Freel, J. *J. Catal.* **25**, 139 (1972).
9. Bugge, P. E., Kerlogue, R. H. and Westwick, F. *Nature* **158**, 28 (1946).
10. Bugge, P. E. and Kerlogue, R. H. *J. Soc. Chem. Ind.* **66**, 377 (1947).
11. Innes, W. B. *Anal. Chem.* **23**, 759 (1951).
12. Lange, K. R. *J. Coll. Sci.* **18**, 65 (1963).
13. Ballou, E. V. and Doolin, O. K. *Anal. Chem.* **32**, 532 (1960).
14. Ballou, E. V. and Barth, R. T. *Advances in Chem. Series* **33**, 133 (1961).
15. Hansen, N. and Littman, W. *Z. Instrumentenk.* **71**, 153 (1963).
16. Nelson, F. M. and Eggertson, F. T. *Anal. Chem.* **30**, 1387 (1958).
17. Loebenstein, W. V. and Deitz, V. R. *J. Res. Nat. Bur. Stand.* **46**, 51 (1951).
18. Haley, A. J. *J. Appl. Chem.* **13**, 392 (1963).
19. Miller, G. A. *J. Phys. Chem.* **67**, 1359 (1963).
20. Partington, J. R. "An Advanced Treatise on Physical Chemistry", Longmans, London, Vol. 1 (1949).
21. Landolt-Bornstein, II Band. I Teil. Springer-Verlag, (1970).
22. Podgurski, H. H. and Davis, F. N. *J. Phys. Chem.* **65**, 1343 (1961).
23. Liang, S. C. *J. Appl. Phys.* **22**, 148 (1951); *J. Phys. Chem.* **56**, 660 (1952); **57**, 910 (1953); *Canad. J. Chem.* **33**, 279 (1955).
24. Joyner, L. G. *In* "Scientific and Industrial Glass Blowing" (W. E. Barr and V. J. Anhorn, eds), Instruments Publishing Co., Pittsburgh (1949).
25. Huttig, G. F. *Monatsh. Chem.* **78**, 177 (1948).
26. Lopez-Gonzalez, J. de D. and Deitz, V. R. *J. Res. Nat. Bur. Stand.* **48**, 325 (1952).
27. Huttig, G. F. and Theimer, O. *Kolloid Z.* **119**, 69 (1950).
28. Gregg, S. J. *J. Chem. Soc.* **1942**, 696.
29. Crawford, V. A. and Tompkins, F. C. *Trans. Faraday Soc.* **46**, 504 (1950).
30. Jura, G. "Physical Methods in Chemical Analysis," Academic Press, N.Y. Vol. 2, (1951), p. 2.
31. Steele, W. A. *Advances in Chem. Series* **33**, 269 (1961).
32. Barker, J. and Everett, D. H. *Trans. Faraday. Soc.* **58**, 1608 (1962).
33. Hanlon, J. F. and Freeman, M. *Can. J. Chem.* **37**, 1575 (1959).
34. Freeman, M. P., and Kolb, K. *J. Phys. Chem.* **67**, 217 (1963).

35. Freeman, M. P. *J. Phys. Chem.* **62**, 723 (1958).
36. Halsey, G. D. *J. Chem. Phys.* **16**, 931 (1948).
37. Freeman, M. P. and Halsey, G. D. *J. Phys. Chem.* **59**, 181 (1955).
38. Wolf, R. and Sams, J. R. *J. Phys. Chem.* **69**, 1129 (1965).
39. Steele, W. A. and Halsey, G. D. *J. Chem. Phys.* **22**, 979 (1954).
40. Steele, W. A. and Halsey, G. D. *J. Phys. Chem.* **59**, 57 (1955).
41. Jura, G. and Harkins, W. D. *J. Chem. Phys.* **11**, 430 (1943).
42. Emmett, P. H. and Brunauer, S. *J. Amer. Chem. Soc.* **59**, 1553 (1937).
43. Roberts, R. W. and Vanderslice, T. A. "Ultrahigh Vacuum and its Applications", Prentice-Hall, Englewood Cliffs, N.J. (1963).
44. Dushman, S. "Scientific Foundations of Vacuum Technique", Wiley, New York (1962).
45. Ehrlich, G. *Advances in Catalysis* **14**, 255 (1963).
46. Benson, J. E. and Garten, R. L. *J. Catal.* **20**, 416 (1971).
47. Hayes, K. E. *J. Catal.* **20**, 414 (1971).
48. Beeck, O. *Advances in Catalysis* **2**, 151 (1950).
49. Rideal, E. K. and Sweett, F. *Proc. Roy. Soc.* **A257**, 291 (1960).
50. Cremer, E. and Huber, H. *Angew. Chem.* **73**, 461 (1965).
51. Brennan, D. and Hayes, F. H. *Trans. Faraday Soc.* **60**, 589 (1964).
52. Schuit, G. C. A. and van Reijen, L. L. *Advances in Catalysis* **10**, 242 (1958).
53. Schuit, G. C. A. and de Boer, N. H. *Rec. Trav. Chim.* **70**, 1067 (1951).
54. Schuit, G. C. A. and de Boer, N. H. *Rec. Trav. Chim.* **72**, 909 (1953).
55. Roberts, M. W. and Sykes, K. W. *Proc. Roy. Soc.* **A242**, 534 (1957).
56. Roberts, M. W. and Sykes, K. W. *Trans. Faraday Soc.* **54**, 548 (1958).
57. Lewis, R. and Gomer, R. *Surface Sci.* **17**, 333 (1969).
58. Carter, J. L., Cusumano, J. A. and Sinfelt, J. H. *J. Phys. Chem.* **70**, 2257 (1966).
59. Sinfelt, J. H., Taylor, W. F. and Yates, D. J. C. *J. Phys. Chem.* **69**, 95 (1965).
60. Cusumano, J. A., Dembinksi, G. W. and Sinfelt, J. H. *J. Catal.* **5**, 471 (1966).
61. Boronin, V. S., Nikulina, V. S. and Poltorak, O. M. *Russ. J. Phys. Chem.* **37**, 626 (1963).
62. Mignolet, J. C. P. *J. Chim. Phys.* **54**, 19 (1957).
63. Lyon, H. B. and Somorjai, G. A. *J. Chem. Phys.* **46**, 2539 (1967).
64. Morgan, A. E. and Somorjai, G. A. *Surface Sci.* **12**, 405 (1968).
65. Lang, B., Joyner, R. W. and Somorjai, G. A. *Surface Sci.* **30**, 454 (1972).
66. Weinberg, W. H., Lambert, R. M., Comrie, C. M. and Linnett, J. W. *In* "Proceedings 5th International Congress on Catalysis" (J. W. Hightower, ed.), North-Holland, Amsterdam (1973), p. 513.
67. Akhtar, M. and Tompkins, F. C. *Trans. Faraday Soc.* **67**, 2454 (1971).
68. Eley, D. D. *In* "Proceedings 5th International Congress on Catalysis" (J. W. Hightower, ed.), North-Holland, Amsterdam (1973), p. 525.
69. Ohlmann, G. *In* "Proceedings 5th International Congress on Catalysis" (J. W. Hightower, ed.), North-Holland, Amsterdam (1973), p. 523.

70. Poltorak, O. M. and Boronin, V. S. *Russ. J. Phys. Chem.* **39**, 871 (1965).
71. Gruber, H. L. *J. Phys. Chem.* **66**, 48 (1962).
72. Anderson, J. R. *Advances in Catalysis* **23**, 1 (1973).
73. Spenadel, L. and Boudart, M. *J. Phys. Chem.* **64**, 204 (1960).
74. Ponec, V., Knor, Z. and Cerny, S. *In* "Proceedings 3rd International Congress on Catalysis" (W. M. H. Sachtler, G. C. A. Schuit and P. Zwietering, eds), North-Holland, Amsterdam (1965), p. 353.
75. Boudart, M., Vannice, M. A. and Benson, J. E. *Z. Phys. Chem.* (N.F.) **64**, 171 (1969).
76. Benson, J. E., Kohn, H. W. and Boudart, M. *J. Catal.* **5**, 307 (1966).
77. Boudart, M. *Advances in Catalysis* **20**, 157 (1969).
78. Vannice, M. A. and Neikam, W. C. *J. Catal.* **20**, 260 (1971).
79. Boudart, M., Aldag, A. W. and Vannice, M. A. *J. Catal.* **18**, 46 (1970).
80. Boudart, M., Aldag, A. W., Ptak, L. D. and Benson, J. E. *J. Catal.* **11**, 35 (1968).
81. Robell, A. J., Ballou, E. V., Boudart, M. *J. Phys. Chem.* **68**, 2748 (1964).
82. Rabo, J. A., Schomaker, V. and Pickert, P. E. *In* "Proceedings 3rd International Congress on Catalysis" (W. M. H. Sachtler, G. C. A. Schuit and P. Zwietering, eds), North-Holland, Amsterdam (1965) p. 1264.
83. L'Homme, G. A., Boudart, M. and D'Or, L. *Bull. Acad. Roy. Belg.* (Cl. Sci.) **52**, 1206, 1249 (1966).
84. Dalla Betta, R. A. and Boudart, M. *In* "Proceedings 5th International Congress on Catalysis" (J. W. Hightower ed.), North-Holland, Amsterdam (1973), p. 1329.
85. Hansen, A. and Gruber, H. L. *J. Catal.* **20**, 97 (1971).
86. Hall, W. K. and Lutinski, F. E. *J. Catal.* **2**, 518 (1963).
87. Bond, G. C. "Catalysis by Metals", Academic Press, London (1962).
88. Wilson, G. R. and Hall, W. K. *J. Catal.* **17**, 190 (1970).
89. Wilson, G. R. and Hall, W. K. *J. Catal.* **24**, 306 (1972).
90. Freel, J., *J. Catal.* **25**, 149 (1972).
91. Poltorak, O. M. and Boronin, V. S. *Russ. J. Phys. Chem.* **40**, 1436 (1966).
92. Adams, C. R., Benesi, H. A., Curtis, R. M. and Meisenheimer, R. G. *J. Catal.*, **1**, 336 (1962).
93. Giordano, N. and Moretti, E. *J. Catal.* **18**, 228 (1970).
94. Benson, J. E. and Boudart, M., *J. Catal.* **4**, 704 (1965).
95. Mears, D. E. and Hansford, R. C. *J. Catal.* **9**, 125 (1967).
96. Dorling, T. A. and Moss, R. L. *J. Catal.* **7**, 378 (1967).
97. Bett, J., Kinoshita, K., Rontis, K. and Stonehart, P. *J. Catal.* **29**, 160 (1973).
98. Aben, P. C. *J. Catal.* **10**, 224 (1968).
99. Sermon, P. A. *J. Catal.* **24**, 460 (1972).
100. Scholten, J. J. F. and van Montfoort, A. *J. Catal.* **1**, 85 (1962).
101. Wicke, E. and Nernst, G. H. *Ber. Bunsenges.* **68**, 224 (1964).
102. Sermon, P. A. *J. Catal.* **24**, 467 (1972).
103. Kral, H. *Z. Phys. Chem.* (N. F.) **48**, 129 (1966).

104. Wanke, S. E. and Dougharty, N. A. *J. Catal.* **24**, 367 (1972).
105. Yates, D. J. C. and Sinfelt, J. H. *J. Catal.* **8**, 348 (1967).
106. Lanyon, M. A. H. and Trapnell, B. M. W. *Proc. Roy. Soc.* **A227**, 387 (1955).
107. Ponec, V., Knor, Z. and Cerny, S. *Coll. Czec. Chem. Comm.* **30**, 208 (1965).
108. Brooks, C. S. *J. Coll. and Interface Sci.* **34**, 419 (1970).
109. Sinfelt, J. H. *J. Catal.* **8**, 82 (1967).
110. Sinfelt, J. H. and Yates, D. J. C. *J. Catal.* **10**, 362 (1968).
111. Hayward, D. O. and Trapnell, B. M. W. "Chemisorption", Butterworths, London (1964).
112. "Chemisorption and Reactions on Metallic Films" (J. R. Anderson, ed.), Academic Press, London, Vols. 1 and 2 (1972).
113. Gregg, S. J., and Stock, R. *In* "Gas Chromatography" (D. H. Desty, ed.), Butterworth, London (1958), p. 90.
114. Barrer, R. M. and Rideal, E. K. *Proc. Roy. Soc.* **A149**, 231, 253 (1935); Barrer, R. M. *Proc. Roy. Soc.* **A149**, 253 (1935); Barrer, R. M. *J. Chem. Soc.* **1936**, 1256.
115. Benson, J. E. and Boudart, M. *J. Catal.* **8**, 93 (1967).
116. Barrer, R. M. and Sutherland, J. W. *Proc. Roy. Soc.* **A237**, 439 (1956).
117. Angell, C. L. and Schaffer, P. C. *J. Phys. Chem.* **70**, 1413 (1966).
118. Ponec, V. and Knor, Z. *In* "Actes du Deuxieme Congres de Catalyse", Editions Technip, Paris (1961), p. 195.
119. Swift, H. E., Lutinski, F. E. and Tobin, H. H. *J. Catal.* **5**, 285 (1966).
120. Nikolajenko, V., Bosacek, V. and Danes, V. *J. Catal.* **2**, 127 (1963).
121. Brooks, C. S. and Christopher, G. L. M. *J. Catal.* **10**, 211 (1968).
122. Sinfelt, J. H. and Taylor, W. F. *Trans. Faraday Soc.* **64**, 3086 (1968).
123. Brennan, D. and Hayes, F. H. *Phil. Trans. Roy. Soc.* **A258**, 347 (1965).
124. Brunauer, S. and Emmett, P. H. *J. Amer. Chem. Soc.* **57**, 1754 (1935).
125. Emmett, P. H. and Brunauer, S. *J. Amer. Chem. Soc.* **59**, 310 (1937).
126. Emmett, P. H. and Brunauer, S. *J. Amer. Chem. Soc.* **59**, 1533 (1937).
127. Bayer, J., Stein, K. C., Hofer, L. J. E. and Anderson, R. B. *J. Catal.* **3**, 145 (1964).
128. Smeltzer, W. W., Tollepon, E. L. and Cambrou, A. *Can. J. Chem.* **34**, 1046 (1956).
129. Scholten, J. J. F. and Konvalinka, J. A. *Trans. Faraday Soc.* **65**, 2465 (1969).
130. Emmett, P. H. and Skau, N. *J. Amer. Chem. Soc.* **65**, 1029 (1943).
131. Kholyavenko, K. M., Rubanik, M. Y. and Chermuklina, N. A. *Kinetics Katal.* **5**, 505 (1964).
132. Sandler, Y. L. and Hickam, W. M. *In* "Proceedings 3rd International Congress on Catalysis" (W. M. H. Sachtler, G. C. A. Schuit and P. Zweitering, eds), North-Holland, Amsterdam (1965), p. 227.
133. Scholten, J. J. F., Konvalinka, J. A. and Beekman, F. W. *J. Catal.* **28**, 209 (1973).
134. Cadenhead, D. A. and Wagner, H. J. *J. Catal.* **27**, 475 (1972).
135. van der Plank, P. and Sachtler, W. M. H. *J. Catal.* **12**, 35 (1968).
136. Sinfelt, J. H., Carter, J. L. and Yates, D. J. C. *J. Catal.* **24**, 283 (1972).

137. Anderson, J. H., Conn, P. J. and Brandenberger, S. G. *J. Catal.* **16**, 326 (1970).
138. Menon, P. G., Sieders, J., Streefkerk, F. J. and van Kenlen, G. J. M. *J. Catal.* **29**, 188 (1973).
139. Bechtold, E. *In* "Gas Chromatography 1962" (M. van Swaay, ed.), Butterworth, London (1963), p. 49.
140. Brennan, D. and Hayward, D. O. *Phil. Trans. Roy. Soc.* **A258**, 375 (1965).
141. Collins, A. C. and Trapnell, B. M. W. *Trans. Faraday Soc.* **53**. 1476 (1957).
142. Bridges, J. M., MacIver, D. S. and Tobin, H. H. *In* "Actes du Deuxieme Congres de Catalyse", Editions Technip, Paris (1961), p. 2161.
143. Honz, J. and Nemec, L. *Coll. Czec. Chem. Comm.* **34**, 2030 (1969).
144. Gilman, S. *In* "Electroanalytical Chemistry" (A. J. Bard, ed.), Arnold, London (1967), p. 111.
145. Ross, S. *J. Phys. Chem.* **53**, 383 (1949).
146. Rand, D. A. J. and Woods, R. *J. Electroanal. Chem.* **31**, 29 (1971).
147. Rand, D. A. J. and Woods, R. *J. Electroanal. Chem.* **44**, 83 (1973).
148. Michri, A. A., Pshenichnikov, A. G. and Burshtein, R. K. *Elektrokhim*, **8**, 364 (1972).
149. Kipling, J. J. and Wright, E. H. M. *J. Chem. Soc.* **1962**, 855.
150. Giles, C. H., MacEwan, T. H., Nakhwa, S. N. and Smith, D. *J. Chem. Soc.* **1960**, 3973.
151. Giles, C. H. and Nakhwa, S. N. *J. Appl. Chem.* **12**, 266 (1962).
152. Langille, R. C., Braid, P. E. and Kenrick, F. B. *Can. J. Res.* **23**, 31 (1945).
153. Kipling, J. J. and Wilson, R. B. *J. Appl. Chem.* **10**, 109 (1960).
154. Smith, H. A. and Fujek, J. F. *J. Amer. Chem. Soc.* **68**, 229 (1946).
155. Lancaster, J. K. and Rouse, R. L. *Research* **4**, 44 (1951).
156. Garner, W. E., Stone, F. S. and Tiley, P. F. *Proc. Roy. Soc.* **A211**, 472 (1952).
157. Rudham, R. and Stone, F. S. *In* "Chemisorption" (W. E. Garner, ed.), Butterworths, London (1957), p. 205.
158. Gregg, S. J. and Sing, K. S. W. "Adsorption, Surface Area and Porosity", Academic Press, London (1967).
159. Brennan, D., Graham, M. J. and Hayes, F. H. *Nature* **199**, 1152 (1963).
160. Brennan, D. and Graham, M. J. *Discussions Faraday Soc.* **41**, 95 (1966).
161. Brennan, D., Hayward, D. O. and Trapnell, B. M. W. *Proc. Roy. Soc.* **A256**, 81 (1960).
162. Smithells, C. J. "Gases and Metals", Chapman and Hall, London (1937).
163. Smith, D. P., Eastwood, L. W., Carney, D. J. and Sims, C. E. "Gases in Metals", American Society for Metals, Cleveland (1953).
164. Anderson, J. R. and Shimoyama, Y. *In* "Proceedings 5th International Congress on Catalysis" (J. W. Hightower, ed.) North-Holland, Amsterdam (1973), p. 695.
165. Shimoyama, Y. Ph.D. Thesis, Flinders University, Adelaide, Australia (1971).
166. Cadenhead, D. A. and Wagner, N. J. *J. Catal.* **21**, 312 (1971).

167. "Ultramicro Weight Determination in Controlled Environments" (S. P. Wolsky and E. J. Zdanuk, eds), Interscience, New York (1969).
168. Westmoreland, J. *Chem. Ind.* **1965**, 2000.
169. Barrett, P. *Bull. Soc. Chim. France* **1958**, 376.
170. Fujii, C. T., Carpenter, C. D. and Meussner, R. A. *Rev. Sci. Inst.* **33**, 362 (1962).
171. Joshi, R. M. *J. Polymer Sci.* **35**, 271 (1959).
172. Moreau, C. *Vac. Microbalance Tech.* **4**, 21 (1965).
173. Gulbransen, E. A. *Rev. Sci. Inst.* **15**, 201 (1944).
174. Jennings, T. J. *In* "The Defect Solid State" (T. J. Gray, ed.), Interscience, New York (1957), p. 487.
175. Whittle, J. E. *J. Sci. Inst.* **43**, 150 (1966).
176. Schwoebel, R. L. *Surface Sci.* **2**, 356 (1964).
177. Boggs, W. E. *Vac. Microbalance Tech.* **6**, 45 (1967).
178. Mayer, H., Niedermayer, R., Schroen, W., Stuenkel, D. and Goehre, H. *Vac. Microbalance Tech.* **3**, 75 (1963).
179. Cahn, L. and Schultz, H. R. *Vac. Microbalance Tech.* **2**, 7 (1962).
180. Cahn, L. *Inst. Control Systems*, **35**, 107 (1962).
181. Cahn, L. and Schultz, H. R. *Vac. Microbalance Tech.* **3**, 29 (1963).
182. Kolenkow, R. J. and Zitewitz, P. W. *Vac. Microbalance Tech.* **4**, 195 (1965).
183. Carrera, N. J. and Walker, R. F. *Vac. Microbalance Tech.* **3**, 153 (1963).
184. Gruber, H. L. and Shipley, C. S. *Vac. Microbalance Tech.* **3**, 131 (1963).
185. Massen, C. H. and Poulis, J. A. *In* "Ultramicro Weight Determination in Controlled Environments" (S. P. Wolsky and E. J. Zdanuk, eds) Interscience New York (1969), p. 107.
186. Lippens, B. C., Linsen, B. G. and deBoer, J. H. *J. Catal.* **3**, 32 (1964).
187. Halsey, G. D. *J. Chem. Phys.* **16**, 931 (1948).
188. Wheeler, A. *In* "Catalysis" (P. H. Emmett, ed.), Reinhold, New York (1955), p. 105.
189. Innes, W. B. *Anal. Chem.* **29**, 1069 (1957).
190. Innes, W. B. *In* "Experimental Methods in Catalytic Research" (R. B. Anderson, ed.), Academic Press, New York (1968), p. 44.
191. Barrett, E. P., Joyner, L. G. and Halenda, P. P. *J. Amer. Chem. Soc.* **73**, 373 (1951).
192. Aylmore, D. W. and Jepson, W. B. *J. Sci. Inst.* **38**, 156 (1961).
193. Clarke, J. T. *J. Phys. Chem.* **68**, 884 (1964).
194. Houtman, J. P. W. and Medema, J. *Ber. Bunsenges.* **70**, 489 (1966).
195. Cochrane, H., Walker, P. L., Diethorn, W. S. and Friedman, H. C. *J. Coll. Interface Sci.* **24**, 405 (1967).
196. Hughes, T. R., Houston, R. J. and Sieg, R. P. *Ind. Eng. Chem. Process Design Devel.* **1**, 96 (1962).
197. Cormack, D. and Moss, R. L. *J. Catal.* **13**, 1 (1969).
198. Ritter, H. L. and Drake, L. C. *Ind. Eng. Chem.* **17**, 782 (1945).
199. Drake, L. C. and Ritter, H. L. *Ind. Eng. Chem.* **17**, 787 (1945).
200. Moss, R. L. *Platinum Metals Rev.* **11**, 141 (1967).

201. Anderson, J. R., Baker, B. G. and Sanders, J. V. *J. Catal.* **1**, 443 (1962).
202. Thomas, G. "Transmission Electron Microscopy of Metals", Wiley, New York (1962).
203. Warren, E. B. *J. Appl. Phys.* **12**, 375 (1941).
204. Williamson, G. K. and Hall, W. H. *Acta Met.* **1**, 22 (1953).
205. Rachinger, W. A. *J. Sci. Inst.* **25**, 254 (1948).
206. Warren, B. E. *Prog. Met. Phys.* **8**, 147 (1959).
207. Pope, D., Smith, W. L., Eastlake, M. J. and Moss, R. L. *J. Catal.* **22**, 72 (1971).
208. Whyte, T. E., Kirklin, P. W., Gould, R. W. and Heinemann, H. *J. Catal.* **25**, 407 (1972).
209. Guinier, A. and Fournet, G. "Small Angle Scattering of X-rays", Wiley, New York (1955).
210. Guinier, A. "X-ray Diffraction", Freeman, San Francisco (1963).
211. Harkness, S. D., Gould, R. W. and Hren, J. J. *Phil Mag.* **19**, 115 (1969).
212. Porod, G. *Kolloid. Z.* **124**, 83 (1951).
213. Porod, G. *Kolloid. Z.* **125**, 51, 109 (1952).
214. Baur, R. and Gerold, V. *Acta. Met* **12**, 1448 (1964).
215. Kratky, O. *Z. Elektrochem.* **58**, 49 (1954).
216. Brennan, D. and Graham, M. J. *Phil. Trans. Roy. Soc.* **A258**, 325 (1965).
217. Martin, G., Blyth, C. E. and Tongue, H. *Trans. Ceramic Soc.* **23**, 61 (1923/24). Martin, G., Bowes, E. A. and Christelow, J. W. *Trans. Ceramic Soc.* **25**, 51 (1925/26). Martin, G. *Trans. Ceramic Soc.* **27**, 285 (1927/28).
218. Endter, F. and Gebauer, H. *Optik* **13**, 87 (1956).
219. Becher, P. *J. Coll. Sci.* **19**, 468 (1964).
220. Klug, H. P. and Alexander, L. E. "X-ray Diffraction Procedures", Wiley, New York (1954).
221. Jones, F. W. *Proc. Roy. Soc.* **A166**, 16 (1938).
222. Shull, C. G. and Roess, L. C. *J. Appl. Phys.* **18**, 295, 308 (1947).
223. Hosemann, R. *Kolloid Z.* **117**, 13 (1950).
224. Yudowitch, K. L. *Rev. Sci. Inst.* **23**, 83 (1952).
225. Jellinek, M. H., Solomon, E. and Fankuchen, I. *Ind. Eng. Chem. Anal. Ed.* **18**, 172 (1946).
226. Freise, E. J., Kelly, A. and Nicholson, R. B. *Acta Met.* **9**, 250 (1961).
227. Kratky, O., Porod, G. and Kahovek, L. *Z. Electrochem.* **55**, 53 (1951).
228. Brumberger, H. "Small Angle X-ray Scattering", Gordon and Breach, New York (1967).
229. Gunn, E. L. *J. Phys. Chem.* **62**, 928 (1958).
230. Lewis, P. H. *J. Phys. Chem.* **66**, 105 (1962).
231. Lewis, P. H. *J. Phys. Chem.* **64**, 1103 (1960).
232. Parratt, L. G., Hempstead, C. F. and Jossem, E. L. *Phys. Rev.* **105**, 1228 (1957).
233. Beckman, O., Axelsson, B. and Bergvall, P. *Arkiv Fysik* **15**, 567 (1959).
234. Lewis, P. H. *J. Phys. Chem.* **67**, 2151 (1963).
235. Lewis, P. H. *J. Catal.* **11**, 162 (1968).

236. van Nordstrand, R. A. *Advances in Catalysis*, **12**, 149 (1960).
237. Selwood, P. W. "Adsorption and Collective Paramagnetism", Academic Press, New York (1962).
238. MacNab, J. I. and Anderson, R. B. *J. Catal.* **29**, 338 (1973).
239. Foner, S. *Rev. Sci. Inst.* **30**, 548 (1959).
240. Carter, J. L. and Sinfelt, J. H. *J. Phys. Chem.* **70**, 3003 (1966).
241. Allen, T. "Particle Size Measurement", Chapman and Hall, London (1968).
242. Basset, J. M., Theolier, A., Primet, M. and Prettre, M. *In* "Proceedings 5th International Congress on Catalysis" (J. W. Hightower, ed.) North-Holland, Amsterdam (1973), p. 915.
243. Benson, J. E., Hwang, H. S. and Boudart, M. *J. Catal.* **30**, 146 (1973).
244. Kirkpatrick, P. *Rev. Sci. Inst.* **10**, 186 (1939).
245. Ogawa, S., Watanabe, D., Ino, S., Kato, T. and Ota, H. *Science Repts. Research Institute, Tohoku Univ. Ser. A.* **18**, (*Suppl.*), 171 (1966).
246. Robins, J. L. and Robinson, V. N. E. *Vacuum*, **18**, 641 (1968).
247. de Boer, J. H. *In* "The Structure and Properties of Porous Materials" (D. H. Everett and F. S. Stone, eds), Butterworths, London (1958), p. 68.
248. de Boer, J. H. and Lippens, B. C. *J. Catal.* **3**, 38 (1964).
249. Lippens, B. C. and de Boer, J. H. *J. Catal.* **3**, 44 (1964).
250. de Boer, J. H., van den Heuvel, A. and Linsen, B. G. *J. Catal.* **3**, 268 (1964).
251. Lippens, B. C. and de Boer, J. H. *J. Catal.* **4**, 319 (1965).
252. de Boer, J. H., Linsen, B. G. and Osinga, Th. J. *J. Catal.* **4**, 643 (1967).
253. de Boer, J. H., Linsen, B. G., van der Plas, Th. and Zondervan, G. J. *J. Catal.* **4**, 649 (1965).
254. de Boer, J. H., Brockhoff, J. C. P., Linsen, B. G. and Meijer, A. L. *J. Catal.* **7**, 135 (1967).
255. Brockhoff, J. C. P. and de Boer, J. H. *J. Catal.* **9**, 8 (1967).
256. Brockhoff, J. C. P. and de Boer, J. H. *J. Catal.* **9**, 15 (1967).
257. Brockhoff, J. C. P. and de Boer, J. H. *J. Catal.* **10**, 153 (1968).
258. Brockhoff, J. C. P. and de Boer, J. H. *J. Catal.* **10**, 368 (1968).
259. Brockhoff, J. C. P. and de Boer, J. H. *J. Catal.* **10**, 377 (1968).
260. Brockhoff, J. C. P. and de Boer, J. H. *J. Catal.* **10**, 391 (1968).
261. de Boer, J. H., Linsen, B. G., Brockhoff, J. C. P. and Osinga, Th. J. *J. Catal.* **11**, 46 (1968).
262. Becker, J. J. *Trans. Am. Inst. Mining Met. Petrol. Eng.* **209**, 59 (1957).
263. Cahn, J. W. *Trans. Am. Inst. Mining. Met. Petrol. Eng.* **209**, 1309 (1957).
264. Heukelom, W., Broeder, J. J. and van Reijen, L. L. *J. Chim. Phys.* **51**, 474 (1954).
265. Kinoshita, K., Lundquist, J., and Stonehart, P. *J. Catal.* **31**, 325 (1973).
266. Adrian, J. C. and Smith, J. M. *J. Catal.* **18**, 57 (1970).
267. Kobayashi, M. and Shirasaki, T. *J. Catal.* **32**, 254 (1974).
268. Bartholomew, C. H. Ph.D. Thesis, Stanford University, Stanford, U.S.A. (1972).

CHAPTER 7

Measurement Techniques: Surface Composition and Structure

	page
1. SURFACE STRUCTURE	398
Low Energy Electron Diffraction	398
2. SURFACE TOPOGRAPHY	403
Scanning Electron Microscopy	403
Transmission Electron Microscopy	404
Low Energy Electron Diffraction	407
Grazing Incidence Electron Diffraction	408
Other Measurements	408
3. SURFACE COMPOSITION	409
Electron Spectroscopy	409
Auger Electron Spectroscopy	411
Photoelectron Spectroscopy	419
Characteristic Loss Spectroscopy	421
Experimental Factors	421
Applicability to Metallic Catalyst Samples	425
Ion Probe Microanalysis	426
Ion Scattering	428
Ion Beam Spectrochemical Analysis	429
Mössbauer Spectroscopy	430
Work-Function Measurement	432
Gas Chemisorption	434
Hydrogen Chemisorption	435
Carbon Monoxide Chemisorption	437
Supplementary Measurements	438

This chapter is mainly devoted to techniques for studying surface structure and composition of metallic catalysts. Nevertheless, since many catalysts and certainly the majority of those of technical and industrial importance consist of more than one component material, catalyst characterization requires information about the structure of the various components and the way they are arranged in relation to one another. It is therefore convenient to begin by briefly summarizing some important ways in which these latter problems may be approached.

In general there is no single experimental method which can provide all the answers with respect to component composition, structure and distribution in a complex multi-component aggregate. It is necessary to build up the complete picture by a combination of techniques. The degree of detail which finally emerges is then dependent on the nature and the resolving power of the probes which are used.

An overall elemental analysis of a component aggregate is assumed to be available by classical chemical analytical methods. Our attention will be restricted to the component materials.

Methods based upon the use of a high energy electron beam probe are the most generally useful for studying the location, composition and structure of components in an aggregate, while laser and ion beam microprobes are also useful. The ion microprobe has in fact a rather greater degree of surface specificity than the other two, and it is described in more detail in a later section. Microprobe techniques have been reviewed in some detail recently.[141]

Provided the specimen can be obtained thin enough (thickness less than about 200 nm for an electron beam energy of 100 keV), it may be examined by transmission electron microscopy and electron diffraction. In crystalline specimens, diffraction effects rather than simple electron absorption are the most important origin of contrast in electron micrographs, and this depends on crystallite orientation. However, if different phases contain atoms of widely differing scattering power for electrons, that is widely differing atomic number, this will also be a significant source of contrast. Diffraction data obtained on the same (or part of the same) specimen field as used for electron microscopy, are an important source of information for component characterization and differentiation, and selected-area diffraction may be obtained on a portion of the specimen field down to about $0 \cdot 5$ μm diameter. Dark-field microscopy uses a selected diffraction beam as an image-forming beam, and this therefore gives information about the distribution of crystallites of a specified orientation of a particular component.

The transmitted electrons which are elastically scattered contain the diffraction information. Electrons are also transmitted with energy loss and the most important loss mechanisms are the generation of interband transitions and plasma oscillations. Interband transitions predominate in the transition metals, whereas plasma oscillations (plasmon losses) predominate with lighter metals. Inasmuch as the magnitude of the energy loss tends to be component specific, loss measurement offers a means of component differentiation and identification. The field has recently been reviewed by Phillips and Lifshin.[142] The dependence of energy loss on sample composition is generally not known in advance, and careful calibration with materials of known composition is therefore essential. The ultimate spacial resolution in the specimen is limited by the physics of the inelastic scattering process to about 10 nm, and this limit has been approached in practice.

X-ray emission analysis may be performed in an electron microscope by the addition of an X-ray spectrometer to analyse the X-rays backscattered from the specimen: in effect the microscope then functions as a microprobe analyser. The spatial resolution at the specimen can be no better than the incident electron beam size which, in a normal electron microscope is $> 0 \cdot 5$ μm. This is much poorer than the best that can be achieved in an electron microprobe or a scanning electron microscope where, due to a rather different electron optical system a beam diameter at the specimen of 2 nm is possible. However, an analytical resolution commensurate with this small beam size cannot be achieved because the X-rays are emitted from a volume of specimen material whose size is dictated by the distance through which the exciting electrons move before being degraded in energy to near or below the critical value for X-ray production. In practice in an electron microprobe or scanning electron microscope, an exciting electron energy in the range 5–30 keV is used, and the X-rays are generated from a volume of a few μm^3 immediately beneath the surface. This volume is minimized at lower electron energies, and it is enlarged if X-rays are generated by fluorescence.

X-rays are analysed by either a wavelength dispersive or energy selective analyser. An approximate elemental analysis may be obtained by comparing the characteristic X-ray line intensities from the unknown specimen with those from standards of known composition. Using raw intensity data, an analytical accuracy of perhaps $\pm 25\%$ may be expected for components present to $> 10\%$. By correcting for X-ray absorption in the specimen and for non-linearity in X-ray production, and by paying careful attention to instrumental stability, constant specimen positioning, surface finish on the specimen (an optically flat surface is needed) and quality of calibration, an analytical accuracy of $\pm 2\%$ can be achieved. If the specimen is an electrical insulator it needs to be coated with an electrically conducting layer to avoid charging problems.

All analytical methods which use a high energy electron beam probe cause some specimen heating. In most metallic specimens, particularly bulk specimens such as may be examined by an electron microprobe or a scanning electron microscope, specimen heating is usually not important;

however, in non-metallic specimens where the thermal conductivity is poorer, the degree of specimen heating may be substantial and may amount to a temperature rise of several hundred K in the electron absorbing volume. This can only be controlled by minimizing the electron beam energy and current density.

The laser microprobe depends upon vaporization of some of the specimen material and its analysis by emission spectroscopy (the most usual method), mass spectrometry, or atomic absorption spectrometry. The technique has a lower spacial resolution than is available from an electron beam probe: the laser beam vaporizes material from an area of diameter typically 10–100 μm, and the crater volume is in the range 10–500 μm^3.

It is obvious that all these analytical techniques deal more or less with the surface region of the specimen. If data are required through the depth of a bulk specimen this can only be obtained insofar as the specimen can be sectioned or otherwise sampled at varying depths. X-rays penetrate solids much more efficiently than do electrons, the penetration depth being of the order of one thousand times greater, and X-ray techniques can be used to obtain composition and phase analysis data averaged over much greater specimen depths than can be done with electrons. However, X-rays cannot be efficiently focused and the degree of spatial resolution of components that can be achieved with them is relatively gross.

In the succeeding sections of this Chapter we shall make the arbitrary distinction of defining the surface region with which we are concerned as roughly the upper 1–10 atomic layers of the solid. We shall mainly be concerned with metals.

It is convenient to approach the nature of a metal surface in three stages. First there is the question of surface structure, by which we mean the arrangement of surface atoms in long-range order. Secondly, there is the question of surface topography which is the geometric shape of the surface as distinct from that dictated by ideal surface crystallography, and thirdly there is the question of the chemical composition of the surface.

1. Surface Structure

One technique, low energy electron diffraction, is dominant. High energy electron diffraction at grazing incidence has mainly been used in studying surface topography or the nature of foreign overlayers.

Low Energy Electron Diffraction

In this technique, a low energy electron beam, typically in the range 20–500 eV, strikes the specimen surface, and the elastically back-reflected electron beams are recorded and interpreted in terms of the crystallography of the metal surface layers. The angle of incidence of the primary beam is often near normal, although this is not necessarily so.

The most commonly used apparatus is of the post-acceleration display type, and this is illustrated schematically in a four grid configuration in Fig. 7.1. In the normal mode of operation, the specimen crystal C, the drift-tube D, and the first grid G_1 are all at ground potential so that the incident and reflected beams move in a field-free region. The two central

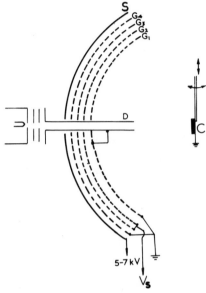

FIG. 7.1 Schematic illustration of four-grid post-acceleration LEED display apparatus.

grids, G_2 and G_3, are strapped together electrically and are set at a negative potential, V_s, with respect to earth to an extent almost equal to the primary beam energy: thus elastically scattered electrons have just enough energy to pass, while electrons which have lost appreciable energy by inelastic processes in the specimen, are rejected. The elastically scattered electrons are finally accelerated through 5–7 kV to the fluorescent display screen S. The grid G_4 acts to shield the suppressors from the field due to S. For LEED operation, only one grid is really needed as a suppressor, but the two grid configuration is often used because it is directly convertible to four-grid Auger electron spectroscopy operation. The electron gun provides a beam which covers an area of $0 \cdot 5$–1 mm^2 at the specimen.

In commercial equipment, the vacuum chamber is made of stainless steel, and is provided with a viewing window, together with ports for the usual pumping lines, pressure measuring devices and other ancillary equipment. The specimen is carried on a manipulator which provides for translation and rotation. Specimen cleaning by ion bombardment, gas reaction and thermal desorption is provided, and the entire equipment must be designed for UHV operation. There are a number of commercial suppliers, the main ones being Varian Associates and Vacuum Generators.

A magnetic deflection postacceleration display type of LEED system is also available, but is less convenient for most purposes than the direct display type.

Because low energy electrons, say < 500 eV, undergo strong interactions with the valence electrons of the solid which renders them likely to lose about 10 eV of energy within 0·5–1 nm of the surface, the elastically scattered electrons originate in the surface region of the specimen. Clearly, if structural information is required about the metal surface itself, some pains are needed to ensure that the surface is devoid of adsorbed material.

Although the interpretation of LEED data is complicated by multiple and dynamical scattering phenomena, most past (and present) interpretations are done on an elementary geometric model which uses only the information contained in the directions of the reflected beams.

Elastic electron scattering from a crystal has a necessary energy conservation

$$E(\mathbf{k}) = E(\mathbf{k}') \qquad (7.1)$$

where $E(\mathbf{k})$ represents the energy as a function of the wave vector $\mathbf{k}$; the prime indicates the wave vector after the scattering event. Moreover, the component of $\mathbf{k}$ parallel to the surface, $\mathbf{k}_{11}$, also must be conserved with a reciprocal lattice vector $\mathbf{g}$ of the surface Bravais net

$$\mathbf{k}'_{11} = \mathbf{k}_{11} + \mathbf{g} \qquad (7.2)$$

The allowed scattered beams are related to the surface Bravais net via 7.2. Thus if the incident beam is normal to the surface, $\mathbf{k}_{11} = 0$, and the directions of the exit beams correspond directly to the various vectors in the reciprocal lattice. The result is that

$$\mathbf{g} = 2\pi(h\mathbf{a}^* + k\mathbf{b}^*) \qquad (7.3)$$

and the reciprocal lattice vectors $\mathbf{a}^*$ and $\mathbf{b}^*$ for the surface unit mesh are, as usual defined by

$$\left. \begin{array}{l} \mathbf{a}^* = \mathbf{a} \times \mathbf{n}/(\mathbf{a}\cdot\mathbf{b}) \\ \mathbf{b}^* = \mathbf{b} \times \mathbf{n}/(\mathbf{a}\cdot\mathbf{b}) \end{array} \right\} \qquad (7.4)$$

in which **n** is the interior surface normal. The indices (hk) may be used to label the beams associated with the (hk) reciprocal lattice vector. Provided the vectors **a** and **b** (and **a*** and **b***) define primitive unit cells, there will be a reflection for every (hk). If the surface net is defined in terms of a non-primitive unit cell, then space-group extinctions will occur; the final result in terms of the allowed beams is of course the same.

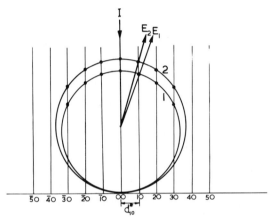

FIG. 7.2 Ewald construction for reflection from two-dimensional lattice, sectioned through (00). The vertical reciprocal lattice rods correspond with the two-dimensional lattice upon which the primary beam I is normally incident, and E_1 and E_2 are the (10) reflected beams for two different primary electron energies which correspond to the Ewald spheres 1 and 2. In each case the radius of the Ewald sphere is $2\pi/\lambda$, and the d^*_{10} distance in reciprocal space equals $2\pi/d_{10}$.

An equivalent representation may be made using an Ewald construction. For an electron beam incident upon a two-dimensional net, the Ewald construction consists of the sphere of reflection intersecting reciprocal lattice rods, each rod being erected on a reciprocal lattice point. The reciprocal lattice rods may be thought of as originating from the lines of intersection of two cones of reflection, in contrast to the three-dimensional case where there are three cones of reflection which intersect at a point. The construction is indicated in Fig. 7.2, for a section on the (h_0) reciprocal lattice row. The wavelength, λ, is related to the electron energy by

$$\lambda = (1 \cdot 504/E)^{\frac{1}{2}} \tag{7.5}$$

where E is the electron energy in eV, and λ in nm.

The Ewald construction makes it clear that, in this approximation, an increase in the size of the sphere of reflection with increasing beam energy only causes new beams to appear as the sphere cuts more reciprocal lattice rods; an established beam is continuously present as λ varies, and only the scale of the diffraction pattern changes.

In practice, however, an examination of the intensity of a reflected beam with varying electron energy shows considerable structure, and there is clearly three-dimensional diffraction occurring due to electron penetration beneath the outermost atomic layer. If the scattering cross-sections for electrons in the solid were low, then full three-dimensional diffraction would result, and a kinematic model would be a good approximation. However, in the LEED energy range the atomic scattering cross-sections are fairly large, multiple electron scattering is not negligible, and the intensity/energy profiles for reflected beams cannot be fully explained without recourse to a dynamical multiple scattering theory. The current situation has been covered by a number of summaries[109-111] based on various approaches.[112-118] Although complex, probably the most straightforward approach is due to Duke and Tucker[112,116] based on the general T-matrix formulation of Beebe.[115] On this basis and using s-wave scattering potentials[112,116] and preferably higher partial waves (cf. ref. 109) good agreement can be achieved between experimental and computed intensity profiles.

However, from the point of view of obtaining some surface structural information with tolerable facility from intensity profile data, the most reasonable procedure seems to be to seek to reduce the experimental data to forms to which a kinematic analysis can be applied. Thus, in the Bragg-envelope concept of Duke and Tucker,[112] it was noted that peaks in the intensity profile which are fundamentally of dynamical multiple scattering origin, are often most intense for energies close to those predicted from kinematics, that is the dynamical intensities behave as if a kinematic scattering cross-section modulates an array of closely spaced multiple scattering peaks which would otherwise exhibit fairly uniform intensities. The envelopes are then treated as the appropriate features for a normal kinematic analysis. The usefulness of this approximation depends on the electron energy range. At energies < about 25 eV, dynamical effects which are strongly dependent on the electronic structure of the scattering atoms, together with discrete inelastic loss processes, are so important that the concept is not useful. At higher energies, up to perhaps 500 eV, the concept is usable by integrating the energy profile over an energy range which includes several modulations, but which is centred on a Bragg peak position, an integrated intensity may be obtained which is proportional to the kinematic intensity. At somewhat higher energies, say from 200–500 eV, the intensity profiles are usually dominated by the Bragg peaks themselves. The data may be analysed by normal kinematics (e.g. ref. 112) or, in principle, by a Patterson-function analysis: the latter has not been often done, but work by Clarke et al.[113,114] gives examples but the accuracy is yet uncertain.

The lateral coherence length, Δx, of the electrons reaching the specimen, that is the distance within which the electron waves can undergo constructive interference, is given by

$$\Delta x = \frac{\lambda}{2\beta(1 + \Delta E/2E)} \qquad (7.6)$$

where β is the half-angle indeterminancy of the beam and ΔE its energy spread. Taking typical values, $\lambda = 0.1$ nm, $E = 150$ eV, $\beta = 10^{-3}$ rad and $\Delta E = 0.2$ eV, Δx is about 50 nm, which is then about the minimum size of the area in the surface over which long range order must exist.

2. Surface Topography

The various techniques of electron microscopy are the most generally powerful methods available for the study of surface topography. Gross features are, of course, resolvable by conventional techniques such as optical microscopy (both normal and stereo) and mechanical tracing devices. Some topographic features also become apparent in electron diffraction results.

Scanning Electron Microscopy

In the SEM, a finely focused electron beam impinges at a point on the specimen surface. The interaction between the electrons and the solid generate a variety of signals, each of which can in principle be collected and amplified. The resulting signal is used to control the brightness on a cathode ray tube, and to display the nature of an area of the surface the electron beam is scanned over the surface, to which is synchronized the cathode ray tube scan.

The electron beam energy is variable, typically, in the range 1–50 keV, and an electron lens system reduces the size of the beam spot from that of the source (some 50 μm) to about 10 nm at the specimen. In the normal mode of operation, it is the secondary electrons emitted at the specimen which are detected using a scintillator photomultiplier. Other possible emissive methods include the use of backscattered electrons, emitted X-rays or emitted light. At the voltages used, the secondary electrons come from within the first 10 nm or so of the specimen. The backscattered electrons, because of their greater energy, can diffuse through greater distances in the solid, that is up to about 100 nm under high voltage SEM conditions. These electron diffusion distances place a practical limit on SEM resolution.

Modes of SEM operation other than emissive are possible. In the absorptive mode, the detected signal is the difference between the primary and secondary electron currents to the specimen. If the specimen is thin enough, the various signals may be collected at the exit surface rather than the entrance surface of the primary beam: this is the transmissive mode for which the specimen thickness is typically < 100 nm. The detection of Auger electrons allows a spacially resolved elemental analysis of the surface to be obtained.[104] Of course, the X-rays may be analysed for analytical information in the usual way, as in electron probe microanalysis, but at high primary beam energies this yields composition data for the specimen below the surface.

Used in the secondary electron emissive mode, the practical optimum resolution currently obtainable with commercial instruments is about 15 nm, although improvements to the order of 5 nm have been produced using a field-emitter tip as the electron source[103] with a commercially available instrument.[150] In an experimental SEM which operated in the transmission mode, which used a field-emitter source and which was designed for minimum spot size and minimum specimen contamination, a point resolution of about $0 \cdot 5$ nm was achieved after optimization of the operating conditions, and individual heavy metal atoms on a carbon support have been distinguished.[150-153]

Surface topography becomes visible because the secondary electron yield depends on the angle of electron incidence. However, compositional changes may also be apparent in the image if they affect the secondary yield. The depth of field is large, about 15 μm at a magnification of $\times 10^4$ is typical, and this is one of the main advantages of the technique.

In normal instruments, specimens a few centimetres in size can be accommodated, and the specimen stage provides for specimen orientation. Non-conductive surfaces are normally coated by metal evaporation to prevent surface charging.

A substantial number of commercial instruments are now available including: Cambridge Instrument Company, Hitachi, JEOL, Philips.

Transmission Electron Microscopy

The operation of the modern transmission electron microscope is so well known and so extensively documented that we shall confine this discussion to outlining the main techniques by means of which it may be applied to problems of surface topography. Massive metal specimens which are too thick to allow the passage of an electron beam must be examined by surface replication methods.

Surface replicas may be made by one-step or two-step processes. Plastic is often used as the material with which the first impression of the surface is made. This may be done by evaporation of a solution of the plastic (e.g. Formvar or Collodion) or a thin layer of solvent-softened plastic is pressed against the surface (e.g. Bexfilm softened with acetone). After drying, the plastic layer is stripped from the surface. In the one-step process, the replica is completed at this stage by shadowing. In the two-step process, the plastic film carrying the surface impression is coated by a thin layer of evaporated carbon to which the impression is transferred; the plastic is now dissolved away, and the carbon replica is then shadowed for examination. A shadowed two-step plastic–carbon replica gives a resolution for surface detail of about 2 nm and is better than a one-step replica in which the relatively thick plastic layer has to transmit the electron beam. Plastic replication has the advantage that the original specimen surface is largely undamaged.

One-stage replicas can also be made by evaporating a thin layer of carbon or silicon monoxide directly onto the specimen surface. Recovery of the replica requires chemical or electrochemical processing so that the specimen surface is destroyed.

Shadowing is best done by evaporative deposition of platinum/carbon. Specimen preparation manuals are available (e.g. refs 105, 106).

If the specimen is thin enough to transmit the electron beam in the microscope, the transmission electron micrograph may contain information about surface topography. Assuming the metal specimen is polycrystalline, it is possible to interpret extinction contours which occur near the edges of the crystal images in terms of the crystal shapes, which implicitly gives information about the surface geometry. An example is to be found in the study of evaporated metal films by Anderson et al.[107]

If a catalyst specimen is in dispersed form, evidence of surface topography will largely come from an examination of the profiles of the metal particles seen in the micrograph. If the specimen consists of metal dispersed on a support, there are several approaches. Moss and co-workers (e.g. ref. 108) have made extensive use of a technique whereby the catalyst is set into "Araldite" resin which, after curing (330–350 K) is thin sectioned using an ultramicrotome. Alternatively, the supported catalyst may first be lightly ground and the ground product dispersed in a liquid (ultrasonic treatment may be helpful). If a substance such as butyl alcohol, which readily wets an oxide support, is used as the liquid, the specimen may be simply prepared by evaporating the liquid after placing a small drop of the suspension on a carbon film carried on a standard electron microscope grid. On the other hand, the suspension may be prepared in a 2% nitrocellulose solution and a drop of this suspension allowed to evaporate on a glass slide: after backing this with a carbon film the duplex layer may be floated onto a water surface and so transferred to a grid. These latter two techniques have the advantage of simplicity, and they do not require access to an ultramicrotome: there is really little chance that the grinding will significantly affect the metal particles but of course only the thin sectioning method retains the support in something approaching its original morphology.

Specimens of ultrathin evaporated metal films on a mica support may be prepared for electron microscope examination by stripping a thin layer of mica carrying the metal particles from the top of the thicker mica sheet, using a piece of "Cellotape". This mica sliver may then be separated from the "Cellotape" by solvent and recovered onto a grid.

Unsupported dispersed metal is best handled by dispersion in a liquid as previously described (non-noble metals are pyrophoric in this form).

Some comments are in order concerning the application of conventional transmission electron microscopy for the examination of very small particles. In the first place amorphous supports, for instance amorphous carbon, have an inherent granularity at high magnification. This is a phase contrast effect and can vary from $0 \cdot 5$ to 5 nm depending on beam coherence, focal conditions and instrument resolution. There are considerable advantages to the use of crystalline supports with a high degree of perfection: cleaved mica, graphite or molybdenite, or extremely small crystals of oxides such as beryllia or magnesia have been used. Stringent control of specimen contamination is essential for optimal resolution.

The resolution of individual metal atoms either as isolated atoms or in atom clusters remains at (or more often just beyond) the accessible limit in transmission electron microscopy. A theoretical analysis by Hashimoto *et al.*[154-156] has shown that for arrays of a few atoms, tilted dark-field images should give better resolution than bright-field images. Of course, with very small aggregates of only a few atoms, contrast arises entirely from phase contrast, unlike the situation with larger particles where diffraction contrast occurs. An analysis by Flynn *et al.*[78] of phase contrast from such atom aggregates demonstrates how sensitive the contrast is to the conditions of focus. In particular, the relation between the geometry of an atom array and the nature of the computed image is very much dependent on the focus conditions: the two are not necessarily even qualitatively similar. It is clear that the interpretation of an image which, at first glance, seems to indicate the presence of a cluster of a few atoms, needs to be undertaken with the greatest circumspection. An interpretation with any degree of confidence demands that the image be studied in detail as a function of focus defect. This analysis also shows that, so far as particle size measurements are concerned, due to dependency on conditions of focus, the relation between true and apparent particle size becomes increasingly uncertain for particles less than about 2 nm in size, and measurements of this sort cease to be worthwhile for particles less than about 1 nm in size.

Low Energy Electron Diffraction

Although LEED is not particularly sensitive to topographical details of the surface some features, particularly facets and steps can become apparent from the diffraction pattern.

Facets become immediately obvious in the diffraction pattern when they expose a small number of particular lattice planes, each type of plane giving rise to its characteristic diffraction features from which the facet planes may be indexed. An aid to distinguishing reflections from facet planes from those due to the nominal surface plane is the way the features move in the pattern, with changing electron energy. We saw previously that with the primary beam normal to the surface plane, increasing the beam energy causes the diffraction pattern to contract symmetrically around the (00) reflection; however, the reflections from facet planes do not follow this simple behaviour and undergo a complex motion depending on the beam energy and the indices of the facet plane. This becomes apparent from the appropriate Ewald construction, as indicated in Fig. 7.3. The exposed facet planes may occur in either hills or valleys, and these diffraction data alone do not allow them to be distinguished. Facets need

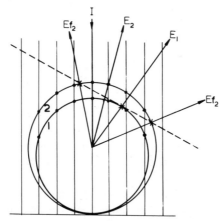

FIG. 7.3 Section through Ewald construction for reflection from two-dimensional lattice with exposed facet plane. E_2 indicates one of the beams reflected from the plane upon which the incident beam impinges normally and corresponding to the sphere of reflection 2. The dotted line is one of the reciprocal lattice rods from a facet plane. If, as shown, this rod is a tangent to sphere 1 a single beam E_{f_1} is reflected, and the figure shows how this reflection is split into two E_{f_2} beams on changing the beam energy to give the sphere of reflection 2.

to be quite large to be observable in LEED, bearing in mind the lateral coherence width of the electrons of the order of 50 nm.

A stepped surface also becomes apparent in LEED, with the diffracted beams split into doublets[119-121] provided the steps are of constant height and spacing. Henzler[120] describes how the step height and spacing may be evaluated. Briefly, the separation of the doublets is proportional to the step spacing, while the step height can be found from

$$V_{00} \text{ (singlet max)} = \frac{1 \cdot 504 \, s^2}{4d^2} \qquad (7.7)$$

where V_{00} are the voltages where a singlet spot of maximum intensity is observed, d is the step height in nm, and s an integer. Houston and Park[122] have discussed a model applicable to a statistical distribution of steps.

Grazing Incidence Electron Diffraction

The nature of the diffraction pattern obtained with high energy electrons (10–50 keV) at grazing incidence can give some indication about surface topography. Thus the ring patterns obtained from a polycrystalline surface must be mainly due to transmission through projecting crystals which the electron beam can traverse without appreciable energy loss, an argument which is substantiated by the fact that the ring positions usually show no marked displacements. On the other hand, if the crystals have comparatively smooth surfaces of small curvature, the beam may enter and leave the crystals by parts of the surface inclined at only a small angle to the beam, so that the diffracted beams will be broadened owing to small penetration, and refraction will cause a displacement and further broadening. A rough surface on a single crystal would, of course, give spots rather than rings. The general technique has been summarized by Bauer.[123]

Other Measurements

Adsorption data including the measurement of infrared absorption spectra of adsorbed molecules, and the nature of some catalytic reactions, have been interpreted in terms of the topography of the metal surface. While measurements of this sort are not primary data for the characterization of surface topography they should not be overlooked because they can give some information under conditions when direct physical methods may be of limited applicability, particularly with supported dispersed catalysts.

van Hardeveld and van Montfoort[124] interpreted the infrared absorption bands at $4 \cdot 54$–$4 \cdot 48$ μm (2202–2230 cm^{-1}) on supported nickel, platinum and palladium catalysts as being due to the perturbation of physically adsorbed molecular nitrogen at special surface sites which have their origin in surface geometry (S_5 sites, cf. p. 253). Cormack and Moss[125] interpreted infrared absorption spectra and temperature programmed desorption data for carbon monoxide on a platinum/silica catalyst in terms of the fraction of the surface containing atoms of low co-ordination, while this same surface feature has been correlated[126] with the extent of low temperature hydrocarbon cyclization on platinum catalysts. These interpretations obviously depend upon specific assumptions about the nature of the adsorption or catalytic processes, and require a good deal of ancillary information to justify their validity.

3. Surface Composition

There are a number of important ways in which one may seek information about the surface composition of metals. Atoms may be removed entirely from the sample and their identity estimated mass spectrometrically. Ion scattering is also sensitive to surface composition. Gas adsorption data may be interpreted in terms of surface composition, at least in favourable cases, and a measurement of spectroscopic or thermodynamic properties relating to the adsorbate–adsorbent interaction may be useful. The chemical identity of a surface atom may be inferred from its electronic structure (controlled by nuclear charge) and electron or photon probes have been widely used: in principle, ion probes may also be used, but in practice their analytical significance is so limited that we shall not consider them further. The chemical environment of a surface atom may also be studied by Mössbauer spectroscopy in certain circumstances.

The applicability of various techniques for studying surface composition may be heavily dependent on the nature of the specimen, in particular on the accessibility of the surface to the probe being used. This becomes very apparent in comparing the utility of the various techniques for massive metal specimens which expose a free surface on the one hand, with supported dispersed metal specimens on the other: in the latter situation the use of electron or photon probes, or the complete removal of a surface atom for mass spectrometric identification, may well be difficult or quite impossible.

Electron Spectroscopy

So far as methods which depend on an identification of some part of the electronic structure of a surface atom are concerned, one may measure the energy required to create a hole in a core electronic state of the atom in question (various forms of ionization spectroscopy, including characteristic loss spectroscopy and photoelectron spectroscopy (ESCA)), or one may measure the energies of the electronic transitions involved in the recombination of the hole with an electron of higher energy (Auger electron emission and X-ray emission spectroscopy). Of these, Auger electron spectroscopy has assumed the greatest importance because it can be carried out relatively simply, and the fact that conventional LEED equipment can be used with but trivial modification has been a major convenience. Furthermore, X-ray emission becomes much less sensitive as the element being examined becomes lighter, and is unfavourable for elements of atomic number (Z) less than about 20: the transition rate for the filling of a vacant K-level hole with the emission of a photon decreases as approximately $(Z-1)^4$ with decreasing Z. In contrast, the Auger transition rate is approximately independent of Z.[82] There are also some other factors which have made Auger electron spectroscopy the more important surface analytical tool. It is much easier to provide a dispersing element of high resolving power for electrons than for X-rays: for the latter a maximum resolution ($\Delta E_{1/2}/E$, the ratio of line width at half height to the line energy) of 1–2% appears to be the best that can currently be achieved. The X-ray emission system is much more complex, and it is more difficult to make it UHV-compatible. The emitted X-rays have a relatively long mean free path in the metal, so in order to make the technique surface sensitive, the penetration depth of the exciting electron beam must be minimized: this can be done by limiting the energy of the primary exciting electron beam to, say, 1 keV or so, or alternatively if higher energy electrons are used, say 10–20 keV, adequate surface sensitivity has to be achieved by using grazing electron beam incidence of 1–2°. The use of lower energy exciting electrons results in only relatively soft X-rays being produced and this poses problems with absorption in window materials and the like. Nevertheless this sort of X-ray emission technique has had some use for surface analytical studies, and typical examples are due to Sewell et al.[46]

In addition to X-ray emission spectroscopy itself there has appeared recently a related technique in which the thresholds, measured in terms of the electron energy just sufficient to excite the various emission lines, are studied. This goes under the name of appearance potential spectroscopy.[61-66] Chemical shifts in the thresholds have been demonstrated, as in the oxidation of chromium[63] and in nickel–titanium alloys[64] and in this respect the technique appears to have considerable promise, particularly for 3d metals. It is relatively insensitive to 4d metals, semiconductors and insulators.[67] Park et al.[61] have described a fairly simple apparatus for this type of measurement in which the X-rays are detected via the photoelectrons they eject from the walls of the stainless steel vacuum chamber. Although the X-ray intensities near the thresholds are extremely low, it is fairly easy to extract the required signal from the noise by standard phase-lock amplification and differentiation techniques provided one is only interested in threshold position. Quantitative surface analysis does not appear to be currently possible.

Auger Electron Spectroscopy

Consider an atom ionized in an inner-shell level X. If the hole is filled by movement of an electron from a higher atomic level Y, an amount of energy $(E_X - E_Y)$ will be released. In the Auger regime, this released energy is accommodated by the ejection of an electron from another level, say Z, so that the energy of the emitted Auger electron, E_A, will be (approximately) given by

$$E_A = E_X - E_Y - E_Z \qquad (7.8)$$

The initial excitation may be produced photoelectrically (and by other means) as well as by electron impact, but in any case the Auger energy is independent of the mode of excitation. Although the ionization cross-sections for photons are some four orders of magnitude greater than for electrons of comparable energy, the intrusion of various instrumental factors, including source intensity and beam collimation, results in electron excitation offering the larger absolute Auger signal. On the other hand, photon excitation gives a rather lower background from valence electron excitation so the intrinsic signal-to-noise ratio is better than with electron excitation. Nevertheless, the great ease and convenience of electron excitation has resulted in its extensive use, particularly in equipment that is adapted from another use (e.g. LEED).

The processes in electron spectroscopy are designated in terms of the electron states involved, and X-ray or orbital term notation is used. These notations are related as follows. The principle quantum numbers 1–6 are labelled by the letters K–P respectively. Further subclassification is done in terms of the quantum numbers l and j, the latter being equal to $|l \pm \tfrac{1}{2}|$. For a given l, the state with the lowest j is designated with the subscript I on the letter symbol: subscripts II, III etc. designate terms in increasing j then increasing l. Thus the state $(n = 2, l = 0, j = \tfrac{1}{2})$ is called L_I, $(n = 2, l = 1, j = \tfrac{1}{2})$ is L_{II}, $(n = 2, l = 1, j = \tfrac{3}{2})$ is L_{III}. Auger transitions are labelled by a triplet of X-ray terms: the first labels the state from which the initial excitation occurred, the second labels the state from which the electron falls to fill the first hole, and the third labels the state from which the Auger electron is ejected. Thus, in the example given previously, if $X \equiv K$, $Y = L_I$ and $Z = L_{II}$, the transition is $KL_I L_{II}$. Since in a metal an electron state in the valence band may be involved, this is labelled V without further specification.

Auger electron spectroscopy is a surface-sensitive analytical tool because both the emitted electrons and the exciting electron have only a limited mean free path in the metal. Experimentally, maximum ionization is found when the incident electron energy is about three times the energy required for the initial excitation, and incident energies in the range 1–1·5 keV are common. Furthermore, the Auger peaks of most interest lie in the range 20–1000 eV, and these energies cover the range of mean free paths in the region 0·5–5 nm. It is the escape depth of the Auger electrons which mainly determines the depth-sensitivity of the technique, and low energy Auger electrons from metals of high atomic number are likely to have the smallest escape depth.

The technique has potential both for element identification and for quantitative estimation.

Auger electron spectra have been observed for many of the elements, and on the whole there is good agreement between observed Auger energies and those calculated using the appropriate electron level energies. For the purposes of calculation the simple expression 7.8 needs correction because after the ionization step, the transitions do not refer to a neutral atom. One simple way of taking this into account[1,2] is to use instead of E_Y the term $[E_{Y(N)} + E_{Y(N+1)}]/2$, and instead of E_Z the term $[E_{Z(N)} + E_{Z(N+1)}]/2$, where $E_{Y(N)}$ and $E_{Y(N+1)}$ etc. are the appropriate electron binding energies for the atom in question with atomic number N, and for the atom with atomic number $N + 1$.

Element identification is thus attempted in terms of the observed Auger energies. These need to be corrected by subtracting the work-function of the analyser material: the work-function of the specimen is not involved because the electron binding energies in a metal are given with reference to the Fermi level. For the heavier elements the total number of possible Auger transitions is quite large, but the situation is simplified somewhat if attention is confined to the more intense lines: Coghlan and Clausing provide a summary of these which is reproduced in Fig. 7.4, and which is adequate for most purposes. These authors also present a comprehensive tabulation of all calculated Auger transitions in the range 10–3000 eV for all the element from atomic number 3 to 92.

Auger peak heights offer, in principle, means for quantitative surface analysis. While this has been achieved, the method is not without its problems with respect to calibration and the significance of the composition–depth profile, the latter being of particular significance in the case of alloys.

So far as the surfaces of metals and alloys are concerned, it is valid to assume that the Auger peak height, either in the form of N(E) vs E or the more usual differentiated form dN(E)/dE vs E (peak-to-peak), is directly

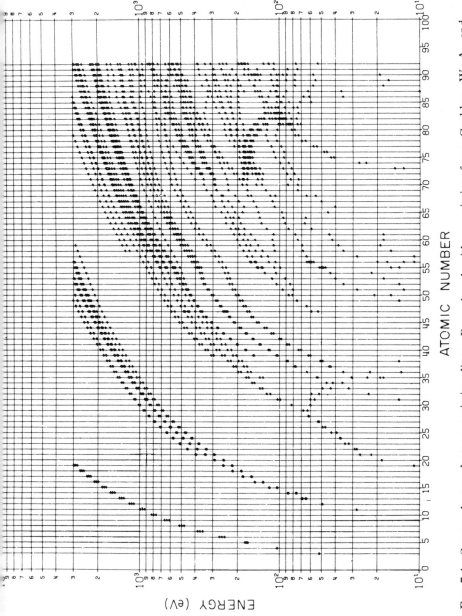

FIG. 7.4 Strong Auger electron emission lines. Reproduced with permission from Coghlan, W. A. and Clausing, R. E. *Atomic Data* **5**, 317 (1973)

proportional to concentration.[3,4,133] This assumes that the peak shape is constant, and while this seems to be true for metals and alloys, transition from a metal to (say) a metal oxide does frequently result in a considerable change in peak shape: in this situation comparison should presumably be made in terms of integrated peak areas (requiring double integration from the differentiated form, $dN(E)/dE$ vs E).

It is useful to consider first some of the general factors controlling the Auger electron current. Imagine the specimen to consist only of atoms A arranged in a sequence, atom layers designated 1, 2, 3, ... i, layer 1 being the surface layer. The i'th layer contains N_{iA} atoms (in this case N_{iA} is independent of i). The Auger electron current I_{iA} originating from the i'th layer will be

$$I_{iA} = kN_{iA}f_{iE_p}r_{iE_p}q_{iE_A}\sigma_{iAE_p} \qquad (7.9)$$

Here k is a constant containing geometric and instrumental factors; f_{iE_p} is the fraction of the primary exciting electron current to reach the i'th layer; q_{iE_A} is the chance that an Auger electron emitted from an A atom in the i'th layer in the direction of the analyser will escape from the specimen without appreciable energy loss. Both f_{iE_p} and q_{iE_A} depend on the distance the electron must travel in the solid and thus both depend on i.

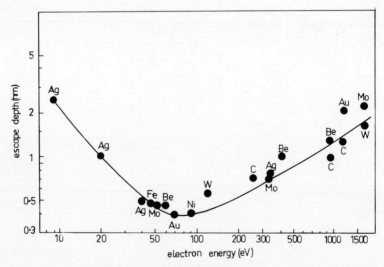

FIG. 7.5 Escape depth for electrons for various materials as a function of electron energy. After Palmberg, P. W. *Anal. Chem.* **45**, 549A (1973), based upon data assembled by J. C. Tracy.

In principle both f_{iE_p} and q_{iE_A} are dependent on composition and electron energy. However, there is evidence that they are dominated by electron energy, and to a modest approximation can be taken as independent of composition. This is illustrated in Fig. 7.5, which shows the escape depth of Auger electrons as a function of energy for various materials.[134] In photoelectron spectroscopy the same conclusion follows from the data of Wagner.[145] Thus it is reasonable to take f_{iE_p} as depending only on i and on the primary exciting electron energy E_p, and q_{iE_A} only on i and on the the Auger electron energy E_A. The parameter σ_{iAE_p} includes the ionization cross-section of the inner level for the Auger process and the transition

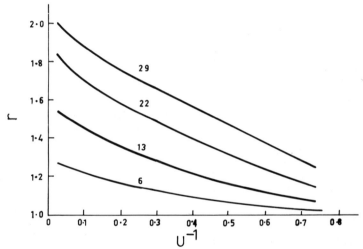

FIG. 7.6 Backscattering factor (r) for normal electron incidence as a function of U^{-1}; where $U = E_p/E_c$, E_p being the incident electron energy and E_c the atomic ionization energy. Calculations for 10 keV electrons. Figures on each line are atomic numbers of target materials. After Bishop H. E. and Rivière, J. C. *J. Appl. Phys.* **40**, 1740 (1969) and Bishop, H. E. *Brit. J. Appl. Phys.* **18**, 703 (1967).

probability for subsequent emission. In principle, σ_{iAE_p} depends on the particular Auger process occurring with a particular type of atom, on E_p and on i, the latter because the energy retained by an exciting electron decreases with increasing depth of penetration. Vrakking and Meyer[136] have discussed the evaluation of effective values for σ_{iAE_p} from experimentally available ionization cross-sections such as those due to Glupe and Mehlhorn.[137] The parameter r_{iE_p} is the backscattering factor by which the exciting current at the i'th layer must be multiplied to take into account the effects of backscattering from layers beneath the i'th: r_{iE_p} is a function of specimen composition and of E_p, as well as of i. Figure 7.6 shows some values calculated by Bishop,[135] and these agree reasonably well with experimental data.[136]

Consider the way in which the Auger electron current depends on specimen thickness, assuming the specimen to consist of only a single constituent. Gallon[143] has given a simple analysis in which it is assumed that:

(i) $f_{iE_p} = F^{i-1}$ where F is the chance of a primary exciting electron penetrating any of the i–1 layers in order to reach the i'th;

(ii) $q_{iE_A} = Q^{i-1}$ where Q is the chance of an Auger electron generated in the i'th layer penetrating any of the i–1 layers above it in order to escape;

(iii) $r_{iE_p} \sigma_{iAE_p} = A$ where A is constant. The total Auger electron current, $_iI_T$, from all of the layers from 1 to i is obtained by summation of the contributions from the layers, giving

$$_iI_T = {_\infty}I_T \left\{ 1 - (1 - I_1/{_\infty}I_T)^i \right\} \tag{7.10}$$

where I_1 and $_\infty I_T$ are the Auger currents from a monolayer and from an infinitely thick specimen respectively, and are introduced in order to eliminate the unknown parameters. Of the assumptions made, (i) and (ii) are clearly reasonable, but (iii) is more questionable. Nevertheless, equation 7.10 well fits the data for silver films of thicknesses ranging from 1 to 16 layers,[12] and it can be used as a basis for the estimation of the thickness of evaporated metal films (cf. ref. 144).

Consider next the problem of quantitative analysis with specimens in which there is a variation in composition with depth beneath the surface. Bearing in mind the finite escape depth of the Auger electrons and the finite penetration depth of the primary exciting electrons, Auger electron analysis of (say) an alloy surface will measure the composition averaged to some depth beneath the surface, but with the average constructed with decreasing weighting factors for layers further below the surface. Moreover, since the escape depth varies with the energy of the Auger electron, different Auger lines will, in principle, give different averages. The contribution of the backscattered electrons increases with the atomic number of the elements in the specimen, and for heavier elements may contribute up to half of the total Auger yield when the incident beam energy is much larger than the Auger threshold energy. In many applications, all that has been done is to treat the technique as only a relatively gross tool, to recognize that Auger sensitivity is heavily weighted to contributions from the outermost one or two layers, and to express the apparent average surface composition using the simplest interpretation of peak height data calibrated by the use of pure component metal specimens (cf. refs 5, 6).

Harris has shown that, by measuring the Auger spectrum at large angles from normal emittance, the surface composition is emphasized with respect to layers lying deeper beneath the surface.[7] This arises, of course, from the decreased escape depth for electrons emerging at large angles to the normal. The Auger yield also varies as csc ϕ, where ϕ is the angle of incidence of the primary exciting beam. The expected increase in yield of a factor of 4 has been verified experimentally[136] on going from normal incidence ($\phi = 90°$) to $\phi = 14\cdot5°$.

Suppose a specimen contains two elements A and B in an alloy. We wish to write down an expression for the total Auger current, I_{TA}, for a line characteristic of A: this follows directly from equation 7.9, there only being a modification of convenience whereby the current is expressed on a relative scale by replacing N_{iA} by the mole fraction of element A in the i'th layer, x_{iA}

$$I_{TA} = \sum_i I_{iA} = k \sum_i x_{iA} f_{iE_p} r_{iE_p} q_{iE_A} \sigma_{iAE_p} \qquad (7.11)$$

with an analogous expression in I_{TB} for B. To a modest approximation, some simplification of 7.11 is possible. In practice E_p is set considerably above the Auger electron energy (typically a factor of three greater), so that the penetration depth of the exciting electrons is considerably greater than the escape depth of the Auger electrons. Thus, over the Auger electron escape depth it is reasonable to assume that f_{iE_p} and σ_{iAE_p} are independent of i, giving

$$I_{TA} = k_A \sum_i x_{iA} r_{iE_p} q_{iE_A} \qquad (7.12)$$

the presummation constant now being dependent on the particular process. Further progress requires further assumptions and/or numerical data for the unknown parameters. Bouwman and Biloen[146] have elected to proceed by assuming that: (i) r_{iE_p} is also independent of i and can thus be lumped into k_A; (ii) the dependence of q_{iE_A} on i can be represented by an analytic function, thus converting the summation in 7.12 to an integral

$$I_{TA} = k'_A \int_0^\infty x_A(z)[\exp(-z/\lambda_{E_A})]dz \qquad (7.13)$$

where z measures the distance into the specimen and λ_{E_A} (a function of the Auger electron energy) is the mean free path of the escaping electron. Values for λ_{E_A} were identified with the escape depths given in Fig. 7.5. It is assumed that the specimen is infinitely thick so far as Auger emission is concerned. An analogous expression to 7.13 exists for component B, and involves $x_B(z)$ and λ_{E_B}. If more than one Auger line is measured for each component, each will generate its own equation of this form.

In principle, k'_A may be eliminated from 7.13 by calibration with a homogeneous sample for which $x_{A(calib)}$ is known, since with $x_{A(calib)}$ independent of z the integral is readily evaluated and one obtains for each of the lines measured for each component an equation of the form

$$\int_0^\infty x_A(z)[\exp(-z/\lambda_{E_A})]dz = \lambda_{E_A}(I_{TA}/I_{TA(calib)})x_{A(calib)} \qquad (7.14)$$

This also contains the implicit assumption that r_{iE_p} is independent of composition. The best-fit functions $x_A(z)$ and $x_B(z)$ for the various equations summarized by 7.14 may be obtained by trial and error. In practice, to obtain any worthwhile data about the depth-composition profiles requires measurement on at least two lines for each constituent, and these should be separated as widely in energy as possible. The resulting profiles are subject to the appropriate mass balance criteria. At any particular value of z (or i), one must have $x_A + x_B = 1$. Furthermore, if it can be assumed that there has been no net removal of either component from the surface region, then the amounts of the two components summed over the specimen depth sampled by the Auger technique must stand in the same ratio as in the bulk (assuming that departure from bulk composition becomes negligible when the maximum sampling depth is reached). However, the possibility of net component removal during specimen preparation should never be neglected, particularly if the surface has been subjected to chemical etching or ion bombardment treatments.

It should be recognized that if pure component specimens are used for calibration, one may well be using a calibration composition much different from that of the unknown. In principle, the errors will be reduced if calibration can be effected with a material nearer in composition to the unknown. However, a calibration alloy of known and strictly homogeneous composition is only likely to be available in occasional exceptional circumstances: one illustration is provided by the use of intermetallic compounds in the platinum–tin system where strictly homogeneous calibration samples could be obtained from specimens fractured *in situ*.[146-148]

Comparison between unknown and calibration specimens is subject to uncertainties due to varying instrumental factors, and to differences in surface morphologies and surface contamination (the latter has been shown to be particularly important in photoelectron spectroscopic systems which do not have UHV capability).[146] For this reason, if strictly homogeneous alloy calibration samples are available, there is an advantage in working with the ratios I_{TA}/I_{TB}, $I_{TA(calib)}/I_{TB(calib)}$ (cf. ref. 146).

Of the assumptions made in the treatment based on equation 7.13, probably the most questionable is that of the constancy of r_{iE_p}. To remove this assumption requires detailed knowledge of the dependence of r_{iE_p} on composition and depth, and this is not generally available. However, when r_{iE_p} values can be estimated, the starting point for the data analysis is best taken at equation 7.12: the composition dependence of r_{iE_p} then requires iteration to the depth–composition profile.

Auger peaks are superimposed on a broad background spectrum which originates from inelastic scattering of both primary exciting electrons and of the various secondary electrons. The analytical detection limit of the method depends on the type of hardware and on the type of atom being detected. However, roughly a system based on a grid analyser will detect down to a surface atom concentration of about 10^{16} atom m^{-2}. Better types of analyser and detector can improve this limit by a factor of perhaps 100.

Auger peaks from metals are broadened by small energy losses suffered by the escaping electrons in the form of interband transitions. Such broadening is typically 1–10 eV, and is the greater for electrons originating deeper in the metal. In addition to this, Auger peaks are also affected by lifetime broadening due to the imposition of the uncertainty principle with the very short transition times. Finally, there will be an inherent width in Auger peaks which involve one or more valence electrons since such electrons can come from within quite a wide band: structure in the destiny of states within the band may well be reflected in structure in the Auger peak.

The energy of an Auger electron is sensitive to the atomic chemical environment, so that chemical shifts occur, although the inherent peak width tends to place a limit on the analytical usefulness of such shifts. Nevertheless, chemical shifts are readily observable, but in a purely analytical sense not much use has yet been made of them. A problem is that since three electron levels are involved in an Auger process, there will be three electron energy level shifts which contribute to the overall chemical shift. From an interpretative point of view then, Auger chemical shifts are at a decided disadvantage compared to those observed in photo-electron spectroscopy, where the shift of only a single level is involved, and the lines are in any case, less broad.

Photoelectron Spectroscopy*

This technique can be divided into two classes, depending on the type of exciting radiation used to produce ionization, soft X-rays (1–20 keV) or

---

* Well-known manufacturers of commercial equipment include AEI, Vacuum Generators, Hewlett Packard, Du Pont, Kokusai Electric Co.

ultraviolet ($\leqslant 60$ eV). The name ESCA (electron spectroscopy for chemical analysis) is frequently used for X-ray photoelectron spectroscopy. In all of these techniques, the absorption of a photon of known energy $h\nu$ is used to eject an electron from some energy level, E_X. If the ejected electron has an energy E_P,

$$E_P = h\nu - E_X \qquad (7.15)$$

so if E_P is measured, E_X can be obtained. The energies E_X relate directly to the individual X-ray terms discussed in the previous section. If the electron is ejected from an atomic core level the value of E_X is diagnostic of the atom identity, while peak height or area is a measurement of concentration. If core electrons are to be excited, X-ray photons are needed, and the most commonly used X-rays which are used are Mg K_α at $1254\cdot6$ eV and Al K_α at $1486\cdot6$ eV. In this situation, electrons ejected from core levels will thus have energies roughly in the range 100–1400 eV, which is the same sort of energy range as for Auger electrons. Thus, although the penetration depth of the exciting X-ray photon into a metal will be quite large, the mean free path for the ejected ESCA electrons will be of the same order as for Auger electrons, and ESCA will thus also be a surface-dominated analytical tool with metals, in much the same way as Auger electron spectroscopy is, that is, the surface region will make its greatest contribution at lower ejected electron energies, and at higher atomic numbers for the metal being examined. This surface sensitivity has been experimentally confirmed for physical adsorption on evaporated films of gold[59] and for carbon monoxide chemisorption on evaporated films of tungsten and molybdenum.[60]

With ultraviolet radiation as the exciting source, the most commonly used photon energies are He II, $40\cdot8$ eV; Ne II, $26\cdot9$ eV; He I, $21\cdot2$ eV and Ne I, $16\cdot8$ eV. Excitation thus cannot occur deep in the atomic core, and the technique is mainly of use for valence electron excitation. The u.v. photon penetration depth in metals is much less than for 1–$1\cdot5$ keV X-ray photons. While this tends to make u.v. excitation more surface specific, the gain is restricted to the extent that ESCA is already surface specific due to electron escape depth limitations. Furthermore, in practice ESCA spectra are more specific in terms of chemical identity because they mostly have their origin in core electron excitations, and u.v. excitation with metals has mainly found use for the elucidation of band structure rather than for chemical analysis.

A spectrum obtained on an ESCA spectrometer will contain Auger lines as well as true ESCA lines. The Auger lines are readily identified since they are invariant with changing photon energy. Other lines, which are usually relatively weak and sometimes occur as satellites to the main lines, arise from various discrete energy loss processes, including interband transitions, and plasmon generation.

Chemical shifts are readily measured for ESCA lines, and give information about the atomic chemical environment. The subject has been reviewed.[8-10]

Wagner[145] has given a compilation of the relative ESCA sensitivities for the most intense emitted lines for a number of the elements, using the F(1s) line and the Na (1s) line as primary and secondary standards respectively. These data were obtained using exciting X-radiation from an aluminium anode, and refer to "infinitely thick" homogeneous specimens. The data are given in Table 7.1.

Characteristic Loss Spectroscopy

Energy losses suffered by electrons making inelastic collisions with a metal contain information about the electronic structure of the metal. However this sort of spectroscopy has not yet been used for analytical purposes as such.

Experimental Factors

At the most elementary level, the arrangement needed for electron spectroscopy with metals is shown in Fig. 7.7. There are now a large number of possible experimental arrangements which mostly depend on

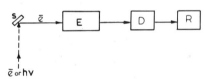

FIG. 7.7 Schematic illustration of basic electron spectroscopy equipment. The sample S is irradiated with electrons or photons, and the ejected electrons pass to the energy analyser E which controls the energy of the electrons which reach the detector D. The output from D, after suitable processing passes to the recorder R.

the design chosen for the electron energy analyser. In the past, this choice has been strongly influenced by the aims of the measurement. Much of the thrust with ESCA has been towards the measurement of chemical shifts where it is important to optimize electron energy resolution. On the other hand, Auger electron spectroscopy has been mainly used for qualitative and quantitative analysis of surface composition, and the equipment has been dominated by the desire to make the technique easily compatible with UHV methods and with existing LEED equipment. Another reason for the importance of Auger electron spectroscopy lies

TABLE 7.1 Relative atomic sensitivity of elements for X-ray photoelectron spectroscopy

$F(1s)$ line as primary standard: aluminium X-radiation

Z	Element	Compound	Strong line	Reference element	Sensitivity by peak height	Sensitivity by peak area
3	Li	LiF	$1s$	F	0·024	0·022
5	B	$NaBF_4$	$1s$	F	0·14	0·14
6	C	$-CF_2-$	$1s$	F	0·27	0·24
7	N	BN	$1s$	B	0·42	0·41
8	O	$NaHCO_3$	$1s$	Na	0·52	0·61
9	F	(std)	$1s$	...	1·00	1·00
11	Na	Various[a]	$1s$	F	2·14	2·09
12	Mg	MgF_2	$1s$	F	2·24	2·24
13	Al	K_3AlF_6	$2s$	F	0·28	0·23
14	Si	Na_2SiF_6	$2p-2p_{3/2}$[c]	F	0·22[c]	0·17[c]
15	P	$NaPO_3$	$2p-2p_{3/2}$[c]	Na	0·40[c]	0·26[c]
16	S	Na_2SO_3	$2p-2p_{3/2}$[c]	Na	0·46[c]	0·33[c]
17	Cl	NaCl	$2p-2p_{3/2}$[c]	Na	0·55[c]	0·46[c]
19	K	Various[b]	$2p_{3/2}$	F	0·94	0·85
20	Ca	CaF_2	$2p_{3/2}$	F	1·09	1·01
22	Ti	K_2TiF_6	$2p_{3/2}$	F	1·10	1·10
24	Cr	Na_2CrO_4	$2p_{3/2}$	Na	1·69	1·53
25	Mn	MnF_2	$2p_{3/2}$	F	1·18	1·55
26	Fe	$K_4Fe(CN)_6$	$2p_{3/2}$	K	1·82	1·76
28	Ni	$K_4Ni(CN)_6$	$2p_{3/2}$	K	4·10	3·68
30	Zn	ZnF_2	$2p_{3/2}$	F	3·73	4·24
32	Ge	Na_2GeF_6	$2p_{3/2}$	F	4·80	5·60
33	As	$NaAsO_2$	$2p_{3/2}$	Na	4·80	5·87
34	Se	Na_2SeO_3	$3p_{3/2}$	Na	0·64	0·86
35	Br	NaBr	$3d-3p_{3/2}$[d]	Na	0·54[d]	0·71[d]
37	Rb	RbF	$3d-3d_{5/2}$[d]	F	1·20[e]	0·95[e]
38	Sr	SrF_2	$3d-3d_{5/2}$[d]	F	1·26	1·03[e]
42	Mo	Na_2MoO_4	$3d_{5/2}$	Na	2·37	2·04
45	Rh	Na_3RhCl_6	$3d_{5/2}$	Na	2·16	2·00
48	Cd	CdF_2	$3d_{5/2}$	F	3·60	3·80
50	Sn	$NaSnF_3$	$3d_{5/2}$	F	5·75	5·75
51	Sb	$KSbF_6$	$3d_{5/2}$	F	6·2	6·7
53	I	NaI	$3d_{5/2}$	Na	5·14	5·03
56	Ba	BaF_2	$3d_{5/2}$	F	6·16	6·63
62	Sm	SmF_3	$3d_{5/2}$	F	2·80	6·90
72	Hf	HfF_4	$4f-4f_{7/2}$[f]	F	1·14[f]	0·85[f]
74	W	Na_2WO_4	$4f_{7/2}$	Na	1·52	1·37
75	Re	$KReO_4$	$4f_{7/2}$	K	2·11	1·77
77	Ir	Na_2IrCl_6	$4f_{7/2}$	Na	1·80	1·72
78	Pt	$K_2Pt(CN)_6$	$4f_{7/2}$	K	2·24	1·93
82	Pb	PbF_2	$4f_{7/2}$	F	3·92	4·10
83	Bi	$NaBiO_3$	$4f_{7/2}$	Na	3·48	4·16
92	U	UO_2F_2	$4f_{7/2}$	F	5·80	6·45

in its high sensitivity relative to photoelectron spectroscopic processes involving the same initial ionization process.

The widely used LEED back-reflection apparatus adapted for the measurement of Auger electron spectroscopy is shown in Fig. 7.8.[11,12] The grid system acts as a high pass electron energy filter, and so the current which is collected on the display screen needs to be differentiated with respect to voltage in order to generate a normal spectrum. In practice, it is often differentiated a second time in order to improve the sensitivity. In the arrangement shown in Fig. 7.8, the grids G_1 and G_4 are grounded, while the two central grids, G_2 and G_3 which are electrically strapped together, act as the retarding elements and their potential is slowly swept from zero up to the primary beam energy. The retarding grid potential has superimposed on it a small modulating potential. The alternating component of the current collected by the screen is detected and either the first or second harmonic when plotted against the retarding voltage gives the Auger-electron energy distribution, or its first derivative respectively. The most important (and essential) refinement is the provision of an external circuit for capacitive decoupling of the grids.[49] If a standard LEED apparatus is being used, the provision of another electron gun (B) to give incident electrons at 10–20° with respect to the surface can give a significant improvement in Auger sensitivity compared to the use of the normal diffraction gun (A) (Fig. 7.8). This benefit accrues both from the low angle of incidence, and from the higher beam currents

---

Notes to Table 7.1 on facing page

a. Height and area ratios for Na–F compounds: NaF 2·00, 2·04; NaBF$_4$ 2·59, 2·59; Na$_2$SiF$_6$ 2·14, 2·04; Na$_2$GeF$_6$ 2·33, 2·33; Na$_3$FeF$_6$ 1·48, 1·60; NaSnF$_3$ 2·30, 1·96. Average 2·14 ± 0·27, 2·09 ± 0·24. Average used where Na is reference element.
b. Height and area ratios for K–F compounds: KF 0·94, 0·94; K$_3$AlF$_6$ 0·80, 0·59; K$_2$TiF$_6$ 0·97, 0·88; KSbF$_6$ 0·67, 0·70. Average 0·84 ± 0·11, 0·78 ± 0·13.
c. Where p doublets are insufficiently resolved to permit valid area measurement of right half of the $2p_{3/2}$ line, overall peak height was used for the peak comparison and $\frac{2}{3}$ of the area for the area comparison.
d. The $3d$ unresolved doublet was more intense than the $3p_{3/2}$ line and was used in the peak height comparison, but the area of the $3p_{3/2}$ line was larger than that of the unresolved $3d_{5/2}$ line and was therefore used in the area comparison.
e. Where d doublets are insufficiently resolved to permit valid area measurement of the right half of the $3d_{5/2}$ line, overall peak height was used for the peak height comparison and $\frac{3}{5}$ of the area for the area comparison.
f. Where f doublets are insufficiently resolved to permit valid area measurement of the right half of the $4f_{7/2}$ line, overall peak height was used for the peak height comparison and $\frac{4}{7}$ of the area for the area comparison.

Reproduced with permission from Wagner, C. D. *Anal. Chem.* **44**, 1050 (1972).

which can easily be used with the auxiliary gun. Instrumental artifacts can occur in Auger spectra obtained with this type of equipment, and the problems have been discussed by Morrison and Lander[13] and by Wei et al.[14]

The use of a four-grid system as shown in Fig. 7.8 offers an appreciable improvement in resolution, compared with a three-grid system, and their relative merits have been analysed by Taylor.[15] With a three-grid system the energy resolution is $2 \cdot 5$–3%, while with a four-grid system in optimum

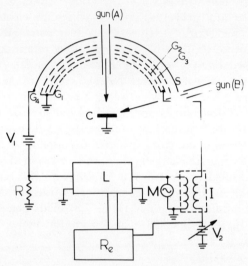

FIG 7.8 Retarding grid analyser for Auger electron spectroscopy. The specimen C may be bombarded with electrons from gun (A) or gun (B). The grids G_1–G_4 act as a high pass filter for the emitted electrons which are collected by the screen S. The collected current is modulated by the voltage generator M via the isolation transformer, I. L is a phase-lock amplifier feeding the recorder Re. V_1 is about 200 V; V_2 0–2000 V sweep; R, 10^6 ohm. The modulation frequency should be a non-integral multiple of mains frequency.

conditions a resolution of about $0 \cdot 3$–$0 \cdot 5\%$ can be achieved. In normal practice a resolution of $0 \cdot 5\%$ with a four-grid system is a reasonable expectation: in other words, in the middle of the typical Auger energy range, say 400 eV, the resolution is around 2 eV. This needs to be compared with the magnitude of Auger chemical shifts which (as with ESCA chemical shifts) lie in the range $\leqslant 20$ eV, and with the Auger peak widths which lie in the range 1–10 eV (or more if the transition involves a valence electron). A reasonable conclusion is that quite a lot of what is possible can be done with a retarding grid analyser, but rather better resolution would be an advantage.

The literature on the various type of electron energy analysers is now quite extensive, including: general discussions;[16-20] parallel plate analysers;[21, 22] curved parallel plate analysers;[23-30] spherical plate analysers;[17, 31-34] cylindrical plate analysers;[17, 35] retarding field analysers;[15, 36-40] retarding field analysers with postmonochromator;[43] retarding field analysers with postfocus grids;[44] combined reflection and transmission filters;[41] time of flight analysers;[42] Mollenstedt analysers.[58] All of the above are electrostatic analysers. Magnetic analysers are also used (e.g. refs 47, 48). For the measurement of chemical shifts in ESCA, the resolution of the analyser should be such that it does not contribute more than a few tenths of an eV to total instrumental line broadening. At ejected electron energies of, say, 1 keV this means that the analyser resolution ($\Delta_{\frac{1}{2}}/E$) should be of the order of a few parts in 10^4. Both single-focusing and double-focusing analysers have been used, and of the electrostatic types those most often used have been the cylindrical plate analyser or the spherical plate analyser. The combined reflection and transmission filter design has also recently been used. Commercial ESCA spectrometers mainly use electrostatic analysers. Palmberg et al.[45] illustrate the application of a cylindrical plate analyser to Auger electron spectroscopy.*

One commercial ESCA instrument (Vacuum Generators Limited, England) is UHV-compatible and is particularly suited to surface work: it can also easily be converted to X-ray, u.v., or electron irradiation sources. In all current commercial equipment however, the overall resolution of the ESCA system contains factors additional to the resolution of the electron energy analyser itself. For instance, unmonochromatized Al K_α or Mg K_α radiation has a natural line width of about 0·8 eV, and in practice an overall resolution of 0·8–1 eV appears typical.

Applicability to Metallic Catalyst Samples

The important point is the extent to which the surface of the metallic specimen is accessible to the exciting radiation, and the facility with which the emitted entity (electrons in the case of Auger or ESCA, photon in the case of X-ray fluorescence) can reach the analyser. Obviously, a massive metal specimen with a free surface offers no problems. However, with catalysts consisting of small metal particles distributed through a high area support, the problem is severe because the amount of metal that can contribute to the spectrum is very limited. Some ESCA spectra from supported platinum and nickel catalysts have been reported[10] but fairly high metal loadings 1–5 wt. % appear to be needed to obtain tolerable sensitivity. Nevertheless, it is possible, at least in favourable cases, to

* A cylindrical plate analyser mounted on a UHV flange is available from Physical Electronics Industries.

follow some of the features of the chemistry by which metal particles are generated by reduction of the precursor. Elemental sensitivity is dependent on the sample morphology.

Attempts to apply electron stimulated Auger electron spectroscopy to supported catalysts have been unrewarding.

Ion Probe Microanalysis

The principle of the method is that the surface to be analysed is bombarded by gas ions, and the ions which are sputtered from the surface are analysed mass spectrometrically. The advantages are high sensitivity, applicability to all elements, and a high degree of spacial resolution (about 1 μm) across the surface possible by the use of a finely focused ion beam. Socha[50] and Kane and Larrabee[51] have given summary reviews. The ion source is a "duoplasmatron."[52,53] This consists of a magnetically constricted gas arc at about 2–3 Pa (about 0·02 Torr) from which ions are extracted through an anode pinhole. The ions strike the specimen after further acceleration and focusing. The specimen ions are sputtered with appreciable kinetic energy, and a double focusing mass spectrometer is normally employed.

Although inert gas bombarding ions have been used, Anderson[54] made the important observation that the decrease in the sputtering yield with time which is observed when rare gas bombardment is used with metals, can

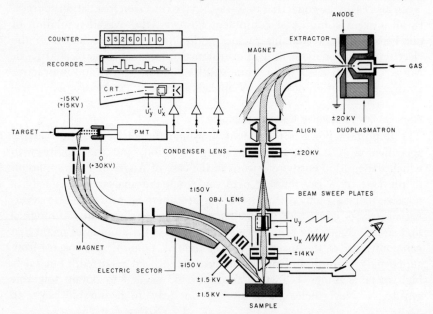

FIG. 7.9 Schematic diagram of ion probe microanalyser system of the Applied Research Laboratories. Reproduced by permission.

be eliminated by bombarding with oxygen ions. The sputtering rate is not strongly dependent on primary ion energy in the range 5–25 keV[53] but the sputtering rate can be controlled via the primary beam intensity and rates in the range $\leqslant 50$ nm s^{-1} are usual. This ability progressively to remove material from the specimen surface makes it possible to generate a depth–composition profile. There are some problems here which are due to material coming from the sides of the "hole", and Socha provides a good account.[50]

Equipment of this type runs to considerable instrumental elaboration, and features which have been described include mass spectrometric purification of the primary ion beam,[55] display of an "ion image" on a CRO and the construction of an "ion microscope".[56,57] In practice, mass spectrometric purification of the primary ion beam is particularly desirable. Figure 7.9 illustrates the apparatus in this configuration.

The ion yields of metals are strongly dependent on the element in question, and there is a rough trend for the yield to decrease with increasing atomic number and increasing cohesive energy. Table 7.2 gives some examples.[53] The differences between the metals are lower the heavier the bombarding ion. This variation in ion yield places a premium on thorough calibration procedures if the technique is to be used for quantita-

TABLE 7.2 Ion yields for some metals resulting from bombardment with 8 keV Xe$^+$ ions

Metal	Ion yield (metal ions per Xe$^+$ ion)
magnesium	20·9
aluminium	7·2
iron	4·2
lead	3·0
nickel	1·68
indium	1·67
cobalt	1·5
tantalum	1·0
zinc	0·95
copper	0·79
tin	0·72
zirconium	0·56
cadmium	0·38
niobium	0·09
silver	0·01
gold	0·006

tive analysis of any precision. This calibration requires the use of surfaces of known composition. However, the calibration problem is not necessarily straightforward, since one cannot assume the sensitivity to a given element is independent of its chemical environment. Nevertheless, the problem of quantitative analysis has been approached in a systematic manner by Anderson and Hinthorne,[83, 84] based on a theoretical model of the sputtering process. In the best tested procedure, internal calibration standards are still needed.

Instruments are available commercially (e.g. Applied Research Laboratory, Cameca, AEI, Hitachi.

Ion Scattering

When an ion makes an inelastic collision with a surface, the energy loss is a function, *inter alia*, of the mass of the surface atoms. This has become the basis of a surface analytical technique which has been reviewed by Smith.[68] Commercial equipment is available (3M Company). The technique uses noble gas ions with primary energies $\leqslant 3$ keV, and the basic arrangement is shown schematically in Fig. 7.10. The energy of the scattered ion relative to that of the incident ion is given to a good approximation by the simple kinematic formula which is strictly applicable to a collision between two isolated atoms: if the scattering angle is 90°,

$$m_2 = m_1(1 + E_1/E_0)/(1 - E_1/E_0) \qquad (7.16)$$

where E_0 is the energy of the primary ion; E_1 the energy of the scattered ion; m_1 the mass of the primary ion; and m_2 the mass of the target atom. An experimental apparatus which is basically similar to the commercial equipment has been described by Goff and Smith.[69] The ions are generated by a relatively standard ion gun, and after focusing irradiate an area of sample about 1 mm diameter. A variety of energy analysers for the scattered ions is possible, but the apparatus described by Goff and Smith uses a simple curved parallel plate analyser with 127° deflection.

The usual primary ion beam is helium ions in the energy range $1 \cdot 5$–2 keV, and is sensitive only to the first atomic layer of the sample.[70] However, according to equation 7.16, the energy spread per unit mass, $d(E_1/E_0)/dm_2$, is increased by using heavier primary ions, and in practice this may dictate the use of heavier ions than helium if improved resolution is needed.

As with ion microprobe analysis, ion bombardment in the present technique results in removal of atoms from the sample surface and variations with depth can be explored. The analytical sensitivity of the method appears to be quite reasonable, the limit of detection being certainly less than 1 % of a monolayer.

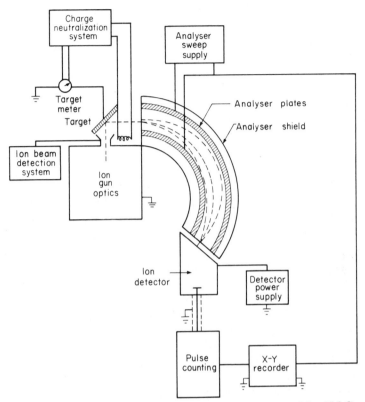

FIG. 7.10 Schematic diagram of ion scattering spectrometer of the 3M Company. Reproduced by permission.

For quantitative surface analysis, calibration with known surfaces is needed, and some comments on this have been made by Smith.[68]

Ion Beam Spectrochemical Analysis

When the surface of a solid is bombarded with energetic ions (or neutral atoms) spontaneous emission of u.v. and visible radiation occurs. When the radiation is analysed by a spectrometer, it is found to contain characteristic lines of the constituent element of the solid.[128, 131] The radiation is due to the decay of excited atoms ejected from the surface. The intensities of the spectral lines are directly proportional to the bombarding ion beam current, but the photon emission process is more efficient with insulating than with metallic targets. Tsong and McLaren describe some instrumental techniques based on argon ion bombardment.[128, 131]

Mössbauer Spectroscopy

Although Mössbauer spectroscopy is not an explicitly surface sensitive technique, we mention it at this juncture because in highly dispersed metals the proportion of surface atoms becomes large enough for them to present a significant Mössbauer component, and at the same time the technique is one of the few which can directly provide information about the nature of the metal particles in supported highly dispersed catalysts. The latter is so because of the intrinsic high sensitivity of the method, and because the support material is readily penetrated by the γ-radiation.

The Mössbauer effect is the result of recoil-free emission and resonant reabsorption of low energy γ-rays by atoms in solids. The most important property of Mössbauer γ-radiation is its very narrow line width, so that very small perturbations due to the environment of the emitting or absorbing atom are measurable.

Mössbauer spectroscopy has been well described for theoretical and experimental detail (e.g. refs 98–102) and it is here only useful to indicate a general outline in the briefest terms.

Line shifts are extremely small on an energy scale, and in practice the spectrum is scanned by making use of a Doppler shift from a small relative velocity between source and absorber: hence spectral parameters are commonly expressed in velocity units (e.g. mm s^{-1}). Experimentally, the basic apparatus consists of an emitter, an absorber and a γ-ray detector: the emitter and absorber are moved with respect to each other repetitively at speeds up to 10^3 mm s^{-1}, and the γ-ray intensity after transmission through the absorber is recorded as a function of relative velocity on a multichannel analyser. The sample being studied may be made either the source or the absorber. Each arrangement has its advantages which have been summarized.[87]

The chemical shift of a Mössbauer line depends, for a given transition, on the electronic charge density at the nucleus, and this comes mainly from s-electrons, but changes may also be produced indirectly through screening effects on s-electrons by p- and d-electrons. Other effects which the atomic environment has on the Mössbauer line are quadrupole splitting and magnetic splitting. Quadrupole splitting arises if an electric field gradient is applied to a nuclear state with a nuclear spin quantum number, I, greater than $\frac{1}{2}$.

For the Mössbauer effect to occur the nuclear transitions must be recoil-free, that is, the recoil momentum must be transmitted to the solid as a whole, and no new phonons excited in the solid. The probability that a recoil-free transition occurs is given by $\exp[-<x^2>/\lambda^2]$ where $<x^2>$ is the mean square displacement of the nucleus in the direction of the γ-ray, and $2\pi\lambda$ the wavelength of the γ-ray. The probability of recoil-free transition increases as the γ-ray energy decreases, and in practice this needs to be less than about 150 keV; it also increases as the strength of binding of the atom to the lattice increases, as the mass of the atom increases, and as the temperature is decreased, the latter because phonon excitation is less likely at the lower temperature.

Application to surface work is well exemplified by the recent work of Bartholomew and Boudart[85] with dispersed platinum–iron catalysts on a graphitized carbon support. Observed spectral line envelopes were decomposed into component lines by computer curve fitting. The room temperature spectra of these samples prepared by hydrogen reduction at 770 K were generally characterized by envelopes which were best-fitted by two quadrupole-split doublets as shown in Fig. 7.11. The outer, less intense doublet having broad line widths was assigned to surface atoms, whereas the inner doublet having smaller line widths, was assigned to bulk atoms. The proportion of iron in the surface relative to the bulk may be estimated from the respective spectral areas, provided it can be

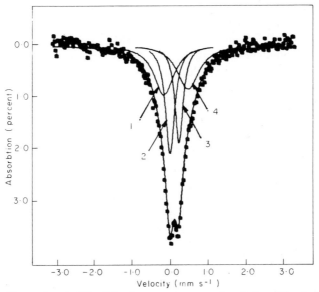

FIG. 7.11 Example of ^{57}Fe Mössbauer spectrum for 9·4 wt.% platinum–iron (90/10) carbon catalyst after hydrogen reduction at 743 K, cooling to 298 K and evacuation, followed by exposure to air at 298 K. Spectrum recorded at room temperature. The figure shows a computer fitted decomposition of the experimental envelope into a doublet due to bulk iron (peaks 2 and 3) and a doublet due to surface iron (peaks 1 and 4). Reproduced with permission from Bartholomew, C. H. and Boudart, M. *J. Catal.* **29**, 278 (1973).

assumed that the probability for recoil-free emission is the same for a surface atom as a bulk atom. In general one would expect this not to be so, because the surface atoms are less strongly bonded to the crystal. Bartholomew and Boudart overcame this by measuring the Mössbauer spectra also as a function of gas adsorption, and making use of the proposition that a surface iron atom at which is adsorbed hydrogen or preferably oxygen effectively behaves like a bulk atom. This different quadrupole splitting arises because of the different electrostatic fields experience by iron atoms in the surface and in the bulk, that in the surface being of lower symmetry. In the case of ^{57}Fe the unsplit Mössbauer line involves transitions between the nuclear levels with $I = \frac{1}{2}$ and $I = \frac{3}{2}$. In a field with an axial component the $I = \frac{3}{2}$ level is split into two levels so this results in a doublet.

If a magnetic field is present at the nucleus, the effect is to remove completely the spin degeneracy of all the nuclear levels. In iron this magnetic hyperfine splitting leads to the splitting of the $I = \frac{3}{2}$ state into four, and the $I = \frac{1}{2}$ state into two. With the application of the appropriate selection rules, six possible transitions are allowed, so the spectrum would have six lines. This magnetic field may be either internal as in ferromagnetic or antiferromagnetic materials, or it may be externally applied. The generation of an internal magnetic field requires magnetic ordering, and this is dependent on temperature and particle size. For instance, in α-Fe$_2$O$_3$ antiferromagnetic ordering with consequent magnetic hyperfine splitting of the Mössbauer spectrum only occurs for a particle diameter greater than about 26 nm; below this size the material exhibits collective paramagnetism and only a quadrupole split doublet is observed.[132] The point is that a sensible decomposition of Mössbauer line envelopes in the manner we have just indicated is not to be undertaken lightly, and the way it is done depends upon a knowledge of the hyperfine splitting possibilities.

This type of technique is highly dependent on the natural line width of the Mössbauer transition being sufficiently narrow, and it becomes more complex if higher values of I are involved. Iron is certainly the only transition element which can be used at room temperature. At liquid helium temperatures other possible candidates are ^{99}Ru and ^{193}Ir.

Work-Function Measurement

The work-function, the energy required to remove an electron from the top of the Fermi sea to a point outside the metal, depends *inter alia* on the composition of the surface. From the present point of view, this method of monitoring surface composition is limited to a free surface of a massive metallic specimen.

Photoelectric and contact potential measuring methods have been generally used. Thermionic emission requires the metal to be at a very high temperature.

The photoelectric method requires the estimation of the threshold photon energy for emission to occur. On a simple free-electron model, Fowler's[91] treatment leads to

$$J = B(kT)^2 f[(h\nu - h\nu_0)/kT] \qquad (7.17)$$

where $h\nu_0 = e\phi$, ϕ being the work-function; $h\nu$ is the photon energy giving the emission current J, B is essentially a constant, and f is a universal function of the variable $[(h\nu - h\nu_0)/kT]$. The original procedure for the evaluation of $h\nu_0$ involved curve fitting, but it follows from an expansion of the function f that $h\nu_0$ is given by the intercept on the $h\nu$ axis of an extrapolation of the plot of $J^{1/2}$ against $h\nu$. With modern equipment the method is experimentally straightforward, and Rivière gives a useful summary.[92] However, the method has some pitfalls. If the free-electron model is too bad an approximation the Fowler theory may not work. In any case a problem exists in the interpretation of the results from a nonuniform surface. This has been analysed by Herring and Nichols[93] in terms of the relative magnitudes of the electric fields applied for electron collection (ε_a), and the field due to the different work-functions of the patches (ε_p). The order of magnitude of ε_p is given by $|\Delta\phi|/\Delta x$, where Δx is a patch dimension and $|\Delta\phi|$ the magnitude of the work-function difference. There are two limiting cases. When $\varepsilon_a \gg \varepsilon_p$, all patches emit independently. On the other hand, if $\varepsilon_a \ll \varepsilon_p$, all patches with $\phi_i < \bar{\emptyset}$ (where $\bar{\emptyset}$ is the area weighted mean work-function of the surface) will be represented as a single patch of work-function $\bar{\emptyset}$, and only patches with $\phi_i > \bar{\phi}$ will be individually resolved. If we let $|\Delta\phi| = 0.5$ eV, then with ε_a taking a typical value of 2 V mm^{-1}, the values of Δx for the two regimes would be; $\varepsilon_a \gg \varepsilon_p$, $\Delta x \gg 0.25$ mm; $\varepsilon_a \ll \varepsilon_p$, $\Delta x \ll 0.25$ mm. A number of cases where patch work-functions have been resolved have been reported,[96, 97] but there remains some doubt if the patch size limits defining the two regimes are accurately estimated by Herring and Nichols theory.[97]

There are a number of methods for the measurement of the contact potential difference between the unknown specimen and a reference surface, but they all, in effect, measure the potential difference which becomes established between the two specimens in electron equilibration in order to make the Fermi levels of the two the same. This contact potential difference equals the difference in work-functions. We shall mention only two methods.

In the Kelvin method, the unknown and the reference surface are formed into a parallel plate condenser across which, in the absence of an applied external potential, there is established the contact potential difference. This may be balanced out by the application of a potential equal in magnitude but opposite in sign in the external circuit. The point of cancellation is determined when a change in the value of the condenser capacitance causes no external current flow.

In the diode method, the specimen surface is made the current collector in a saturated diode, so the emitter is the reference surface. The current to the anode is made so small that space-charge effects are negligible, then in the region of retarding potentials on the anode (negative values of $(V + V_c)$), the current i is given by

$$i = i_s \exp[e(V + V_c)/kT] \qquad (7.18)$$

while with $(V + V_c)$ positive the current becomes constant at i_s: here V is the applied potential and V_c is the contact potential difference. A plot of log i vs V gives two straight lines which intersect at a point where $V = -V_c$. In practice, special precautions have to be taken to obtain a sharp intersection point, and the electron focusing method of Shelton[94] appears to be the most satisfactory.

If one is only interested in *changes* in the work-function of the specimen surface, a diode measurement becomes much easier, since one only has to measure the displacement of the current–applied voltage curve along the voltage axis as a result of the change, and measurements can be conveniently made in the space-charge-limited region. A large number of geometric arrangements of filament and collector are then possible. One has to ensure that the curves before and after displacement are parallel for the measurement to be accurate.

Rivière gives a detailed summary of contact potential difference methods, as well as other methods of measuring the work-function itself.[92]

The contact potential difference refers to an area-weighted mean average work-function of non-uniform surfaces.

Sachtler and co-workers[95,77] have made extensive use of photoelectric measurements for following the variation in work-function with composition for a number of alloy evaporated film catalysts. The data are useful in giving an indication of the way the components are distributed, particularly when a uniform solution is not formed.

Gas Chemisorption

The use of gas chemisorption studies for the characterization of the composition of a metal surface relies upon using differences in surface reactivity. While variations in chemisorption capability can, of course, arise from the presence of preadsorbed material, we propose to limit the present discussion to composition defined in terms of metallic constituents. The discussion in Chapter 6 has already dealt with the way in which chemisorption differences between the metals themselves may be utilized, particularly when more than one pure metal phase is present. Here we are mostly concerned with the use of gas chemisorption for the quantitative characterization of alloy surfaces, together with some comments about supplementary information which may be obtained by studying spectroscopic or thermodynamic properties of the system which relate to the nature of the adsorbate–adsorbent interaction.

Hydrogen Chemisorption

A number of workers have attempted to use hydrogen chemisorption to titrate the amount of adsorbing component present in an alloy surface: that is, if an alloy surface contains atoms of metals A and B, where it is known that pure A chemisorbs hydrogen but B does not, it is proposed that the number of hydrogen atoms chemisorbed on the alloy surface counts the number of A atoms present in the surface. The best examples of the application of this idea are to be found with evaporated films of nickel–copper alloy by Sachtler and co-workers,[71, 72] and Sinfelt and co-workers with dispersed group VIII metal–copper catalysts.[73, 74] In the case of Sachtler's alloys films, the hydrogen uptake was measured at room temperature and at about $0 \cdot 2$–$0 \cdot 3$ Pa (a few Torr) where the uptake was approximately independent of hydrogen pressure. With Sinfelt's dispersed catalysts, adsorption measurements were made under much the same conditions as are usually employed for normal monometallic dispersed transition metal catalysts, that is at room temperature and at hydrogen pressures in the range 2×10^2–4×10^3 Pa (about 2–30 Torr). As would be expected from the behaviour of analogous monometallic catalysts, there are both weak and strong components to the total hydrogen uptake, the latter being, for the present purpose, defined as the uptake left after pumping the catalyst to about 10^{-2} Pa (about 10^{-4} Torr) at room temperature for ten minutes. Figure 7.12 shows an example of the strong hydrogen uptake relative to the total for an unsupported dispersed nickel–copper catalyst.[73] It is obvious that the proportion of weakly chemisorbed hydrogen increases strongly as the proportion of copper in the catalyst increases.

In interpreting this augmented weak adsorption, it is reasonable to propose that most of it is due to chemisorption on surface copper atoms. A clean copper surface does not chemisorb hydrogen to an appreciable extent under these conditions. However, it is also known that the thermodynamic limitation on this occurring can be overcome if hydrogen atoms rather than molecules are presented to the copper, and it appears that

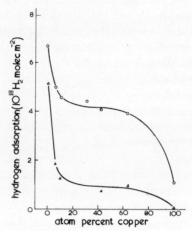

FIG. 7.12 The adsorption of hydrogen on unsupported dispersed nickel–copper catalysts as a function of composition. The upper curve (○) represents the total amount of hydrogen adsorbed at room temperature at 13·3 kPa (100 Torr); the lower curve (△) represents the amount of strongly adsorbed hydrogen which is not removed by evacuation at room temperature following the completion of the initial isotherm. Reproduced with permission from Sinfelt, J. H., Carter, J. L. and Yates, D. J. C. *J. Catal.* **24**, 283 (1972).

nickel sites adjacent to the copper can function in this way: that is, there is spillover of chemisorbed hydrogen atoms from nickel sites to the copper. However, the binding energy of a chemisorbed hydrogen atom on copper is low, and desorption as molecular hydrogen occurs when the pressure is low enough. On this basis then, it is the strong hydrogen chemisorption component which should give the best estimate of the proportion of nickel atoms in the surface,[73] and the same conclusion should be extended to other group VIII metal–copper alloys.[74] Inasmuch as Sachtler used much lower hydrogen pressures for his uptake measurements, his technique is also consistent with this general idea.

However, it is very important not to read more accuracy into these measurements than the model warrants. There is a weak chemisorption component with all dispersed monometallic transition metal catalysts to which hydrogen adsorption measurements can usefully be made, and the significance of this has been discussed in Chapter 6. It is true that this is normally a good deal less than the size of the weak uptake which occurs with the bimetallic catalysts, but it is still appreciable, and we have seen in Chapter 6 that it needs to be included when using hydrogen chemisorption for surface area estimation. In view of this, it is likely that using the strong hydrogen uptake somewhat underestimates the proportion of group VIII metal in the surface. There is also the problem of the extent to which a transition metal atom in the alloy surface retains its chemisorptive properties. To take the nickel–copper system as an example, it would not seem unreasonable to suppose that a group of, say, ten nickel atoms all adjacent to one another in the surface could behave with respect to hydrogen chemisorption not too unlike a small bit of nickel surface; on the other hand, how reasonable is it to suppose that a single nickel atom completely surrounded in the surface by copper atoms retains its essential nickel-like behaviour? Some earlier measurements[75] of hydrogen chemisorption on platinum-copper/silica catalysts have been interpreted to show that the copper does not simply dilute the platinum atoms without changing their properties; however, total hydrogen uptakes were used to elicit this conclusion. Certainly we must say that in a transition metal–group IB alloy surface, some modification of the chemical properties of the transition metal atoms must occur relative to the properties of the pure metal. This modification is presumably greatest for a single transition metal atom with only group IB atom neighbours, and becomes less for transition metal clusters of larger size. The situation badly needs illumination by making comparative measurements with chemisorption and Auger electron spectroscopic analysis.

Carbon Monoxide Chemisorption

This has been used[74] with dispersed ruthenium–copper and osmium–copper catalysts as an alternative to hydrogen chemisorption for the estimation of the proportion of group VIII metal atoms in the surface. The uptake was measured at room temperature and at pressures in the range $2 \times 10^2 - 4 \times 10^3$ Pa (about 2–30 Torr). As with the use of hydrogen, the strong chemisorption component was taken as a measure of the number of group VIII metal atoms, and this was obtained as the difference between an initial isotherm and a second isotherm obtained after ten minutes evacuation at room temperature. For a given surface, the ratio of the number of carbon monoxide molecules adsorbed to the number of hydrogen molecules adsorbed is mostly somewhat greater than unity, the ratio ranging up to about 1.5. In this sort of situation where one of the metallic components is a group IB metal, there is not sufficient knowledge of the adsorption chemistry to make a detailed assessment of the relative merits of hydrogen or carbon monoxide adsorption for the estimation of the group VIII metal component in the surface. Nevertheless, on general grounds related to the known variability of carbon monoxide chemisorption stoichiometry, and the somewhat greater propensity of carbon monoxide than hydrogen for chemisorption on group IB metals, we incline to the view that the use of hydrogen is preferable.

Supplementary Measurements

We are concerned here with the measurement of properties which can depend on the nature of the adsorbate–adsorbent interaction, and which can give information about the nature of the metal atom to which the adsorbate species is bonded. This is a matter of particular importance when we are dealing with dispersed bimetallic catalysts the two components of which do not show large quantitative differences in chemisorption behaviour. For instance, one cannot hope to study a catalyst containing two group VIII metals, both of which strongly chemisorb gases like hydrogen or carbon monoxide, simply by gas uptake measurements in the manner that can be attempted with transition metal–group IB catalysts.

What is possible here is a function of the physical form of the catalyst, since this strongly influences the accessibility of the metal surface.

Because so little has yet been attempted in this direction, we shall mainly confine our remarks to suggest ways in which the problem can be approached. This type of measurement should aim towards specificity.

Infrared Absorption Spectrophotometry

With dispersed catalysts on supports which have sufficient infrared transparency, such as silica or alumina, the measurement of the infrared absorption spectrum of some convenient adsorbate such as carbon monoxide or nitric oxide should prove to be a generally useful diagnostic tool, since it is known that the main features of the absorption spectrum show some sensitivity to the nature of the metal. For example, the bands assigned to the carbon–oxygen stretching frequencies for the linear form of adsorbed carbon monoxide on some reduced metal/silica catalysts fall in the range 4·72–5·05 μm (2120 − 1980 cm^{-1})[79-81, 89] while bands due to the bridged form fall in the range 5·10–5·46 μm (1960–1830 cm^{-1}). At least in favourable cases then, the detailed structure of the absorption bands should reflect some of the features of surface composition. The general technique for infrared absorption spectrophotometry with supported dispersed metal specimens has been summarized by Little.[86] With the free surface of a massive metal specimen, the measurement of an infrared absorption spectrum is much more difficult and Little[87] and Pritchard[88] give recent summaries.

A recent study[149] with carbon monoxide on palladium–silver/silica catalysts illustrates some of the potential of this technique. Three carbon monoxide absorption bands were observed on supported palladium and on supported palladium–silver catalysts at about 4·85 μm (2060 cm^{-1}), 5·10 μm (1960 cm^{-1}) and 5·21 μm (1920 cm^{-1}). The 4·85 μm band was ascribed to linear adsorbed carbon monoxide and, although relatively weak on palladium, this became the dominant one with the bimetallic catalysts. The other two bands which were ascribed to bridged carbon monoxide were very faint with the bimetallic catalysts. The clear interpretation is that, as expected, the two metals form alloy particles: thus as the surface becomes more dilute in palladium as the silver content is increased, the proportion of palladium atoms as adjacent pairs falls (bridged carbon monoxide) much more rapidly than the surface palladium concentration itself (linear carbon monoxide). The behaviour is obviously quite different from that which would be expected if alloy particles were not formed.

Electron Spectroscopy
Chemical shifts in the ESCA carbon ($1s$) line of carbon monoxide adsorbed on evaporated films of tungsten and molybdenum have been measured[60] and further work with other metals will very probably show a carbon chemical shift dependent on the nature of the metal.

Differential Hydrogen Analysis
This technique, described by Hall and Lutinskii[90] involves making use of the difference in reactivity of hydrogen for exchange with deuterium, depending on the nature of the surface to which it is attached. Although only so far applied to differentiate hydrogen on the metal from that on the oxide of supported platinum, it may prove possible to find differences in reactivity of different metal surfaces if the reaction temperature is made low enough. The method has recently been used to distinguish between different hydrogen adsorption modes on a dispersed platinum specimen (platinum black) which had been identified by temperature programmed desorption.[140]

Temperature Programmed Desorption
The temperature required to desorb a gas from a metal surface depends upon its binding energy to the surface. With pure metal specimens the resolution of peaks in the desorption spectrum has often been interpreted in terms of different types of adsorption sites on the surface. A recent summary is given by Hayward.[127] Temperature programmed desorption of hydrogen has recently been studied from a dispersed platinum catalyst (platinum black),[138] and the technique for use with this sort of sample, desorption into a stream of carrier gas, has been reviewed.[139] It may prove possible to observe modifications in desorption spectra with gases such as carbon monoxide, hydrogen or nitrogen which could give information about the surface composition of alloy catalysts. Although most often

practised with unsupported metal specimens, work with supported metal is possible in favourable circumstances (e.g. ref. 125) and combination with infrared absorption may then be very helpful.

Change in Work-function

The change in work-function consequent upon gas adsorption can give information related to surface composition if the change is known to be considerably different for the two pure components of a bimetallic system. It can be used most reliably for obtaining insight into component distribution. Sachtler and co-workers[95, 96] have used a photoelectric method with carbon monoxide adsorption on a variety of metal films, while Whalley et al.[76] have used a diode method with carbon monoxide adsorption on palladium–silver films.

References

1. Chung, M. F. and Jenkins, L. H. *Surface Sci.* **22**, 479 (1970).
2. Coghlan, W. A. and Clausing, R. E. *Atomic Data* **5**, 317 (1973).
3. Perdereau, M. *Surface Sci.* **24**, 239 (1971).
4. Palmberg, P. W. and Rhodin, T. N. *J. Appl. Phys.* **39**, 2425 (1968).
5. Christmann, K. and Ertl, G. *Surface Sci.* **33**, 254 (1972).
6. Ertl, G. and Kuppers, J. *J. Vac. Sci. Tech.* **9**, 829 (1972).
7. Harris, L. A. *Surface Sci.* **15**, 17 (1969).
8. Siegbahn, K., Nordling, C., Fahlman, A., Nordberg, R., Hamrin, K., Hedman, J., Johansson, G. and Bergmark, T., Karlsson, S.-E., Lindgren, I. and Lindberg, B. "ESCA–Atomic Molecular and Solid State Structure Studied by Means of Electron Spectroscopy" *Nov. Act. Reg. Sci.Upsaliensis* (*IV*), **20**, Almqvist and Wiksells, Uppsala (1967).
9. Turner, D. W. *In* "Physical Methods in Advanced Inorganic Chemistry" (H. A. O. Hill and P. Day, eds), Interscience, London (1968), p. 74.
10. Delgass, W. N., Hughes, T. R. and Fadley, C. S. *Catal. Rev.* **4**, 179 (1971).
11. Weber, R. E. and Peria, W. T. *J. Appl. Phys.* **38**, 4355 (1967).
12. Palmberg, P. W. and Rhodin, T. N. *J. Appl. Phys.* **39**, 2425 (1968).
13. Morrison, J. and Lander, J. J. *Bull. Am. Phys. Soc. II*, **13**, 945 (1968).
14. Wei, P. S. P., Cho, A. Y. and Caldwell, C. W. *Rev. Sci. Inst.* **40**, 1075 (1969).
15. Taylor, N. J. *Rev. Sci. Inst.* **40**, 792 (1969).
16. Kuyatt, C. E. and Simpson, J. A. *Rev. Sci. Inst.* **38**, 103 (1967).
17. Hafner, H., Simpson, J. A. and Kuyatt, C. E. *Rev. Sci. Inst.* **39**, 33 (1968).
18. Simpson, J. A. and Kuyatt, C. E. *J. Appl. Phys.* **37**, 3805 (1966).
19. Simpson, J. A. *Rev. Sci. Inst.* **35**, 1698 (1964).
20. Klemperer, O. *Reports Prog. Phys.* **28**, 77 (1965).
21. Harrower, G. A. *Rev. Sci. Inst.* **26**, 850 (1955).

22. Yarnold, G. D. and Bolton, H. C. *J. Sci. Inst.* **26**, 38 (1949).
23. Hughes, A. L. and Rojansky, V. *Phys. Rev.* **34**, 284 (1929).
24. Theodoridis, G. C. and Paolini, F. R. *Rev. Sci. Inst.* **39**, 326 (1968).
25. Hughes, A. L., McMillen, J. H. *Phys. Rev.* **34**, 291 (1929).
26. Rogers, F. T. and Horton, C. W. *Rev. Sci. Inst.* **14**, 216 (1943).
27. Schultz, G. J. *Phys. Rev.* **125**, 229 (1962).
28. Marmet, P. and Kerwin, L. *Can. J. Phys.* **38**, 787 (1960).
29. Clarke, E. M. *Can. J. Phys.* **32**, 764 (1954).
30. Marmet, P., Morrison, J. D. and Swingler, D. L. *Rev. Sci. Inst.* **33**, 239 (1962).
31. Reynolds, R. and Scherb, F. *Rev. Sci. Inst.* **38**, 348 (1967).
32. Paolini, F. R. and Theodoridis, G. C. *Rev. Sci. Inst.* **38**, 579 (1967).
33. Rogers, F. T. *Rev. Sci. Inst.* **22**, 723 (1951).
34. Purcell, E. M. *Phys. Rev.* **54**, 818 (1938).
35. Sar-el, N. Z. *Rev. Sci. Inst.* **38**, 1210 (1967).
36. Simpson, J. A. *Rev. Sci. Inst.* **32**, 1283 (1961).
37. Coulton, M. *RCA Rev.* **26**, 217 (1955).
38. Schultz, G. J. *J. Appl. Phys.* **31**, 1134 (1960).
39. Fox, R. E., Hickam, W. M., Grove, D. J. and Kjeldaar, T. *Rev. Sci. Inst.* **26**, 1101 (1955).
40. Hickam, W. M. and Fox, R. E. *J. Chem. Phys.* **25**, 642 (1956).
41. Lee, J. D. *Rev. Sci. Inst.* **44**, 893 (1973).
42. Baldwin, G. C. *Rev. Sci. Inst.* **38**, 519 (1967).
43. Huchital, D. A. and Rigden, J. D. *Appl. Phys. Letters.* **16**, 348 (1970).
44. Staib, P. *J. Phys.* (C), **5**, 484 (1972).
45. Palmberg, P. W., Bohn, G. K. and Tracy, J. D. *Appl. Phys. Letters* **15**, 254 (1969).
46. Sewell, P. B., Mitchell, D. F. and Cohen, M. *Developments in Appl. Spect.* **7A**, 61 (1969).
47. Helmer, J. C. and Weichert, N. H. *Appl. Phys. Letters* **13**, 266 (1968).
48. Fadley, C. S., Miner, C. E. and Hollander, J. M. *Appl. Phys. Letters* **15**, 223 (1969).
49. Gerlach, R. L., Houston, J. E. and Park, R. L. *Appl. Phys. Letters* **16**, 179 (1970).
50. Socha, A. J. *Surface Sci.* **25**, 147 (1971).
51. Kane, P. F. and Larrabee, G. B. *Ann. Rev. Mat. Sci.* **2**, 33 (1972).
52. Liebl, H. J. and Herzog, R. F. K. *J. Appl. Phys.* **34**, 2893 (1963).
53. Barrington, A. E., Herzog, R. F. K. and Poschenrieder, W. P. *In* "Progress in Nuclear Energy" Series IX, Anal. Chem., Vol. 7, Pergamon, Oxford (1966), p. 243.
54. Anderson, C. A. *Internat. J. Mass. Spect. Ion. Phys.* **2**, 61 (1969).
55. Liebl, H. *J. Appl. Phys.* **38**, 5277 (1967).
56. Castaing, R. and Slodzian, G. *J. Microscopie* **1**, 395 (1962).
57. Slodzian, G. *Ann. Phys.* **9**, 591 (1964).
58. Metherell, A. J. F. and Cook, R. F. *Optik* **34**, 535 (1972).

59. Brundle, C. R. and Roberts, M. W. *Proc. Roy. Soc.* **A331**, 383 (1972).
60. Atkinson, S. J., Brundle, C. R. and Roberts, M. W. *J. Electron Spect. Related Phenom.* **2**, 105 (1973).
61. Park, R. L., Houston, J. E. and Schreiner, D. G. *Rev. Sci. Inst.* **41**, 1810 (1970).
62. Tracy, J. C. *Appl. Phys. Letters* **19**, 353 (1971).
63. Houston, J. E. and Park, R. L. *J. Chem. Phys.* **55**, 4601 (1971).
64. Houston, J. E. and Park, R. L. *J. Vac. Sci. Tech.* **9**, 579 (1972).
65. Houston, J. E. and Park, R. L. *J. Vac. Sci. Tech.* **8**, 91 (1971).
66. Houston, J. E. and Park, R. L. *Solid State Comm.* **10**, 91 (1972).
67. Tracy, J. C. *J. Appl. Phys.* **43**, 4164 (1972).
68. Smith, D. P. *Surface Sci.* **25**, 171 (1971).
69. Goff, R. F. and Smith, D. P. *J. Vac. Sci. Tech.* **7**, 72 (1970).
70. Strehlow, W. H. and Smith, D. P. *Appl. Phys. Letters* **13**, 34 (1968).
71. van der Plank, P. and Sachtler, W. M. H. *J. Catal.* **12**, 35 (1968).
72. Ponec, V. and Sachtler, W. M. H. *J. Catal.* **24**, 250 (1972).
73. Sinfelt, J. H., Carter, J. L. and Yates, D. J. C. *J. Catal.* **24**, 283 (1972).
74. Sinfelt, J. H. *J. Catal.* **29**, 308 (1973).
75. Anderson, J. H., Conn, P. J. and Brandenberger, S. G. *J. Catal.* **16**, 326 (1970).
76. Whalley, L., Thomas, D. H. and Moss, R. L. *J. Catal.* **22**, 302 (1971).
77. Bouwman, R. and Sachtler, W. M. H. *J. Catal.* **26**, 63 (1972).
78. Flynn, P. C., Wanke, S. E. and Turner, P. S. *J. Catal.* **33**, 233 (1974).
79. Eischens, R. P., Francis, S. A. and Pliskin, W. A. *J. Phys. Chem.* **60**, 194 (1956).
80. Blyholder, G. and Neff, L. O. *J. Phys. Chem.* **66**, 1464 (1962).
81. Smith, A. W. and Quets, J. M. *J. Catal.* **4**, 163, 172 (1965).
82. Burhop, E. H. S. "The Auger Effect and Other Radiationless Transitions", Cambridge University Press (1952).
83. Anderson, C. A. and Hinthorne, J. R. *Science* **175**, 853 (1972).
84. Anderson, C. A. and Hinthorne, J. R. *Anal. Chem.* **45**, 1421 (1973).
85. Bartholomew, C. H. and Boudart, M. *J. Catal.* **29**, 278 (1973).
86. Little, L. H. "Infrared Spectra of Adsorbed Species", Academic Press, London (1966).
87. Little, L. H. *In* "Chemisorption and Reactions on Metallic Films" (J. R. Anderson, ed.), Vol. 1, Academic Press, London (1971), p. 490.
88. Pritchard, J. *Specialist Periodical Reports Chem. Soc.* **1**, 222, (1972).
89. Guerra, C. R. and Schulman, J. H. *Surface Sci.* **7**, 229 (1967).
90. Hall, W. K. and Lutinskii, F. E. *J. Catal.* **2**, 518 (1963).
91. Fowler, R. H. *Phys. Rev.* **38**, 45 (1931).
92. Rivière, J. C. *Solid State Surface Science*, **1**, 179 (1969).
93. Herring, C. and Nichols, M. H. *Rev. Mod. Phys.* **21**, 185 (1949).
94. Shelton, H. *Phys. Rev.* **107**, 1553 (1957).
95. Sachtler, W. M. H. and Dorgelo, G. J. H. *J. Catal.* **4**, 654 (1965).
96. Bouwman, R. Ph.D. Thesis, University of Leiden, 1970.

97. Maire, G., Anderson, J. R. and Johnson, B. B. *Proc. Roy. Soc.* **A320**, 227 (1970).
98. Frauenfelder, H. "The Mössbauer Effect", Benjamin, New York (1963).
99. Wertheim, G. K. "Mössbauer Effect: Principles and Applications", Academic Press, New York (1964).
100. Delgass, W. N. and Boudart, M. *Catal. Rev.* **2**, 129 (1969).
101. Hobson, M. C. *Advances in Coll. and Interface Sci.* **3**, 1 (1971).
102. "Mössbauer Effect Methodology" (I. J. Gruverman, ed.), Plenum Press, New York, Vols. 1–7 (1965–1971).
103. Crewe, A. V., Wall, J. and Welter, L. M. *J. Appl. Phys.* **39**, 5861 (1968).
104. MacDonald, N. C. *Appl. Phys. Letters* **16**, 76 (1970).
105. Brammer, I. S. and Dewey, M. A. P. "Specimen Preparation for Electron Microscopy", Blackwell, Oxford (1966).
106. Kay, D. "Techniques for Electron Microscopy", Blackwell, Oxford (1961).
107. Anderson, J. R., Baker, B. G. and Sanders, J. V. *J. Catal.* **1**, 443 (1962).
108. Moss, R. L. *Platinum Metals Rev.* **11**, 141 (1967).
109. Joyner, R. W. and Somorjai, G. A. *Specialist Periodical Reports Chem. Soc.* **2**, 1 (1973).
110. Duke, C. B. In "LEED: Surface Structure of Solids" (M. Laznicka, ed.), Prague Vol. 2, (1972), p. 125.
111. Pendry, J. B. In "LEED: Surface Structure of Solids" (M. Laznicka, ed.), Prague, Vol. 2 (1972), p. 305.
112. Duke, C. B. and Tucker, C. W. *Surface Sci.* **24**, 31 (1971).
113. Pendry, J. B. *J. Phys.* (*C*), **2**, 2273, 2283 (1969).
114. Strozier, J. A. and Jones, R. O. *Phys. Rev.* (*B*) **3**, 3228 (1971).
115. Beeby, J. L. *J. Phys.* (*C*) **1**, 82 (1968).
116. Duke, C. B., Anderson, J. R. and Tucker, C. W. *Surface Sci.* **19**, 117 (1970).
117. Jepsen, D. W., Marcus, P. M. and Jona, F. *Phys. Rev. Letters* **26**, 1365 (1971); *Phys. Rev.* (*B*) **5**, 3933 (1972).
118. Tong, S. Y. and Rhodin, T. N. *Phys. Rev. Letters* **26**, 711 (1971).
119. Ellis, W. P. and Schwoebel, R. L. *Surface Sci.* **11**, 82 (1968).
120. Henzler, M. *Surface Sci.* **19**, 159 (1970); **22**, 12 (1970).
121. Perdereau, J. and Rhead, G. E. *Surface Sci.* **24**, 555 (1971).
122. Houston, J. E. and Park, R. L. *Surface Sci.* **26**, 269 (1971).
123. Bauer, E. In "Techniques of Metals Research, Vol. 2, Part 2" (R. F. Bunshah, ed.), Interscience, New York (1969), p. 502.
124. van Hardeveld, R. and van Montfoort, A. *Surface Sci.* **4**, 396 (1966).
125. Cormack, D. and Moss, R. L. *J. Catal.* **13**, 1 (1969).
126. Anderson, J. R. *Advances in Catalysis* **23**, 1 (1973).
127. Hayward, D. O. In "Chemisorption and Reactions on Metallic Films" (J. R. Anderson, ed.), Vol. 1, Academic Press, London (1971), p. 225.
128. Tsong, I. S. T. *Phys. Stat. Sol.* (a) **7**, 451 (1971).
129. White, C. W. and Tolk, N. H. *Phys. Rev. Letters* **26**, 486 (1971).
130. White, C. W., Sims, D. L. and Tolk, N. H. *Science* **177**, 481 (1972).
131. Tsong, I. S. T. and McLaren, A. C. *Nature* **248**, 43 (1974).

132. Kundig, W., Bommel, H., Constabaris, G. and Lundquist, R. H. *Phys. Rev.* **142**, 327 (1966).
133. Weber, R. E. and Johnson, A. L. *J. Appl. Phys.* **40**, 314 (1969).
134. Palmberg, P. W. *Anal. Chem.* **45**, 549A (1973).
135. Bishop, H. E. *Brit. J. Appl. Phys.* **18**, 703 (1967).
136. Vrakking, J. J. and Meyer, F. *Surface Sci.* **35**, 34 (1973).
137. Glupe, G. and Mehlhorn, W. *Phys. Letters* **25A**, 274 (1967). *J. Phys.* (Paris), Colloq. **4**, 40 (1971).
138. Tsuchiya, S., Amenomiya, Y. and Cvetanovic, R. J. *J. Catal.* **19**, 245 (1970).
139. Cvetanovic, R. J. and Amenomiya, Y. *Advances in Catalysis* **17**, 103 (1967).
140. Tsuchiya, S., Amenomiya, Y. and Cvetanovic, R. T. *J. Catal.* **20**, 1 (1971).
141. "Microprobe Analysis" (C. A. Anderson, ed.), Wiley, New York (1973).
142. Phillips, V. A. and Lifshin, E. *Ann. Rev. Mat. Sci.* **1**, 1 (1971).
143. Gallon, T. E. *Surface Sci.* **17**, 486 (1969).
144. Jackson, D. C., Gallon, T. E. and Chambers, A. *Surface Sci.* **36**, 381 (1973).
145. Wagner, C. D. *Anal. Chem.* **44**, 1050 (1972).
146. Bouwman, R. and Biloen, P. *Surface Sci.* **41**, 384 (1974).
147. Bouwman, R., Toneman, L. H. and Holscher, A. A. *Surface Sci.* **35**, 8 (1973).
148. Bouwman, R., Toneman, L. H. and Holscher, A. A. *Vacuum* **23**, 163 (1973).
149. Soma-Noto, Y., and Sachtler, W. M. H. *J. Catal.* **32**, 315 (1974).
150. Crewe, A. V. and Wall, J. *Optik* **5**, 461 (1970).
151. Crewe, A. V. and Lim, P. S. D. *In* "Eighth International Congress on Electron Microscopy" (J. V. Sanders and D. J. Goodchild, eds.). The Australian Academy of Science, Canberra, Vol. 1, 1974 p. 38.
152. Crewe, A. V. *Progress in Optics* **11**, 223 (1973).
153. Crewe, A. V., Wall, J. and Langmore, J. *Science* **168**, 1338 (1970).
154. Hashimoto, H., Kumao, A., Hino, K., Yotsumoto, H. and Ono, A. *Japan J. Appl. Phys.* **10**, 1115 (1971).
155. Hashimoto, H., Kumao, A., Hino, K., Endoh, H., Yotsumoto, H. and Ono, A. *J. Electron Microscopy* **22**, 123 (1973).
156. Hashimoto, H., Kumao, A., Endoh, H. and Ono, A. *In* "Eighth International Congress on Electron Microscopy" (J. V. Sanders and D. J. Goodchild, eds.). The Australian Academy of Science, Canberra, Vol. 1, 1974, p. 244.

APPENDIX 1

TABLE A.1. Data for the more common metals

Metal	Atomic weight †	Crystal structure ‡	Lattice parameters ‡ (nm)	Nearest neighbour distance ‡ (nm)*	Melting point (K)	Boiling point (K)	Density (293 K) † (10^3 kg m^{-3})	Heat of atomization (298 K) (kJ mol^{-1})
aluminium	26.98	f.c.c.	0.404	0.286	1033	2740	2.70	314
barium	137.34	b.c.c.	0.501	0.434	998	1413	3.51	176
beryllium	9.01	h.c.p.	0.228; 0.357	0.225	1551	3243	1.85	321
cadmium	112.40	h.c.p.	0.297; 0.561	0.297; 0.329	594	1038	8.65	113
calcium	40.08	f.c.c.	0.557	0.393	1118	1760	1.55	193
cerium	140.12	h.c.p.	0.365; 0.596	0.363	1068	3741	6.78	356
		f.c.c.	0.516					
cesium	132.91	b.c.c.	0.613	0.524	302	963	1.87	79
chromium	52.00	b.c.c.	0.289	0.249	2163	2755	7.19	338
cobalt	48.93	f.c.c.	0.355	0.251	1768	3173	8.90	440
copper	63.54	f.c.c.	0.361	0.255	1356	2868	8.96	343
gold	196.97	f.c.c.	0.407	0.288	1336	3239	19.32	345
hafnium	178.49	h.c.p.	0.320; 0.508	0.316	2420	5670	13.29	(680)
iridium	192.2	f.c.c.	0.383	0.271	2683	4800	22.42	692
iron	55.85	b.c.c.	0.286	0.248	1808	3273	7.87	405
lanthanum	138.91	h.c.p.	0.372; 0.606	0.371	1193	3742	6.19	369
lead	207.19	f.c.c.	0.495	0.349	601	2017	11.35	155
lithium	6.94	b.c.c.	0.350	0.303	452	1590	0.534	156
magnesium	24.31	h.c.p.	0.321; 0.521	0.320	924	1380	1.74	151
manganese	54.94	complex			1517	2370	7.20	286

APPENDIX 1   447

molybdenum	95.94	b.c.c.		0.314	2883	5833	10.22	655
nickel	58.71	f.c.c.		0.352	1726	3005	8.90	426
niobium	92.91	b.c.c.		0.329	2741	5200	8.57	775
osmium	190.2	h.c.p.	0.273; 0.431	0.270	3273	5273	22.57	730
palladium	106.4	f.c.c.		0.388	1825	3200	12.02	390
platinum	195.09	f.c.c.		0.392	2042	4100	21.45	511
potassium	39.10	b.c.c.		0.531	337	1047	0.862	90
rhenium	186.2	h.c.p.	0.276; 0.445	0.274	3453	5900	21.02	792
rhodium	102.91	f.c.c.		0.380	2239	4000	12.41	578
rubidium	85.47	b.c.c.		0.570	312	961	1.53	186
ruthenium	101.07	h.c.p.	0.270; 0.427	0.267	2523	4173	12.41	670
silver	107.87	f.c.c.		0.408	1234	2485	10.50	290
sodium	22.99	b.c.c.		0.428	371	1165	0.971	109
strontium	87.62	f.c.c.		0.605	1042	1657	2.54	164
tantalum	180.95	b.c.c.		0.330	3269	5698	16.60	775
thorium	232.04	f.c.c.		0.508	2053	4273	11.66	(560)
tin	118.69	tetr. double b.c.	0.582; 0.317	0.316	505	2539	7.28	302
titanium	47.96	h.c.p.	0.295; 0.468	0.293	1948	3533	4.54	470
tungsten	183.85	b.c.c.	0.316	0.274	3683	6200	19.30	846
uranium	238.03	complex			1405	4091	18.95	525
vanadium	50.94	b.c.c.		0.302	2163	3273	6.12	504
zinc	65.37	h.c.p.	0.266; 0.494	0.266; 0.291	692	1180	7.12	131
zirconium	91.22	h.c.p.	0.322; 0.512	0.319	2125	3851	6.53	524

* For h.c.p. metals, if the axial ratio is within about 3% of the ideal value, an average figure is quoted for the nearest neighbour distance.
† Based on Carbon-12.
‡ At about room temperature.

TABLE A.2 Work-functions and first atomic ionization potentials of some metals

	Work-function (kJ mol^{-1})		First ionization potential (kJ mol^{-1})	
aluminium	400	(4·15)*	577	(5·98)*
barium	239	(2·48)	502	(5·20)
beryllium	377	(3·91)	890	(9·22)
cadmium	398	(4·13)	866	(8·98)
calcium	261	(2·71)	586	(6·07)
cerium	425	(4·40)	632	(6·55)
cesium	175	(1·81)	372	(3·86)
chromium	440	(4·56)	653	(6·77)
cobalt	450	(4·97)	757	(7·85)
copper	439	(4·55)	745	(7·72)
gold	513	(5·32)	887	(9·19)
hafnium	349	(3·62)		
iridium	480	(5·0)		
iron	444	(4·60)	757	(7·85)
lanthanum	318	(3·30)	540	(5·60)
lead	383	(3·97)	711	(7·37)
lithium	232	(2·40)	519	(5·38)
magnesium	351	(3·64)	736	(7·63)
manganese	386	(4·00)	715	(7·41)
molybdenum	406	(4·21)	703	(7·29)
nickel	488	(5·06)	736	(7·63)
niobium	422	(4·37)	653	(6·77)
osmium	439	(4·55)	841	(8·72)
palladium	480	(4·98)	803	(8·32)
platinum	523	(5·42)	858	(8·89)
potassium	215	(2·23)	418	(4·33)
rhenium	475	(4·92)		
rhodium	459	(4·76)	741	(7·68)
rubidium	205	(2·12)	397	(4·11)
ruthenium	450	(4·66)	732	(7·59)
silver	433	(4·49)	732	(7·59)
sodium	220	(2·28)	494	(5·12)
strontium	264	(2·74)	548	(5·68)
tantalum	406	(4·21)		
thorium	334	(3·46)		
tin	425	(4·40)	707	(7·33)
titanium	391	(4·05)	657	(6·81)
tungsten	439	(4·55)	782	(8·10)
uranium	331	(3·43)		
vanadium	413	(4·28)	649	(6·73)
zinc	407	(4·22)	908	(9·41)
zirconium	396	(4·10)	669	(6·93)

* Values in parentheses are in eV.

TABLE A.3 Surface energy of some metals at or near the melting point

Metal	Surface energy, γ (J m^{-2})	$d\gamma/dT$ (mJ m^{-2}K^{-1})
aluminium	0·91	−0·135
barium	0·224	−0·095
beryllium	1·10	
cadmium	0·56	
calcium	0·337	−0·068
chromium	1·70	
cobalt	1·89	
copper	1·35	−0·18
gold	1·13	−0·10
hafnium	1·65	
iridium	2·25	
iron	1·85	
lead	0·48	−0·26
lithium	0·40	−0·30
magnesium	0·57	
manganese	1·10	0
molybdenum	2·25	
nickel	1·80	
niobium	1·90	
osmium	2·50	
palladium	1·50	
platinum	1·80	
potassium	0·101	−0·06
rhodium	2·00	
rubidium	0·078	
ruthenium	2·25	
silver	0·93	−0·13
sodium	0·200	−0·05
strontium	0·290	−0·106
tin	0·575	−0·075
titanium	1·60	
tungsten	2·30	
vanadium	1·75	
zinc	0·81	−0·25
zirconium	1·40	

After Wilson, J. R. *Metall. Rev.* **10**, 381 (1964).

TABLE A.4 Magnetic susceptibility and electrical resistivity of some metals

Metal	MKS magnetic susceptibility (about room temperature) $10^{-7} m^3 (kg\ mol)^{-1}$*	Electrical resistivity (298 K) ($n\Omega\ m$)
aluminium	+ 2.07	26.77
barium	+ 2.59	
beryllium	− 1.13	41.3
cadmium	− 2.49	69.4
calcium	+ 5.03	40.2
cerium	+305.5	750
cesium	+ 3.65	200
chromium	+ 22.63	129.8
cobalt	ferromagnetic	62.7
copper	− 0.686	16.76
gold	− 3.52	23.7
hafnium	+ 9.43	351
iridium	+ 3.22	53.2
iron	ferromagnetic	97.4
lanthanum	+ 14.83	57.0
lead	− 2.89	206.6
lithium	+ 1.78	86
magnesium	+ 1.65	45.3
manganese	~+ 66.6	1850
molybdenum	+ 11.19	52
nickel	ferromagnetic	68.7
niobium	+ 24.51	125
osmium	+ 1.24	95.2
palladium	+ 71.3	108.2
platinum	+ 25.38	106.2
potassium	+ 2.62	61.5
rhenium	+ 8.50	193.2
rhodium	+ 13.96	45.3
rubidium	+ 2.14	125
ruthenium	+ 5.43	76
silver	− 2.45	16.1
sodium	+ 2.01	42
strontium	+ 11.56	230
tantalum	+ 19.36	124.5
thorium	+ 16.59	130.8
tin	+ 0.390	110.9
titanium	+ 19.23	420
tungsten	+ 7.42	56.5
uranium	+ 50.3	300
vanadium	+ 32.1	250
zinc	− 1.43	59.4
zirconium	+ 15.34	400.2

* Assuming $\mathbf{B} = \mu_0(\mathbf{H} + \mathbf{M})$: to recover values in cgs emu ($cm^3\ mol^{-1}$), divide values by $4\pi \times 10^{-3}$.

APPENDIX 2

Illustrative Recipes for the Preparation of Metallic Catalysts

1. Platinum–metals powder

(After Adams, R. and Shriner, R. L. *J. Amer. Chem. Soc.* **45**, 2171 (1923) see also McKee, D. W. *J. Catal.* **8**, 240 (1967))

A solution of 4·2 g of chloroplatinic acid in 10 ml of water is mixed with 40 g of C.P. sodium nitrate and evaporated to dryness in a casserole or beaker. The mass is then heated with an ordinary Bunsen or Meker burner until fusion takes place. The mixture and melt are stirred continuously with a thermocouple encased in a Pyrex glass tube and the temperature read on a pyrometer. After the fusion is complete the melt is allowed to cool and treated with water until the filtrates are free from nitrates and nitrites. The oxide is then dried in a desiccator and weighed in order to determine the yield.

The fusion temperature which is used may vary from the lowest possible, about 580 K to about 950 K, but the catalytic activity of the product decreases with increasing fusion temperature. For most purposes a fusion temperature of about 700 K is convenient and satisfactory. The material initially produced may contain small lumps or aggregates, and the particle size may be thus reduced by grinding.

The platinum oxide product from this preparative procedure is often used directly as a hydrogenation catalyst when *in situ* reduction occurs. Reduction to platinum metal powder may be carried out as a separate step by flowing pure hydrogen at atmospheric pressure through a tube packed with the oxide powder. Because the reduction is highly exothermic, the reduction should be initially carried out slowly: this means slowly increasing the reduction temperature from room temperature to the final temperature over a period of some hours (e.g. 8 h). Additionally, dilution of the hydrogen with an inert gas such as argon or helium is helpful in controlling the reduction in the initial stages. Final reduction is commonly effected at temperatures in the region 470–570 K. Some platinum particle sintering occurs during reduction. Higher reduction temperatures may be used at the expense of more extensive sintering.

The method has been used with other platinum metal compounds, e.g. rhodium trichloride (hydrate), ruthenium trichloride (hydrate), chloroiridic acid (hydrate).

2. Platinum–metals powder

(After Carter, J. L., Cusumano, J. A. and Sinfelt, J. H. *J. Catal.* **20**, 223 (1971): McKee, D. W. and Norton, F. J. *J. Phys. Chem.* **68**, 481 (1964))

Aqueous solutions of platinum–metal compounds may be reduced by treatment with an aqueous solution of sodium borohydride or hydrazine, with the precipitation of finely divided metal. The following have been successfully used: ammonium tetrachloropalladate; rhodium trichloride (hydrate); ruthenium trichloride (hydrate); chloroplatinic acid hexahydrate.

The reduction is carried out by adding a 5 wt.% solution of sodium borohydride dropwise to the stirred solution of the platinum–metal compound. Alternatively, an 85 wt.% solution of hydrazine hydrate may be used, but in this case the reaction is more vigorous and coagulation of the precipitated metal does not occur so readily. The precipated metal is washed with water thoroughly until the washings are free of impurity and dried at 370 K. The metal particles are heavily oxide covered: the reduction comments given under preparation 1 apply.

The method may be used for the preparation of platinum–ruthenium alloy powder, starting with a mixed solution.

N.B. See warnings on p. 197 and p. 219.

3. Nickel–copper powder

(After Best, R. J. and Russell, W. W. *J. Amer. Chem. Soc.* **76**, 838 (1954))

The catalysts are prepared by dissolving the calculated amounts of analytical reagent grade $Cu(NO_3)_2 \cdot 3H_2O$ and $Ni(NO_3)_2 \cdot 6H_2O$ in water then diluting to correspond to about 5g of NiO and CuO per 100 ml. To this solution, while rapidly stirred at room temperature, powdered reagent grade ammonium bicarbonate is added until a permanent turbidity just forms and then there is added fairly rapidly 2.2 mol of the bicarbonate per mol of metal ion.

After stirring for 10 min the precipitate is settled overnight, and then washed by decantation with just enough hot water for the washings to be colourless and the nitrate ion largely removed. Having decanted the last washing as cleanly as possible, the precipitate is evaporated over a steam-bath to a pasty solid, then dried in an oven at 380 K for 24 h. The dried precipitate is broken into small lumps and sintered for exactly 4 h at a carefully controlled temperature of 670 K. After crushing in an agate mortar the product is screened to 4–5 mm diameter.

The catalysts are reduced in a stream of purified dry hydrogen at a flow rate of about 22 ml min^{-1}, while the temperature is slowly raised over a period of some 8 h to 770 K. At 770 K the hydrogen flow rate is doubled, and reduction is continued at this temperature until less than 0·1 mg of water is collected in an Anhydrone tube in 15 min.

The basic procedure may be equally well applied to the production of nickel or copper powder.

The impurity level may be improved by working in Monel apparatus and by using the highest purity water (Hall, W. K. *J. Phys. Chem.* **62**, 816 (1958)).

4. Skeletal nickel (Raney nickel)

(After Adkins, H. and Billica, H. R. *J. Amer. Chem. Soc.* **70**, 695 (1948). See also Billica, H. R. and Adkins, H. *In* "Organic Syntheses Collective Vol. 3" (E. C. Horning, ed.), Wiley, New York (1955), p. 176.
Freel, J., Pieters, W. J. M. and Anderson, R. B., *J. Catal.* **14**, 247 (1969))

In a 2l Erlenmeyer flask equipped with a thermometer and a stainless steel stirrer, is placed 160 g of sodium hydroxide in 600 ml of distilled water. The rapidly stirred solution is allowed to cool to 323 K in an ice-bath equipped with an overflow siphon. Then 125 g of Raney nickel–aluminium alloy is added in small portions over a period of 25–30 min. The temperature is maintained at 323 ± 2 K by controlling the rate of addition of alloy to the sodium hydroxide and of ice to the cooling bath. When all the alloy has been added, the suspension is digested at 323 K for 50 min with gentle stirring. It is usually necessary to remove the ice-bath and replace it with a hot water-bath. The catalyst, after digestion, is washed three times by decantation and then transferred immediately to the washing tube for further washing.

A glass tube, approximately 5 cm in diameter and 38 cm in length, with a side arm 6 cm from the top, is used as the container in washing the catalyst. The tube is equipped with a snugly fitting rubber stopper which is held in place by a suitable device. The stopper carries a gas-tight bushing through which the 6 mm shaft of a stainless steel stirrer projects to the bottom of the washing tube. A 5 l reservoir for distilled water is so placed that water will flow from it through a stopcock down a glass tube, 8 mm in diameter, which passes through the stopper and down the side of the washing tube to its bottom. The side arm of the test-tube is connected by rubber tubing to a 5 l overflow bottle from which the water may be allowed to flow through a stopcock to the drain. A connection from a source of distilled water is made to the reservoir. All connections of rubber and glass tubes should be so fastened that they will withstand the pressures used.

The catalyst sludge is immediately transferred to the washing tube after the third decantation. The last portions are rinsed from the flask into the tube with distilled water and the tube and reservoir nearly filled with distilled water. The apparatus is then rapidly assembled, and hydrogen introduced so that the water in the reservoir, washing tube and overflow bottle is under a pressure about 0·5 atm above that of the outside atmosphere. The stirrer is operated at such a speed that the catalyst is suspended to a height of 18–20 cm. Distilled water from the reservoir is allowed to flow through the suspended catalyst at a rate of about 250 ml min^{-1}. When the reservoir is nearly empty and the overflow bottle full, the drain cock and distilled water inlet are simultaneously opened to an equal rate of flow such that, as the overflow bottle empties, the reservoir is filled, while the pressure in the system remains constant.

After about 15 l of water has passed through the catalyst, the stirrer and the water are stopped, the pressure released and the apparatus disassembled. The water is decanted from the settled sludge, which is then transferred to a 250 ml centrifuge bottle with 95% ethanol. The catalyst is washed three times by stirring, not shaking, with 150 ml of 95% ethanol, centrifuging after each addition. In the same manner, the catalyst is washed three times with absolute ethanol; 1–2 min centrifugation at 1500–200 r.p.m. is sufficient to separate the catalyst. All operations should be carried out as rapidly as possible if a catalyst of the maximum activity is desired. The catalyst should be stored in a refrigerator in a closed bottle filled with absolute ethanol. The total elapsed time from beginning of the addition to the alloy to the completion of the preparation, with the catalyst in the refrigerator, should be not more than about three hours.

The volume of the settled catalyst in ethanol is about 75–80 ml containing about 62 g of nickel and 7–8 g of aluminium.

See also: (Raney iron) Thomson, A. F. and Wyatt, S. B. *J. Amer. Chem. Soc.* **62**, 2555 (1940); (Raney copper) Stanfield, J. A. and Robbins, P. E. *In* "Actes du Deuxieme Congres de Catalyse" Editions Technip, Paris (1961), p. 2579.

5. Stabilized Porous Iron from Magnetite Fusion

(After Larson, A. T. and Richardson, C. N. *Ind. Eng. Chem.* **17**, 971 (1925): Ciapetta, F. G. and Plank, C. J. *In* "Catalysis" (P. H. Emmett, ed.), Reinhold, New York, Vol. 1 (1954), p. 315.)

A mixture is prepared from the following dry powdered materials: 187 g Fe_3O_4, 9·2 g MgO, 1·3 g Cr_2O_3, 0·21 g $KMnO_4$, 1·65 g K_2CO_3, 1·2 g SiO_2. Although fusion may be effected in an induction furnace (1820–1870 K) using a crucible of a relatively inert refractory such as alundum, arc-melting is preferable since some attack on the crucible always occurs with a consequent alteration in the composition of the melt.

With an arc-melting method (cf. Larson, A. T. and Richardson, C. N. reference above) the melt is in contact with a protecting bed of unmelted material. Water-cooled iron electrodes are used which contribute negligibly to the catalyst composition. The melt is cast by pouring into an iron mould, ground and screened to size. Reduction may be effected in flowing hydrogen at, typically, 700–800 K for several hours.

This fusion method may, of course, be used with other formulations involving different amounts and types of stabilizers and chemical promoters.

6. Cobalt Fischer-Tropsch Catalyst

(After Ciapetta, F. G. and Plank, C. J. *In* "Catalysis" (P. H. Emmett, ed.), Reinhold, New York, Vol. 1 (1954, p. 315: Storch, H. H., Golumbic, N. and Anderson, R. B. "Fischer-Tropsch and Related Syntheses", Wiley, New York (1951).

100 g of kieselguhr and 6 g of magnesia are dispersed in 500 ml of water and heated to boiling (A); 246 g of $Co(NO_3)_2·6H_2O$ and 6·3 g of $Th(NO_3)_4·4H_2O$ are dissolved in 1000 ml of water, made up to 1300 ml and heated to boiling (B). 92 g of Na_2CO_3 is dissolved in 500 ml of water and heated to boiling (C). A and C are added to B with vigorous stirring, the product filtered, washed thoroughly until free of sodium ions, and dried at 380 K. The dried cake is crushed and screened to a suitable size (e.g. 8–12 mesh).

Reduction of the catalyst if effected in a flow of hydrogen (about 3000 ml of hydrogen per ml of catalyst per hour). The reduction temperature is raised from room temperature to 670 K during 2 h, and the temperature held at 670 K for 2 h.

7. Platinum/Silica Gel (impregnation)

(After Adams, C. R., Benesi, H. A., Curtis, R. M. and Meisenheimer, R. G. *J. Catal.* **1**, 336 (1962).)

Enough 0.2 M chloroplatinic acid (19 ml) is added to 30 g of silica gel (Davison Grade 70) to give roughly 2.5 wt. % platinum in the ultimate product. Enough distilled water is added to make a viscous slurry, the slurry is evaporated to near-dryness (with constant stirring) on a hot-plate, and the product is dried for 16 h at 390 K. The final product is reduced in a stream of hydrogen for 2 h at 480 K; subsequent analysis showed that the reduced product contained less than 0.03 wt% chlorine and did indeed contain 2.5 wt% platinum.

8. Platinum/Aluminosilicate (coprecipitation)

(After Weisz, P. B., Frilette, V. J., Maatman, R. W. and Mower, E. G. *J. Catal*, **1**, 307 (1962).)

78 g of $NaAlO_2$ (actual assay 75% $NaAlO_2$, 10–15% $NaOH$, 10–15% H_2O) is dissolved in 275 ml water (A). Sodium metasilicate, 113 g is dissolved in 275 ml water (B). Tetrammine platinous chloride monohydrate, 0.55 g, is dissolved in 70 ml water (C). Solutions A and C are combined, and then solution B is added at room temperature, stirred under reflux for 7 h, and the solution then filtered. Without prior drying the solid is washed in a copious excess of $CaCl_2$ solution to convert it from the Na- to Ca-form and dried. Its X-ray diffraction pattern is that of 5A calcium aluminosilicate The Na content was 0.23%, showing that nearly complete (98.6%) conversion to the calcium form had been accomplished.

The material contained 0.31 wt% of platinum.

9. Platinum/Silica Gel (adsorption)

(After Benesi, H. A., Curtis, R. M. and Studer, H. P. *J. Catal.* **10**, 328 (1968).)

A reagent (A) is prepared which is an aqueous solution 0.01 M with respect to tetrammine platinous hydroxide and 0.001 M with respect to tetrammine platinous chloride. Tetrammine platinous hydroxide may be prepared from a solution of tetrammine platinous chloride by exchange on the hydroxyl form of an anion exchange resin such as Amberlite IRA-400.

Reagent A is added dropwise to a suspension of silica gel (Davison Grade 70) in 0·001 M tetrammine platinous chloride solution until the desired pH is reached and is constant for 10–20 min. At pH 9·0 the equilibrium adsorption is 0·30 mmol $[Pt(NH_3)]^{2+}$ per g SiO_2, while at pH 8·0 the figure is 0·12 mmol.

After equilibration the product is filtered, dried at least 2 h at 390 K and then reduced in flowing hydrogen at 770 K.

10. Platinum/Zeolite (adsorption)

(After Dalla Betta, R. A. and Boudart, M. *In* "Proceedings 5th International Congress on Catalysis" (J. W. Hightower, ed.), North-Holland, Amsterdam (1973), p. 1329.)

A CaY zeolite is prepared by repeated exchange of a Linde NaY zeolite with a $Ca(NO_3)_2$ aqueous solution to yield 0·8 equivalent fraction exchange.

Platinum is exchanged into the CaY zeolite from a $Pt(NH_3)_4Cl_2$ aqueous solution to yield a platinum content of 0·12 equivalent fraction or 4·97 wt. % Pt, as determined spectrophotometrically.

After washing, the exchanged zeolite is dried in air, first at 370 K, then at 620 K and finally reduced in flowing hydrogen at 670 K.

11. Nickel/Kieselguhr (impregnation)

(After Covert, L. W., Connor, R. and Adkins, H. *J. Amer. Chem. Soc.* **54**, 1651 (1932).)

Ammonium carbonate is in general the most satisfactory precipitant in the preparation of nickel catalysts.

Fifty-eight grams of C.P. nickel nitrate hexahydrate $(Ni(NO_3)_2 \cdot 6H_2O)$, dissolved in 80 ml of distilled water, is ground for 30–60 min in a mortar with 50 g of acid-washed kieselguhr until the mixture is apparently homogeneous and flows as freely as a heavy lubricating oil. It is then slowly added to a solution prepared from 34 g of C.P. ammonium carbonate monohydrate $(NH_4)_2CO_3 \cdot H_2O))$ and 200 ml of distilled water. The resulting mixture is filtered with suction, washed with 100 ml of water in two portions, and dried overnight at 380 K. The yield is 66 g. The product so obtained is reduced for 1 h at 720 K in a stream of hydrogen which passes over the catalyst at a rate of 10–15 ml min^{-1}.

12. Nickel/Magnesia (impregnation)

(After Salvi, G., Fiumari, A. and Riganti, V. *J. Catal.* **10**, 307 (1968).)

The catalyst is prepared by impregnating MgO pellets (previously shaped and fired in air for 48 hours at 1720 K) with a 20 wt.% aqueous solution of $Ni(NO_3)_2 \cdot 6H_2O$. Impregnation in vacuum by barely wetting the MgO support with solution is followed by drying overnight at 390 K, and then calcining in air at 670 K for 2–4 hours, followed by reduction in a stream of flowing hydrogen.

13. Nickel/Alumina (coprecipitation)

(After Zelinskii, N. and Komorewsky, W. *Ber.* **57**, 667 (1924): Ciapetta, F. G. and Plank, C. J. *In* "Catalysis" (P. H. Emmett, ed.), Reinhold, New York, Vol. 1 (1954), p. 315.)

454 g of $Al(NO_3)_3 \cdot 9H_2O$ is dissolved in 3 l of water and the solution cooled to 278–283 K in an ice-bath. 200 g of NaOH is dissolved in 1 l of water and the solution cooled to 278–283 K. The sodium hydroxide solution is added dropwise to the aluminium nitrate solution while stirring vigorously. The time of addition should be between 1·5 and and 2·0 h. 101 g of $Ni(NO_3)_2 \cdot 6H_2O$ is dissolved in 600 ml of water, 45 ml of concentrated nitric acid added and the solution cooled to 278–283 K in an ice bath. The nickel nitrate solution is added with vigorous stirring. After filtration, the precipitate is suspended in 2 l of water and stirred for 15 min and again filtered. This washing procedure is repeated six times. The final washed cake is cut into cubes and dried in an air oven for 16 h at 480 K. The dried catalyst is crushed and screened to about 8–12 mesh. Reduction is effected in a stream of hydrogen (2–5 l h^{-1}) for 16 h at 620 K, the temperature being raised to 620 K from room temperature over a 3 h period.

14. Palladium Colloid

(After Rampino, L. D. and Norg, F. F. *J. Amer. Chem. Soc.* **63**, 2745 (1941).)

To 12·5 ml of a 2 wt.% aqueous solution of polyvinyl alcohol is added 11 ml of water. One ml of a palladium chloride solution (1 wt.%Pd) is introduced followed by the addition drop by drop of 0·5 ml of a 4 wt.% solution of sodium carbonate, which is sufficient to convert the palladium to the hydroxide and to neutralize the hydrochloric acid present in the palladium chloride solution. Absolute alcohol is now added to give a 50% alcohol–water mixture or alternatively more water is introduced and the catalyst used in a purely aqueous medium. The colloid is reduced by shaking with hydrogen.

A platinum–polyvinyl alcohol catalyst is prepared in a similar manner. The platinum is incorporated in the form of a K_2PtCl_4 solution (0·5 wt. % Pt). In order to convert the platinum to the hydroxide it is necessary to boil the solution for a few minutes with the required amount of sodium hydroxide solution.

Index

A

Active sites, 25–26
Adsorption of chloroplatinic acid, 181
Adsorption of metal ions
  aquocations, 35–37, 203–207, 212–214
  effect of pH, 36, 175
  $[Pt(NH_3)_4]^{2+}$, 177–180
  on silica, 35–37, 180, 203–207
Adsorption on carbon
  chloroplatinic acid, 181
Adsorption on oxides
  alumina, 177
  aquocations, 203
  corrosion of support, 176, 183
  effect of pH, 36, 175
  metal ion incorporation, 206–207
  nature of adsorbed species, 204–205
  non-noble metals, 203–206, 214
  $[Pt(NH_3)_4]^{2+}$, 177–180
  silica, 35–37, 180
Adsorption on porous supports
  component separation, 176
  control by competition, 175
  sorption–diffusion, 173–174
Adsorption from solution, 353
Alkali halide, 92–93
  evaporated rocksalt, 93
  thermal etching, 92
Alloys
  catalytic activity, 19
  d-band holes, 19
  surface atom ensembles, 19
Alloy, evaporated films, 144–153
  diffusion, 145–146
  formation, 144–146
  nickel–copper, 146–149
  palladium–silver, 152–153
  platinum–gold, 149
  platinum–ruthenium, 150
  rhodium–palladium, 150–151
  structure, 146–153
Alloy, phase diagrams
  nickel–copper, 147
  palladium–silver, 152
  platinum–gold, 150
  rhodium–palladium, 151
Alloy, surfaces
  effect of chemisorbed gas, 157
  grain boundary, 156
  gold–silver, 155
  surface enrichment, 154–157
Alumina, 46–54
  active alumina structure, 47
  active alumina pore structure, 48
  catalytic activity, 53
  dehydration, 48–49
  formation of active alumina, 46–47
  irradiation, 53
  surface acidity, 50–53
  surface hydroxyl, 49–51
Appearance potential spectroscopy, 410
Asbestos, 87
Atomic weights of metals, 446–447
Auger electron spectroscopy, 410–419
  backscattering factor, 415
  chemical shift, 419
  dependence on specimen thickness, 416
  electron escape depth, 414
  line energies, 412–413
  nomenclature, 411
  quantitative analysis, 416–419
  theory, 414–416, 421–425

B

Bimetallic catalysts
 (*see also* Alloy)
 carbon monoxide chemisorption, 437
 chemisorption, 324–325
 hydrogen chemisorption, 435–437
 metal powders, 231–235
 skeletal metals, 233
 supported, 232–234
Bimetallic particles, 263–266
 surface energy effect, 265–266
 phase separation processes, 263–265
Binders, 39
BET isotherm, 291–293
 area per adsorbed molecule, 293
Boiling point of metals, 446–447

C

Calcium aluminate, 87
Calorimetry, 354
Capillary pressure, 171–172
Capillary transport, 172
Carbonates, 66–67
Carbon, 81–86
 graphitization, 82–84
 heat of physical adsorption, 83
 pore structure, 81–82
 surface, 38
 surface groups, 85–86
Carbon dioxide chemisorption on iron, 326
Carbon molecular sieves, 201
Carbon monoxide chemisorption, 310–317, 437
 on alumina, 313
 on bimetallic catalysts, 437
 on carbon, 315
 on copper, 323
 on iron, 323
 on osmium–copper, 437
 on platinum, 311
 on ruthenium–copper, 437
 on silica, 312
 on zeolite, 314
Chemisorption, 8–18, 295–328
 bonding on metals, 8–15
 heats of chemisorption, 11–12, 12–15
 on metals, 8–18, 295–326
 on various oxides, 326–328
 surface atoms per unit area, 296
Chloroplatinic acid
 adsorption, 181
 decomposition, 182–183
 hydrolysis, 182
Chlorides, 66–67
Chromia, 54–56
 catalytic activity, 56
 oxidation, 55
 pore structure, 55
 recrystallization, 55
 surface hydroxyl, 54
Clays, 80–81
Cobalt, skeletal, 229
Cobalt, stabilized porous, 227–228
Cobalt, supported catalysts
 cobalt/alumina, 213
 cobalt/carbon, 215
 cobalt/silica, 213
 cobalt/silica–alumina, 215
Collective paramagnetism, 371–373
 (*see also* Magnetic measurements)
Colloidal metals, 221
Copper, skeletal, 229
Copper, supported catalysts
 copper/magnesia, 215
 copper/magnesium silicate, 215
 copper/zeolite, 214
 nitrous oxide chemisorption, 323
Corrosion of support, 176, 183
Crystal structure of metals, 446–447

D

Deactivation of catalysts, 164–165
Decomposition, noble metal chloride, 182, 199
Density of metals, 446–447
Differential hydrogen analysis, 439
Diffusion, interdiffusion data for alloys, 146
Dispersion, *see* Metal dispersion

E

Electrical resistivity of metals, 450
Electrochemical measurements, 354–357
Electron energy analysis, 421–425
Electron microprobe analysis, 397
Electron microscopy, 396, 403–407
  energy loss spectroscopy, 397
  measurement of particle size, 363
  replication, 405
  resolution, 406–407
  scanning electron microscopy, 403–404
Electronic structure of metals
  band filling in iron, cobalt, nickel and copper, 4
  band theory, 2–7
  density of states, 5–6
  percentage d-character, 8
  valence bond theory, 7–8
Epitaxial film growth, 137–143
  on mica, 137–139
  on rocksalt, 140–143
Evaporated metal films, 130–153
  alloy films, 144–153
  alloy formation, 144–146
  average crystal size, 133
  epitaxy, 137–143
  intercrystal gaps, 131
  preferred crystal orientation, 137
  roughness factor, 131–133
  sintering, 134–135
  surface planes exposed, 133–135
  surface structure, 136, 139, 140, 142
  work-function data, 133–135
Exchange with $[Pt(NH_3)_4]^{2+}$
  on silica, 177
  on zeolite, 178–180

G

Gas adsorption measurements
  evaporated metal films, 342
  flow apparatus, 344–347
  gravimetric apparatus, 347–352
  radiochemical methods, 352–353
  sample transfer device, 340–341
  sample chambers, 337–338
  static apparatus, 332–338
  UHV apparatus, 338–343
Glass, 45
Gravimetric measurement, 347–352
Grazing incidence electron diffraction, 408

H

Harkins-Jura isotherm, 294
Heat of atomization of metals, 446–447
Heat of chemisorption
  calculation, 11–12
  data for metals, 12–15
Hydrogen absorption, 297–299, 306
Hydrogen chemisorption
  bimetallic catalysts, 435–437
  effect of impurity, 302–304
  on alumina, 313, 314
  on carbon, 315
  on copper, 323
  on iridium, 321
  on nickel, 301, 304, 322
  on nickel–copper, 435–436
  on osmium, 321
  on palladium, 319–320
  on platinum, 299–300, 307, 312–319
  on rhodium, 321
  on ruthenium, 321
  on silica, 312
  on various non-noble metals, 322–323
  on zeolite, 314
Hydrogen spillover, 305

I

Impregnation of porous supports
  chloroplatinic acid 182–183
  component separation, 176
  corrosion of support, 176, 183
  distribution of metal, 173
  effect of adsorption, 172–177
  effect of capillary pressure, 171–172
  effect of solvent removal, 175
  mechanisms, 171–172

Impregnation of porous supports—*contd.*
  non-noble metals, 203–206
  rate of capillary transport, 172
Infrared spectroscopy, 408, 409, 438–439
  adsorbed carbon monoxide, 15, 17, 438–439
  adsorbed nitrogen, 18
  bimetallic catalysts, 438–439
Ion beam spectrochemical analysis, 429
Ion bombardment, 117–119
  damage and annealing, 119
Ion probe microanalysis, 426–428
  ion yields, 427
Ion scattering analysis, 428–429
Ionization potential of metals, 448
Iridium, supported catalysts, 197–199, 202
  hydrogen chemisorption, 321
Iron, skeletal, 229
Iron, stabilized porous, 222–217
Iron, supported catalysts
  iron/alumina, 213
  iron/silica, 213
  iron/zeolite, 214

K
Kelvin equation, 379
Kieselguhr, 44

L
Laser microprobe analysis, 398
Ligand exchange, 181, 204
Low energy electron diffraction 398–403
  clean metal surfaces, 102–103
  electron coherence length, 403
  Ewald construction, 401, 407
  from facets, 407–408
  geometric theory, 400
  post-acceleration display, 399
  theories, 402

M
Magnesia, 63–65
  catalytic activity, 65
  dehydration, 63
  irradiation, 65
  pore structure, 64
  surface hydroxyl, 63–64
Magnetic measurements, 370–375
  collective paramagnetism, 371–373
  Faraday method, 373–374
  particle size distribution, 373
  vibrating magnetometer, 375
Magnetic susceptiblity of metals, 450
Melting point of metals, 446–447
Mercury porosimetry, 384–385
Metals, active sites, 25–26
Metals, catalytic activity, 21–24
Metal clusters
  energy, 244–245
  molecular compounds, 270–275
Metal crystallites
  bimetallic, 263–266
  curved surfaces, 263
  electronic properties, 260–269
  magnetic properties, 269–270
  multiple twinning, 256–259
  optical absorption, 269
  particle growth, 280–286
  particle migration, 280–282
  S_5 sites, 253
  shapes, 246–260
  surface atom statistics, 249–253
  surface atoms of low co-ordination, 260–263
Metal dispersion, 360–363
  (*see also* Metal particle size)
  bimetallic systems, 361–362
  platinum/alumina, 190–194
  platinum/silica, 180, 184–190
  platinum/zeolite, 179, 195–196
  relation to particle size, 361
Metal–oxide interface energy, 246
Metal oxide morphology, 169
Metal particle shape, 218
Metal particle size
  cobalt/alumina, 215

Metal particle size—*contd.*
  cobalt/carbon, 215
  cobalt/silica, 215
  cobalt/silica–alumina, 215
  copper/magnesia, 215
  copper/magnesium silicate, 215
  gold/magnesia, 202
  iridium/alumina, 202
  irrdium/silica, 202
  nickel/alumina, 212, 213
  nickel/silica, 209, 210, 215
  nickel/silica–alumina, 213, 215
  nickel/zeolite, 214
  osmium/silica, 202
  palladium/alumina, 199
  palladium black, 220
  palladium/carbon, 200–201
  palladium/silica, 199–200
  platinum/alumina, 191–193
  platinum black, 219 220
  platinum/carbon, 196
  platinum/silica, 185–190
  platinum/silica–alumina, 194
  platinum/zeolite, 195
  rhenium/silica, 215
  rhodium/carbon, 202
  rhodium/silica, 202
  ruthenium black, 220
  ruthenium/carbon, 202
  ruthenium/silica, 202
  silver/alumina, 215
  silver/silica, 215
  silver/zeolite, 214
  skeletal metals, 230
  stabilized porous metal, 227
  ultrathin platinum film, 218
Metal powders, 218–222
  palladium black 220, 222
  platinum black, 219–222
  ruthenium black, 220
  sintering, 222
  surface cleaning, 221–222
Metal structure, 1–8
  band theory, 2–7
  electronic structure, 2–8
  f.c.c., b.c.c. and h.c.p. structures, 3
  interatomic distances, 2
  valence bond theory, 7–8
Metal–support interactions, 275–280
  carbon supports, 277
  charge transfer, 277–280
  effect of oxygen, 275
  interaction energy, 275–277
  oxide supports, 275–276
  work of adhesion, 276
Metal surfaces
  atoms per unit cell, 111–113
  cleaning, 116–120
  crystallography b.c.c., 104–105
  crystallography f.c.c., 106–107
  crystallography h.c.p., 108–110
  dangling orbitals, 9–10
  impurities, 116
  preparation, 114, 116–120
  rearranged surfaces, 103
  surface bonds, 8–15
  surface structure, 102–103
Mica, 90–91
  surface structure, 90–91
Monolithic porous ceramic, 87–88
Mössbauer spectroscopy, 430–432
  platinum–iron catalysts, 234
Mullite, 87

N

Nickel–copper, evaporated films
  hydrogen chemisorption, 435
Nickel–copper, powder
  hydrogen chemisorption, 235, 435–436
  magnetic properties, 231
Nickel, skeletal, 229–230
Nickel, stabilized porous, 227–228
Nickel, supported catalysts
  hydrogen chemisorption, 301, 322
  hydrogen reduction, 208, 212
  nickel/alumina, 211–212
  nickel aluminate, 211–212
  nickel/carbon, 213
  nickel incorporation, 207–208, 211–212
  nickel/silica, 207–210

Nickel, supported catalysts—*contd.*
  nickel/silica–alumina, 213
  nickel silicate, 207–208
  nickel/zeolite, 214
  oxygen chemisorption, 310
  sintering, 213
Nitrate decomposition, 167–168
Nitrous oxide chemisorption
  on copper, 323
  on silver, 324
Non-ideal gases, 331–332

O

Ortho–para hydrogen conversion, 19
Osmium–copper, supported catalysts, carbon monoxide chemisorption, 437
Osmium, supported catalysts, 197–199, 202, 321
  hydrogen chemisorption, 321
Oxides, mixed, 68–74
  alumina–chromia, 73
  silica–alumina, 68–72
  silica–zirconia, 72–73
  surface acidity, 70–71, 74
  various systems, 74
Oxide surfaces
  ion exchange, 35–37
  pH for zero charge, 34
  reaction with water, 33–35
  surface structure, 32–33
Oxygen chemisorption, 307–310, 312–317
  on alumina, 313
  on carbon, 315
  on copper, 323
  on nickel, 310
  on platinum, 307–310
  on silica, 312
  on silver, 323
  on various metals, 308–310
  on various non-noble metals, 322
  on zeolite, 314

P

Palladium, powder, 220, 222

Palladium, supported catalysts, 197–201
  hydrogen chemisorption, 319–320
  sintering, 201
Particle size measurement, 358–376
  electron microscopy, 363
  magnetic methods, 370–375
  X-ray diffraction line broadening, 365–368
  X-ray scattering, small angle, 368–370
Phosphates, 66–67
Photoelectron spectroscopy, 419–426, 439
  and adsorption, 439
  elemental sensitivities, 422–423
  excitation sources, 420
  surface sensitivity, 420
Physical adsorption, 290–295
  BET isotherm, 291–293
  effect of pore shape, 378
  evaluation of pore structure, 379–384
  Harkins-Jura isotherm, 294
  Henry's Law, 294–295
  hysteresis, 376–382
Platinum, powder, 219–222
Platinum, supported catalysts
  carbon monoxide chemisorption, 311
  chlorine content, 190
  effect of pore structure, 188, 193
  from adsorbed $[Pt(NH_3)_4]^{2+}$, 177–180
  hydrogen chemisorption, 300, 316–319
  platinum/alumina, 190–194
  platinum/carbon, 196
  platinum distribution, 192–193
  platinum/silica, 184–190
  platinum/silica–alumina, 194
  platinum/zeolite, 179, 194–195
  sintering, 188–189
  platinum solubility, 194
Polymers, 89
Pore classification, 42

Pore measurement
  mercury porosimetry, 384–385
  physical adsorption, 376–384
  t-method, 382–384
Pore shape, effect on
  physical adsorption, 378
Porous solids, density, 386
Preparation of catalysts, 451–459

R

Raney metal catalysts, *see* Skeletal catalysts
Reaction of metal oxide and support, 166, 183
Redispersion of platinum, 165
Reduction of metal chlorides, thermodynamics, 166–167
Reduction of exchanged metal cations, 170
Reduction of metal oxides
  effect of water vapour, 169
  mechanism, 169
  product morphology, 170–171
  thermodynamics, 166–167
Resins, 89
Rhenium, skeletal, 229
Rhenium, supported catalysts
  rhenium/silica, 215
Rhodium, supported catalysts, 197–199, 202
  hydrogen chemisorption, 321
Ruthenium–copper, supported catalysts
  carbon monoxide chemisorption, 437
Ruthenium, powder, 220
Ruthenium, supported catalysts, 197–199, 202
  hydrogen chemisorption, 321

S

Second virial coefficients, 331
Silica, 39–44
  aerogel, 42
  flocculate, 41
  gel dehydration, 42
  gel structure, 41–42
  microspheroidal powder, 42
  surface hydroxyl, 43
  surface structure, 40
  xerogel, 42
Silica–alumina, 68–72
  irradiation, 71–72
  structure, 68–69
  surface acidity, 69–73
  surface hydroxyl, 70–71
Silicon carbide, 87
Silver, skeletal, 229
Silver, supported catalysts
  nitrous oxide chemisorption, 324
  silver/alumina, 215
  silver/silica, 215
  silver/zeolite, 214
Sintering, dispersed catalysts, 280–286
  interparticle atom transport, 282
  metal particle migration, 280–282
  nickel, 213
  palladium, 201
  platinum, 188–189
  propensity of various metals, 285–286
  rate laws, 282–285
Sintering, evaporated metal films, 134–135
Sintering, metal powders, 222
Skeletal metals, 228–230, 233
  bimetallic, 233
  metal particle size, 230
  pore structure, 230
  structure and composition, 229–230
Stabilized porous metals, 222–228
  cobalt, 227–228
  chemical promoters, 223
  iron, 223–227
  nickel, 227–228
  stabilizers, 224–227
  structure, 223–227
Sulphates, 66–67
Surface area measurement by chemisorption, 295–328

Surface area measurement by chemisorption—*contd*.
  copper, 323
  iridium, 321
  nickel, 322
  osmium, 321
  palladium, 319–320
  platinum, 317
  rhodium, 321
  ruthenium, 321
  silver, 324
  various non-noble metals, 322
Surface area measurement by physical adsorption, 290–295
Surface atoms per unit, 296
Surface atom statistics, 249–253
  S_5 sites, 253
Surface cleaning, 116–120
  damage and annealing, 119
  ion bombardment, 117–119
  thermal and chemical, 116–117
Surface energy of metals, 449
Surface faceting, 125–130
  evaporation faceting, 125
  impurity faceting, 126
  reactive faceting, 127–130
  silver, 126, 129
  tungsten oxidation, 128
Surface imperfections, 120–130
  effect of temperature, 121–122
  equilibrium theory, 121–122
  screw dislocations, 123
  steps, kinks and terraces, 120–122
Surface roughness, 113–114
  (*see also* Surface imperfections)
  faceting, 125–130
  high index surfaces, 123–124

T

Temperature programmed desorption, 16–18, 439
  carbon monoxide on tungsten, 17
  hydrogen on tungsten, 16, 18
  nitrogen on tungsten, 16, 18
Thermomolecular flow, 329–330

Thoria, 65–66
  catalytic activity, 66
  pore structure, 65
  surface hydroxyl, 66
Titania, 56–61
  catalytic activity, 61
  dehydration, 59
  pore structure, 57
  surface acidity, 60
  surface hydroxyl, 57–59
Titration of adsorbed oxygen and hydrogen, 309

U

Ultrathin evaporated metal films, 217–218

V

Volumetric measurement, 328–344

W

Work-function
  evaporated nickel films, 133, 135
  evaporated platinum films, 134
  measurement, 432–434, 440
    contact potential, 433–434
    photoelectric, 433
  nickel crystal planes, 133
  of metals, 448

X

X-ray absorption edge spectroscopy, 370
X-ray diffraction line broadening, 365–368
  Scherrer equation, 365
  strain broadening, 367
X-ray emission spectroscopy, 410
X-ray scattering, small angle, 368–370
X-ray wavelengths, 367

Z

Zeolite, 74–81
  decationized, 80–81
  exchangeable cation positions, 79
  surface acidity, 79–81
  type A, 74, 76, 77
  types X and Y, 74, 76, 79
  various types, 75, 77
Zircon, 87
Zirconia, 61–63
  catalytic activity, 63
  dehydration, 61–62
  structure, 61–62
  surface hydroxyl, 62